Proceedings of the 15th Asia Pacific Physics Conference

Hyoung Joon Choi · Takhee Lee · Woo-Sung Jung
Editors

Proceedings of the 15th Asia Pacific Physics Conference

Springer

Editors
Hyoung Joon Choi
Department of Physics
Yonsei University
Seoul, Korea (Republic of)

Takhee Lee
Department of Physics and Astronomy
Seoul National University
Seoul, Korea (Republic of)

Woo-Sung Jung
Department of Physics
Pohang University of Science
and Technology
Pohang, Korea (Republic of)

ISBN 978-981-96-0190-5 ISBN 978-981-96-0191-2 (eBook)
https://doi.org/10.1007/978-981-96-0191-2

This Springer imprint is published by the registered company Springer Nature Singapore Pte Ltd.
The registered company address is: 152 Beach Road, #21-01/04 Gateway East, Singapore 189721, Singapore

Preface

This publication is the Proceedings of the 15th Asia Pacific Physics Conference (APPC15) which was held online from August 21 to 26, 2022, at Yonsei University, Seoul, Korea.

APPC is the series of triennial conferences of the Association of Asia Pacific Physical Societies (AAPPS), covering all areas of physics. APPC provides a forum to share and discuss physics researches and educations, promoting collaborations and friendship in Asia Pacific regions.

APPC15 was organized by AAPPS and the Korean Physical Society (KPS), co-organized by Asia Pacific Center for Theoretical Physics (APCTP) and Department of Physics, Yonsei University, and sponsored by The Korean Federation of Science and Technology Societies (KOFST). The scientific program of APPC15 consisted of plenary talks, invited talks, contributed talks, and poster presentations, organized by AAPPS together with AAPPS Division of Plasma Physics (DPP), AAPPS Division of Astrophysics, Cosmology, and Gravitation (DACG), AAPPS Division of Nuclear Physics (DNP), and AAPPS Division of Condensed Matter Physics (DCMP).

APPC15 was held in a fully online format (except for very few hybrid sessions), considering uncertainties and difficulties of the pandemic situation. The Zoom platform was used for all talks, providing live talks and live questions and answers, and the ZEP metaverse platform was for poster sessions, exhibition rooms, and discussion rooms. APPC15 covered all areas of physics: (1) Applied Physics, (2) Astrophysics, Cosmology, and Gravitation, (3) Atomic and Molecular Physics, (4) Biological Physics, (5) Condensed Matter Physics, (6) Nuclear Physics, (7) Optics and Quantum Electronics, (8) Particles and Fields, (9) Physics Education, (10) Plasma Physics, (11) Quantum Information, (12) Semiconductor Physics, (13) Statistical Physics, and (14) Women in Physics. The AAPPS-APCTP C. N. Yang Award Ceremony and AAPPS-APCTP C. N. Yang Award Session were held on August 22, 2022, and Editor Session was held on August 24, 2022.

The year, 2022, was the 70th anniversary of KPS, and it was also the Centenary of International Union of Pure and Applied Physics (IUPAP) and the International Year of Basic Sciences for Sustainable Development (IYBSSD 2022). On the occasion of these, a special forum "Global Physics Summit" and a special session "The Role of Physics in the Green Economy" were held on August 22, 2022. The opening session of APPC15, the first Plenary Session on August 22, Global Physics Summit, AAPPS-APCTP C. N. Yang Award Ceremony, and the Role of Physics in the Green Economy were held in a hybrid format using both online and onsite in which the onsite venue was the Grand Ballroom at the Commons "Baekyang Nuri" at Yonsei University, Seoul, Korea.

APPC15 was very successful with more than 1000 participants from more than 20 countries and a total of 963 presentations: 15 plenary talks, 470 invited talks in topical parallel sessions, 298 contributed talks in topical parallel sessions, 12 invited talks in special sessions, and 168 poster presentations in the metaverse poster sessions.

The organizers of APPC15 would like to express sincere thanks to all participants including committee members, session chairs, plenary speakers, invited speakers, special-session invited speakers, contributed-talk speakers, and poster presenters. The organizers would also like to express sincere thanks to APCTP and KOFST for great support and to staff members for their time and great effort.

This conference proceedings contains 26 papers from various fields of physics which were presented at APPC15, submitted, and accepted.

The editors would like to express sincere thanks to all authors of papers and all reviewers.

December 2022

Hyoung Joon Choi
Takhee Lee
Woo-Sung Jung

Contents

Condensed Matter Physics

Nuclear Physics

Statistical Physics

Applied Physics

Absorbance Measurement of a Scattering Sample Placed Inside an Integrating Sphere and Analysis by Monte Carlo Simulation

Ayaka Mori, Kyohei Yamashita, Keisuke Seto, and Eiji Tokunaga(✉)

Department of Physics, Faculty of Science, Tokyo University of Science, 1 3 Kagurazaka, Shinjuku ku, Tokyo 162 8601, Japan
eiji@rs.tus.ac.jp

Abstract. Absorption measurements of scattering samples overestimate the absorption because the transmitted light is attenuated by scattering. In order to avoid this, we placed a sample inside an integrating sphere (IS) and performed absorption spectroscopy. The absorbance measured by this method is non-linear with respect to the absorption coefficient of the sample because light can take different paths IS inside the sample. Therefore, we attempted to estimate the relationship between the absorption coefficient and the measured absorbance by simulating the behavior of photons inside the IS and sample by Monte Carlo simulation(MC method). Using an aqueous nickel nitrate solution which is a non-scattering sample and a scattering sample prepared by mixing an aqueous nickel nitrate solution and polystyrene beads, it was verified whether the measurement and the simulation results agree with each other. As a result, in the scattering sample, the measurement results could be reproduced by the MC method, while in the adjusted scattering sample, a discrepancy was observed between the simulation and the measurement results in the high absorption region. From the DLVO theory, this can be explained to be caused by particle aggregation at high concentrations because the aqueous nickel nitrate solution is an electrolyte solution.

Keywords: absorption · scattering · integrating sphere · Monte Carlo simulation · absorption spectrophotometer

1 Introduction

Absorption spectroscopy in ultraviolet/visible light/near-infrared is safe and easy to handle, so it is used to evaluate physical properties of various substances. Since the sample can be measured in a non-destructive manner, it is suitable for measuring biological samples, and is used in the medical field such as blood glucose measurement [1] and food quality evaluation [2].

In absorption spectroscopy, the absorption coefficient is generally obtained from the transmittance of the sample based on Beer-Lambert's law. However, the transmitted light is attenuated not only by absorption but also by scattering, so this method overestimates the absorption [3]. To prevent this absorption overestimation, it is necessary to measure not only the transmitted light intensity but also the scattered light intensity.

H. J. Choi et al. (Eds.): APPC 2022, *Proceedings of the 15th Asia Pacific Physics Conference*, pp. 3–14, 2024.
https://doi.org/10.1007/978-981-96-0191-2_1

Overestimation of absorption due to scattering can be reduced by incorporating an integrating sphere (IS) into absorption spectroscopy (Fig. 1). Since the inner wall of the IS is coated with a highly reflective material such as barium sulfate, the light entering the IS is repeatedly reflected by the inner wall and detected by the IS built-in detector. In absorption spectroscopy using IS, the sample is generally placed at the opening of the integrating sphere (Fig. 1(a)) [4,5]. The collection of transmitted light and forward scattered light by the IS enables absorption spectroscopy measurements that take forward scattered light into account. However, the arrangement in Fig. 1(a) cannot detect backscattered light, so the absorption is overestimated by the amount of backscattering [6]. Therefore, we installed a sample inside the IS (Fig. 1(b)) and performed absorption spectroscopy measurements considering the scattered light of all solid angles. This arrangement is mainly used in fluorescence quantum yield measurements [7,8]. It has also been used for accurate absorbance measurements of scattering samples [9,10], but is still uncommon. A possible reason for this is that the absorbance measured in the configuration of Fig. 1(b) does not match the true absorbance (absorption coefficient multiplied by sample thickness). When a sample is placed inside the IS, there is light that passes through the sample multiple times, such as being repeatedly scattered by the inner wall of the IS after passing through the sample, and then passing through the sample again. The resulting increase in the optical path length must be estimated. Reflected light from the sample surface is also a factor of absorbance error. Because the IS collects light exiting from the sample at all angles, it also detects light reflected from the sample surface. As a result, the higher the absorption of the sample, the more the transmitted light containing the absorption information of the sample is buried in this reflected light, increasing the absorbance error [11–14]. The fact that the reflected light becomes stray light is also a factor in lowering the saturated absorbance [15]. Various absorbance correction methods have been proposed in order to eliminate absorbance errors that occur when a sample is placed inside the IS [11,16]. However, to the best of our knowledge there have been no studies in which a sample was placed inside the IS, the sample concentration dependence of absorbance was measured from low absorbance to the higher limit of measurement, and the correction of absorbance was applied over the entire absorbance range.

Therefore, in this study, for the purpose of measuring the absorbance of a scattering sample over a wide dynamic range, we investigated a method to obtain an accurate absorption coefficient from the absorbance measured with the arrangement shown in Fig. 1(b) that can be applied to wide range of absorbance of any sample. For this purpose, the concentration dependence of the absorbance measured in the arrangement shown in Fig. 1(b) was obtained from low absorbance to the higher measurement limit of absorbance. We have derived a above-mentioned correction formula that considers the error factors in the absorbance measurement with the sample placed inside the IS, and have confirmed that the measurement results can be reproduced by this correction formula [15]. However, this correction formula has the following problems. i) The reflectance of the inner wall of the IS is assumed to be 100%, and slight absorption loss by the inner wall is not considered. ii) The loss of light from the openings of the IS is ignored. iii) Since the single scattering regime is assumed in the derivation, the multiple scattering regime cannot be accounted for and the results for highly scattering samples are unreliable. As a method to solve these problems, we attempted correction using Monte Carlo simulation (MC method) [17–19], which simulates the behavior of individual photons inside the IS and inside the sample. In this paper, we show the results of reproducing the absorbance measured in the configuration of Fig. 1(b) for a sample with known absorption, scattering, and anisotropy coefficients from these optical coefficients using MC method.

Our report reveals the relation between absorbance measured by placing a sample inside the IS and the absorption coefficient of the sample, from low absorbance to the limit of measurement. This is important in establishing a measurement method to quantitatively evaluate the absorption properties of scattering samples. The accurate quantitative measurement of the absorption properties of scattering samples will lead to a wide range of applications, including the evaluation of composition and physical properties of biological samples such as cell density measurement [20], protein concentration [21] measurement, photosynthetic efficiency measurement [22], and in the materials field such as size effects on optical properties of functional nanocrystals and solar cell [23] property evaluation [24].

2 Sample Preparation

An aqueous nickel nitrate solution (non-scattering in the measurement wavelength region) was prepared as an absorber, and a suspension of polystyrene microspheres (Polysciences) with a diameter of 500 nm (non-absorption in the measurement wavelength region) was prepared as a scatterer. A pseudo-scattering sample (Sample M) was prepared by mixing these at a ratio of 3.64×10^{10} spheres/mL of polystyrene microspheres to 2 M of nickel nitrate solution. A set of sample M was prepared in which this was diluted at 10 different ratios ranging from 1 to 60%. For comparison with the non-scattering sample, we also prepared nickel nitrate aqueous solutions adjusted to 14 different concentrations in the range of 0.02 to 2.0M. Since the absorption coefficient α_a of sample M is proportional to the concentration of nickel nitrate ($\alpha_a = kc_a$), it was obtained from c_a. The proportionality constant k was obtained from the absorbance measurement results of nickel nitrate aqueous solutions with 14 different concentrations in the range of 0.02 to 2.0M in the arrangement of Fig. 1(a), which is a conventional method, and linearly fitting the relationship between concentration and absorbance. Since the nickel nitrate aqueous solution is non-scattering, even in the arrangement of Fig. 1(a), the absorbance can be measured accurately without being affected by backscattering loss. The scattering coefficient and anisotropy coefficient of sample M were obtained from calculations based on the Mie scattering theory of polystyrene microspheres. Calculations were performed using the online Mie scattering calculation tool [25].

3 Experimental Setup

The IS built into a double-beam spectrophotometer (SolidSpec-3700DUV, Shimadzu) was used to measure the absorbance spectrum. Measurements were performed in the configuration of Fig. 1(b). A top view is shown in Fig. 2. A sample placed inside the IS

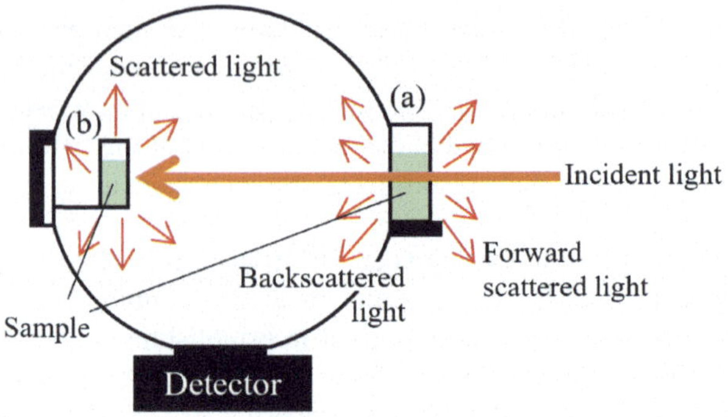

Fig. 1. Schematic diagram of IS absorption spectroscopy setup. (a) shows the case where the sample is placed at the IS opening, and (b) shows the case where the sample is placed inside the IS.

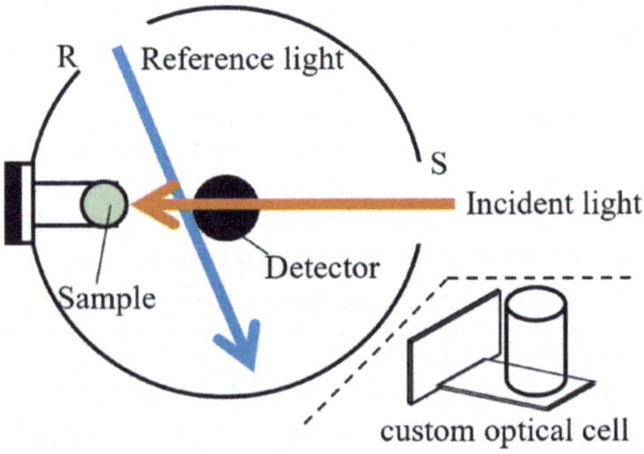

Fig. 2. (a) Absorption spectroscopy setup with the sample inside the IS, top view of the arrangement in Fig. 1(b). Because of the double-beam system, there is an aperture R through which the reference light enters in addition to the aperture S through which the light for measuring the absorption of the sample enters. A detector is attached to the bottom of the sphere and measure the intensity of light exiting through the opening at the bottom of the IS. (b) A custom-made fused silica optical cell containing the sample. A cell containing a sample is inserted into the IS through the port for mounting the diffuse reflection sample on the opposite side of the opening S, and is fixed to the IS by covering the port with a diffuse reflector dedicated to the instrument. The diffuse reflector is the same material as the IS inner wall.

is illuminated by the light incident from the opening S, and the light exiting from the sample at all angles is collected by the IS. A sample placed inside the IS is illuminated by the light incident from the opening S, and the light emitted from the sample at all angles is collected by the IS. A detector is attached to the bottom of the sphere, and the intensity of light incident on the detector is measured. A custom-made cylindrical cell shown in Fig. 2 was filled with 600 μl of liquid sample and placed in the IS as shown in Figs. 1(b) and Figs. 2. Reference light is incident from the opening R. As the intensity of the detected light, a value obtained by dividing the detected light intensity when the light is incident from the opening S by the detected light intensity when the reference light is incident from the opening R is used. Thereby, the influence of the fluctuation of the light source can be reduced. The absorbance A is given by $A = -\log(I/I_0)$, where I is the detected light intensity when the sample is placed in IS, and I_0 is the detected light intensity when the sample is not placed in IS.

4 Monte Carlo Simulation

By Monte Carlo simulation (MC method), we estimated the absorbance $A_{measure}$ measured in this arrangement by simulating the behavior of photons inside the IS and the sample in the arrangement of Fig. 1(b). The execution program of the MC method when the sample is put inside the IS was written in Fortran with reference to prior literature [17–19,26]. The photon's current position $\boldsymbol{r} = (x, y, z)$ and direction of movement $\boldsymbol{d} = (d_x, d_y, d_z)$ are expressed in xyz coordinates with the center of the IS at the origin. $\boldsymbol{d}$ is normalized to $|\boldsymbol{d}| = 1$. The behavior of each photon inside the IS was simulated three-dimensionally by setting the shape and arrangement of the sample cell, the shape and openings of the IS, and the size and position of the detector to the same as in the measurement (Fig. 3). Furthermore, the behavior of photons inside the sample was also three-dimensionally simulated by substituting the absorption coefficient of the medium of the sample, the cross-sectional area, the concentration, the scattering efficiency, and the anisotropy coefficient of the scatterer as input parameters. The incident position of the photon on the IS and direction scattered by the sample, the reflection direction on the inner wall of the IS, etc. are determined randomly according to the probability density distribution. We referred to [26] for the method of randomly determining values based on the probability density distribution. The simulation was performed for $N = 1.0 \times 10^5$ photons according to the flow chart in Fig. 4. A photon is initially given a weight of $W = 1$, and W is reduced each time there is absorption by the inner wall of the IS or by the medium. Therefore, the W of the detected photon is the intensity ratio of the detected light to the incident light ($W = 1$), that is, the transmittance measured. From the average value of W of detected light calculated from the total value W_total, that is, the average value of transmittance $\bar{T} = W_total/N$, measured absorbance $A_{measure} = -\log\bar{T}$ can be estimated. Since the spectrophotometer used this time is a double-beam system (see Sect. 3), $\bar{T} = \bar{T}_s/\bar{T}_r$, where $\bar{T}_s$ is the average transmittance obtained by the simulation of the measurement light incident from the aperture S, and $\bar{T}_r$ is the average transmittance obtained by the simulation of the reference light incident from the aperture R. The initial position of the photon is set at the opening S for the measurement light and at the opening R for the reference light. The intensity distribution of incident light at the aperture was assumed to follow a normal distribution, and the initial positions of a photon were randomly determined based on this distribution function. Beam divergence (dispersion of normal distribution) was determined from the saturated absorbance of non-scattering samples (see Sect. 5).

When a photon is present inside an IS, it is made to travel until it reaches an interface (inner wall, opeing, detector of IS and cell). Specifically, the distance s_0 to the inner wall of the IS and the distance $[s_1, s_2, s_3]$ to each face of the cell [side, top, bottom of the cylinder], respectively, are calculated from the photon's current position and direction of movement, and the photon is moved to the face with the smallest distance ($\boldsymbol{r} \rightarrow$

$\boldsymbol{r} + s_i \boldsymbol{d}$, s_i being the minimum value of $s_0 \sim s_3$). From the arrangement shown in Fig. 3, $s_0 \sim s_3$ are calculated as follows respectively.
Distance s_0 to the IS inner wall:

$$(\boldsymbol{r} + s_0 \boldsymbol{d})^2 = R^2, \tag{1}$$

$$s_0 = -\boldsymbol{d} \cdot \boldsymbol{r} \sqrt{(\boldsymbol{d} \cdot \boldsymbol{r})^2 + R^2 - |\boldsymbol{r}|^2}. \tag{2}$$

Distance s_1 to the sides of the cell:

$$(x + s_1 d_x - x_c ell)^2 + (y + s_1 d_y - y_c ell)^2 = R_{cell}^2, \tag{3}$$

$$s_1 = -\frac{d_x x_\delta + d_y y_\delta + \sqrt{D}}{d_x^2 + d_y^2}, \tag{4}$$

where $x_\delta = (x - x_c ell)$, $y_\delta = (y - y_c ell)$, $D = (d_x x_\delta + d_y y_\delta)^2 - (x_\delta^2 - y_\delta^2 - R_c ell^2)(d_x^2 + d_y^2)$ is defined. If any of conditions $D < 0$, $s_1 < 0$, or the z-coordinate $z' = z + s_1 d_z$ after moving is $z > z_{cell}^+$ or $z < z_{cell}^-$ are met, it will not intersect this face.
Distances s_2, s_3 to the top and bottom of the cell:

$$z + s_{2,3} d_z = z_{cell}^{\pm}, \tag{5}$$

$$s_{2,3} = \frac{z - z_{cell}^{\pm}}{d_z} (z_{cell}^{\pm} \text{ is } \mathrm{z}_{\mathrm{cell}}^{+} \text{ for } \mathrm{s}_2, \mathrm{z}_{\mathrm{cell}}^{-} \text{ for } \mathrm{s}_3). \tag{6}$$

If any of conditions $s_{2,3} < 0$ or $(x' - x_{cell})^2 + (y' - y_{cell})^2 < R_{cell}^2$ at the coordinate $(x', y', z') = \boldsymbol{r} + s_{2,3}\boldsymbol{d}$ after moving are met, it will not intersect this face.
When s_0 is the minimum value, it is determined whether the coordinate $\boldsymbol{r} + s_i \boldsymbol{d}$ after movement is included in the range of the aperture or detector.

When the photon reach the inner wall, they are diffusely reflected by the wall. At this time, the photon weight is changed to $W \rightarrow R_i$ W because there is a slight absorption on the inner wall, where R_i is the reflectance of the inner wall. The direction of reflection is randomly determined assuming that the light is reflected with equal probability in all directions within the range of $0 \leq \phi_r < 2\pi$ for azimuth angles ϕ_r and $0 \leq \theta_r < \pi/2$ for polar angle θ_r. The polar angle is defined as the angle from the normal of the inner wall. When the photon reaches the opening S or R, it is assumed that the photon has left the system, and the processing of this photon ends. When the photon reaches the detector, it is assumed that the photon has been detected, and the W at that time is added to the total value W_total of W of the detected photon, and the processing of this photon ends. When it reaches a cell, it randomly determined whether it should be reflected on the cell surface. The reflectance which is the probability of reflection is calculated from the angle of incidence on the cell and the refractive index of the cell, 1.46 (refractive index of quartz glass [27]), using Fresnel equations. The reflected angle is taken to be the same as the incidence angle.

In the simulation inside the sample, the input parameters of the absorption coefficient α_a of the medium, the cross-sectional area σ_s of the scatterer, the concentration c_s, the scattering efficiency Q_s, and the anisotropy coefficient g are used. Each time the

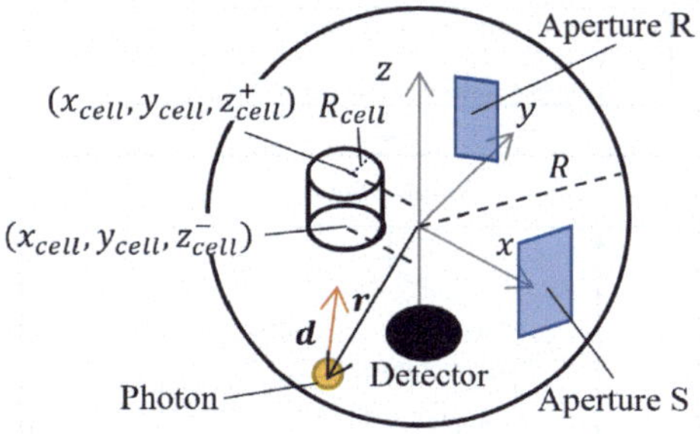

Fig. 3. Arrangement of optical system set in MC. A cylindrical cell with radius $R_{cell} = 0.8$ cm is placed inside IS with radius $R = 3.0$ cm. The inner wall of the IS contains apertures S, R and detectors, as shown in Figs. 1 and 2. A photon travels straight until it hits an interface (the surfaces of the cell or the inner wall of the IS). A photon collides with the nearest surface when moving from its current position $\boldsymbol{r}$ in the direction of movement $\boldsymbol{d}$.

photon travels through the sample, depending on the distance l traveled, absorption by the medium occurs, changing the weight of the photon to $W \rightarrow e^{(-\alpha_a l)}$ according to Beer-Lambert's law. A photon at position $\boldsymbol{r}_0$ inside the sample travels straight until it collides with the scatterer and is scattered. Therefore, first, the distance s traveled by the photon until it hits the scatterer is randomly determined. If the photon collides with the wall of the cell while traveling the distance s, first move the photon from $\boldsymbol{r}_0$ to the collision point (the distance traveled is s_c), and randomly decide whether it will be reflected on the cell surface. If the photon transmitted, it exits the sample and the processing outside the sample is resumed. If the photon is reflected, change the direction and change to $s \rightarrow s - s_c$ which is the distance traveled until the photon hits the scatterer minus the distance it traveled to the cell. If the photon does not reach the cell while traveling the distance s, the photon travels straight for the distance s and is processed in the case when it hits a scatterer. A photon striking a scatterer is scattered by the scatterer with a probability of scattering efficiency Q_s. Whether or not to scatter is randomly determined. When scattered, the scattering direction is randomly determined according to the probability density distribution $p(\cos\theta)$ of the scattering angle θ of the scatterer. $p(\cos\theta)$ is given by the Henyey-Greenstein function, which is an approximation of the Mie scattering theory, as follows [26].

$$p(\cos\theta) = \frac{1 - g^2}{2(1 + g^2 - 2g\cos\theta)^{\frac{3}{2}}} \tag{7}$$

The conditions for terminating the photon simulation are that the photon exits the system through the opening, or is detected by the detector, or that W is sufficiently small to have no effect on the results. The lower limit of W, W_{end}, is taken to be 1.0×10^{-9}.

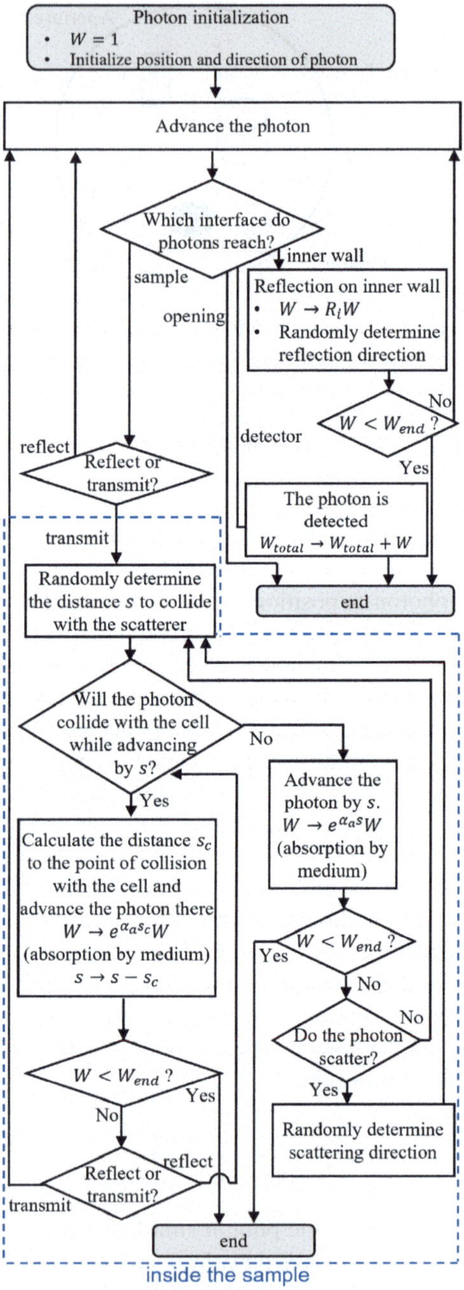

Fig. 4. Flowchart of simulation for one photon of absorbance measurement in the configuration of Fig. 1(b) by the MC method. This processing is performed with 1.0×10^5 photons in the measurement light incident from the opening S and the reference light incident from the opening R, and the light intensity ratio of detection light to the incident light is obtained from the average of the weights of the detected photons. The process enclosed by the dashed line is one when the photon is inside the sample.

5 Results

First, the results of measurements and Monte Carlo simulations for non-scattering samples are shown. Figure 5(a) shows the absorbance spectrum at each concentration of the nickel nitrate aqueous solution which is a non-scattering sample measured in the arrangement of Fig. 1(b). Absorption peaks at 658 and 720 nm originating from the d-d transition ($^3A_{2g}(F) \rightarrow {}^3T_{1g}(F)$) of Ni^{2+} due to the spin-orbit interaction with the mixture of $^3T_{1g}$ and ^1E_g can be confirmed [28]. Figures 5(b) and 5(c) show the relationship between the absorption coefficient and the measured absorbance at absorption peak wavelengths of 658 nm and 720 nm, respectively. The circles are the measurement results, and the triangles are the Monte Carlo simulation results. The measurement results are reproduced by the MC method. The saturated absorbance is about 1.5, which is smaller than the measurement limit of the device, 5.0. As described in Sect. 1, the reduction in the measurement limit value is caused by the light reflected by the

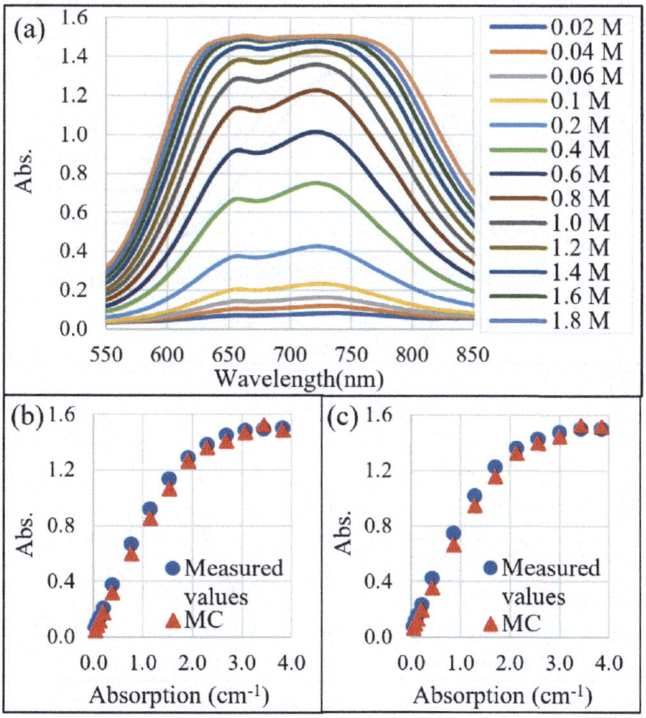

Fig. 5. Absorbance measurement result of nickel nitrate aqueous solution in the arrangement of Fig. 1(b). showing the absorption spectrum (a), and showing the relationship between the absorption coefficient of the nickel nitrate aqueous solution and the measured absorbance at absorption peak wavelengths of (b) 658 nm and (c) 720 nm, respectively. In (b), (c), circles are measurement results, and triangles are simulation results. The absorption coefficients in (b) and (c) were obtained from the absorbance measurements of the aqueous nickel nitrate solution in the configuration of Fig. 1(a) and the concentration of nickel nitrate using the method described in Sect. 2.

cell surface becoming stray light and depends on the spread of the incident light. The spread of the incident light, which is one of the input parameters of the MC method, was determined so that the simulation result would be a value that reproduces this saturated absorbance.

Next, the measurement and simulation results for scattering samples are shown. Figure 6(a) is the absorbance spectrum at each dilution of sample M measured in the arrangement of Fig. 1(b). Absorption peaks at 658 and 720 nm derived from nickel nitrate can be confirmed as in the case without scatterers [Fig. 5(a)]. Figures 6(b) and 6(c) show the relationship between the absorption coefficient and the measured absorbance at absorption peak wavelengths of 658 nm and 720 nm, respectively. In the case of scattering samples, the measurement results are reproduced by the MC method in the low absorption region, but the simulation results are smaller than the measurement values in the high absorption region.

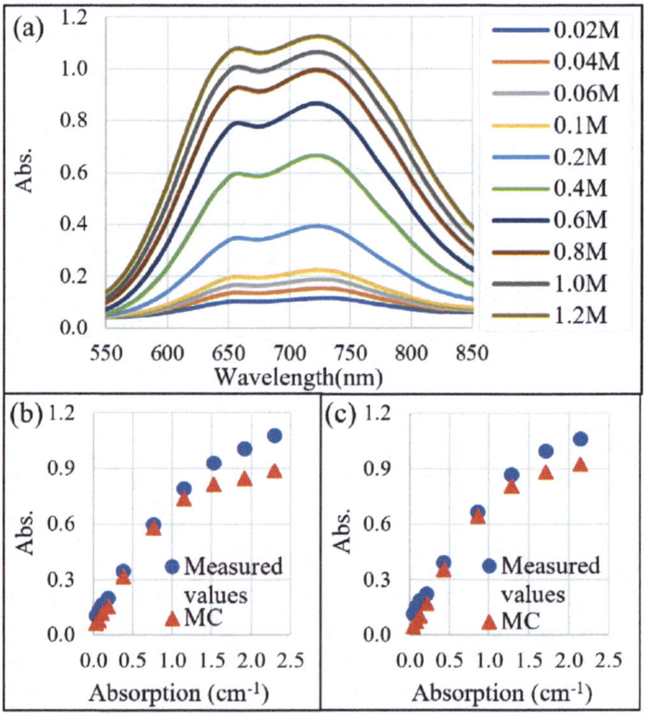

Fig. 6. Absorbance measurement result of sample M in the arrangement of Fig. 1(b). showing the absorption spectrum (a), and showing the relationship between the absorption coefficient of sample M and the measured absorbance at absorption peak wavelengths of (b) 658 nm and (c) 720 nm, respectively. In (b), (c), circles and triangles show measurement and simulation results, respectively. The absorption coefficients in (b) and (c) were obtained from the absorbance measurements of the aqueous nickel nitrate solution in the configuration of Fig. 1(a) and the concentration of nickel nitrate using the method described in Sect. 2.

It is considered that the discrepancy between the simulation and measurement results for sample M was due to the aggregation of polystyrene microspheres in the high-concentration nickel nitrate aqueous solution. The polystyrene microspheres used to prepare sample M are slightly negatively charged, and electrostatic repulsion acts between the particles, enabling stable dispersion even in water. When they are mixed with an electrolyte solution, the charged particles are surrounded by ions, forming an electric double layer around the particles. The interaction between particles is mainly attractive force due to van der Waals force and repulsive force due to electric double layer, both of which become stronger as the distance between particles decreases. When the ion concentration of the medium increases, the thickness of the electric double layer decreases, so the distance over which the repulsive force of the electric double layer acts decreases, causing aggregation. This is explained by the DLVO theory [29,30]. In sample M, slightly negatively charged polystyrene microspheres are mixed with an aqueous solution of nickel nitrate, which is an electrolyte solution. Therefore, it is most likely that polystyrene microspheres aggregate in the high absorption region where nickel nitrate is highly concentrated, and the their scattering characteristics of polystyrene microspheres change from those of the dispersed state. Since the scattering properties of the polystyrene microspheres used in the MC method are those in the dispersed state, an error occurs between the MC calculation and measurement results due to aggregation. To avoid this, it is necessary to use non-electrolytic absorbers.

6 Conclusions

By placing the sample inside the IS and collecting the transmitted light and the scattered light at all solid angles by the IS, we performed absorption spectroscopy measurements that prevented overestimation of the absorbance due to scattering. In this method, light can follow various paths inside the IS and the sample, so the measured absorbance behaves non-linearly with respect to the absorption coefficient of the sample with respect to the absorption coefficient (proportional to solution concentration). As a method to obtain the absorption coefficient of the sample from the measured absorbance, the simulation of the behavior of photons inside the IS and sample was simulated by the Monte Carlo method. For this purpose, we first attempted to reproduce the absorbance measured when a sample with known optical properties was placed inside the IS by Monte Carlo simulation. The cell containing the sample, the opening of the IS, and the position and size of the detector were set to be the same as in the actual measurement, and the three-dimensional movement of photons was simulated. Substituting the optical coefficient of the sample as an input parameter, the measured absorbance was estimated. A pseudo-scattering sample M prepared by mixing polystyrene spheres as a scatterer in an aqueous solution of nickel nitrate as an absorbing medium was placed inside the IS, and absorbance was measured to verify whether the measured and the simulated results agree. For the nickel nitrate aqueous solution, the measurement results were reproduced by simulation using the MC method. On the other hand, in the case of sample M, the low absorption region was reproduced by simulation, but the simulation result was smaller than the measured value in the high absorption region. This is probably because in the high absorption region, that is, when the concentration of nickel nitrate is high, the polystyrene microspheres aggregate due to the high electrolyte concentration, resulting in a change in the scattering characteristics. This discrepancy was mitigated by replacing the nickel nitrate aqueous solution with a non-electrolyte absorbing medium, specifically an aqueous solution of Trypan Blue. Using this modified sample, the simulation results agreed well with the absorbance measurements, as detailed in a subsequent study [31]. In the future, it is one of important targets to acquire the absorption coefficient for samples with unknown optical properties by inverse Monte Carlo from the absorbance measurement results when the sample is placed inside the IS.

References

1. Kasahara, R., Kino, S., Soyama, S., Matsuura, Y.: Biomed. Opt. Express, BOE **9**, 289 (2018)
2. Aleixandre-Tudo, J.L., Buica, A., Nieuwoudt, H., Aleixandre, J.L., du Toit, W.: J. Agric. Food Chem. **65**, 4009 (2017)
3. Gaigalas, A.K., He, H.-J., Wang, L.: J. Res. Natl. Inst. Stand. Technol. **114**, 69 (2009)
4. Dazzi, A., Deniset-Besseau, A., Lasch, P.: Analyst **138**, 4191 (2013)
5. M. N. Merzlyak and K. Razi Naqvi, Journal of Photochemistry and Photobiology B: Biology **58**, 123 (2000)
6. R. Röttgers and S. Gehnke, Appl. Opt., AO **51**, 1336 (2012)
7. Würth, C., Pauli, J., Lochmann, C., Spieles, M., Resch-Genger, U.: Anal. Chem. **84**, 1345 (2012)
8. Leyre, S., et al.: Rev. Sci. Instrum. **85**, 123115 (2014)
9. Wrigglesworth, E.G., Johnston, J.H.: Nanoscale Adv. **3**, 3530 (2021)
10. Darby, B.L., Auguié, B., Meyer, M., Pantoja, A.E., Le Ru, E.C.: Nat. Photonics **10**, 1 (2016)
11. Tassan, S., Ferrari, G.M.: Appl. Opt., AO **42**, 4802 (2003)
12. Chess, D.J.: Anal. Biochem. **439**, 161 (2013)
13. Haardt, H., Maske, H.: Limnol. Oceanogr. **32**, 608 (1987)
14. Aduev, B.P., et al.: Tech. Phys. **59**, 1387 (2014)
15. Mori, A., Yamashita, K., Tabata, Y., Seto, K., Tokunaga, E.: Rev. Sci. Instrum. **92**, 123103 (2021)
16. Sun, L., Bolton, J.R.: J. Phys. Chem. **100**, 4127 (1996)
17. Nilsson, A.M.K., Berg, R., Andersson-Engels, S.: Appl. Opt., AO **34**, 4609 (1995)
18. Nelson, N.B., Prézelin, B.B.: Appl. Opt., AO **32**, 6710 (1993)
19. Villanueva, Y., Veenstra, C., Steenbergen, W.: Appl. Opt., AO **55**, 3030 (2016)
20. Drieschner, T., Ostertag, E., Boldrini, B., Lorenz, A., Brecht, M., Rebner, K.: Anal. Bioanal. Chem. **412**(14), 3359 (2020)
21. McKechnie, W.S., Tugcu, N., Kandula, S.: Biotechnol. Prog. **34**(5), 1234 (2018)
22. Kishino, M., Takahashi, M., Okami, N., Ichimura, S.: Bull. Mar. Sci. **37**(2), 634 (1985)
23. Numata, Y., Nair, S.V., Nakagawa, K., Ishino, H., Kobayashi, T., Tokunaga, E.: Chem. Phys. Lett. **601**, 128 (2014)
24. Adimule, V., Suryavanshi, A., Bc, Y., Nandi, S.S.: Macromol. Symp. **392**(1), 2000001 (2020)
25. Mie Scattering Calculator. https://omlc.org/calc/mie_calc.html
26. Welch, A.J.: and Gemert M. Optical-Thermal Response of Laser-Irradiated Tissue. Plenum Press (1995)
27. https://www.global-optosigma.com/jp/category/opt_d/opt_d01.html
28. Qiu, S., Zhang, Q., Lv, W., Wang, T., Zhang, Q., Ma, L.: RSC Adv. **7**, 24551 (2017)
29. Derjaguin, B., Landau, L.: Prog. Surf. Sci. **43**, 30 (1993)
30. Verwey, E.J.W.: J. Phys. Chem. **51**, 631 (1947)
31. Mori, A., Yamashita, K., Tokunaga, E.: Photonics. **10**, 754 (2023)

Large Anomalous Nernst Effect in the Ferromagnetic Fe_3Si Polycrystal

Yangming Wang[1,2], Susumu Minami[1], Akito Sakai[1,2], Taishi Chen[1,2,3], Zili Feng[1,2], Daisuke Nishio-Hamane[2], and Satoru Nakatsuji[1,2,4,5,6](✉)

[1] Department of Physics, University of Tokyo, Bunkyo-ku, Tokyo 113-0033, Japan
[2] Institute for Solid State Physics, University of Tokyo, Kashiwa, Chiba 277-8581, Japan
[3] School of Physics, Southeast University, Nanjing 211189, China
[4] Institute for Quantum Matter and Department of Physics and Astronomy, Johns Hopkins University, Baltimore, MD 21218, USA
[5] CREST, Japan Science and Technology Agency (JST), 4-1-8 Honcho Kawaguchi, Saitama 332-0012, Japan
[6] Trans-scale Quantum Science Institute, University of Tokyo, Bunkyo-ku, Tokyo 113-8654, Japan

satoru@phys.s.u-tokyo.ac.jp

Abstract. The high-throughput calculation predicts that the Fe-based cubic ferromagnet Fe_3Si may exhibit a large anomalous Nernst effect (ANE). Here, we report our experimental observation of the large Nernst coefficient S_{yx} ~2 μV/K and the transverse thermoelectric coefficient $-\alpha_{yx} \sim 3$ $Am^{-1}K^{-1}$ for Fe_3Si polycrystal at room temperature. The large $-\alpha_{yx}$ indicates that the large ANE originates from the intrinsic Berry curvature mechanism. The high Curie temperature of 840 K and the most abundant raw elements of Fe and Si make Fe_3Si a competitive candidate for Nernst thermoelectric generations.

Keywords: Anomalous Nernst effect · Thermoelectrics · Ferromagnet · Polycrystal

1 Introduction

The thermoelectric (TE) effect, converting the heat current into electric energy directly, has a great potential for energy harvesting and heat flow sensors for advanced Internet of Things (IoT) society [1–3]. Ferromagnets can induce transverse thermoelectric voltages, so-called anomalous Nernst effect (ANE), which appears perpendicular to heat flow and magnetization. Recently, the ANE has attracted wide attention owing to its unique advantages for large-area and flexible thermoelectric devices [4,5]. Thanks to recent developments in topological physics, giant ANE and anomalous Hall effect (AHE) enhanced by the large Berry curvature have been discovered in various ferromagnets and even antiferromagnets [6–10]. One of the most attractive candidates is Fe_3X (X = Ga, Al) where the Nernst coefficient S_{yx} reaches up to 6 and 4 μV/K at room temperature in Fe_3Ga and Fe_3Al, respectively [7,8]. The theoretical analysis indicates that such a giant ANE originated from the topological nodal web structure around the Fermi energy E_F.

H. J. Choi et al. (Eds.): APPC 2022, *Proceedings of the 15th Asia Pacific Physics Conference*, pp. 15–22, 2024.
https://doi.org/10.1007/978-981-96-0191-2_2

Here, we focus on a sister compound, cubic $D0_3$ Fe_3Si (Fig. 1(a)) [12]. The large ANE in Fe_3Si is predicted by the high throughput calculation [7]. The Curie temperature (T_C) of Fe_3Si $\sim$ 840 K is higher than that for Fe_3Ga (720 K) and Fe_3Al (600 K) [13–15], which is beneficial for TE application at high temperature. Besides, silicon is the most abundant element in the earth's crust and widely used for industry. However, the experimental report of ANE for Fe_3Si is limited only for thin films and its value is small $|S_{yx}| <\sim 1$ μV/K [16].

In this paper, we report the temperature (T) and magnetic field (B) dependence of the electric and thermoelectric properties for the bulk polycrystalline Fe_3Si. We find a large room-temperature Nernst coefficient S_{yx} $\sim$ 2 μV/K, which is twice larger than the previous report using thin films [16]. We also estimate the transverse thermoelectric coefficient $|\alpha_{yx}|$ and found a large room-temperature value $\sim$ 3 $Am^{-1}K^{-1}$, suggesting the dominant intrinsic contribution. The analysis of the band structure indicates dispersion-less flat bands on the Γ-X line might have significant contribution to the large ANE response.

2 Experiments and Discussion

Polycrystalline Fe_3Si samples were synthesized by the melt cooling method in a mono-arc furnace. As-grown samples were used for all characterization and measurements. The powder X-ray diffraction (XRD) result shows the single phase of the $D0_3$ Fe_3Si with a lattice constant of 5.65 Å. The scanning Electron Microscope-Energy Dispersive X-ray Spectrometry (SEM-EDX) method shows our Fe_3Si is stoichiometric within a few percent resolutions. The bar-shaped samples were used for all the transport properties, including Hall and longitudinal resistivity (ρ_{yx} and ρ_{xx}), Nernst and Seebeck coefficients (S_{yx} and S_{xx}) in a physical properties measurement system (PPMS, Quantum Design) with a thermal transport option (TTO). To remove the longitudinal contributions, ρ_{xx} and S_{xx}, the temperature dependence of ρ_{yx} and S_{xx} were evaluated by symmetrization of the data with the positive and negative field sweeps. Magnetization was measured by a commercial magnetic properties measurement system (MPMS, Quantum Design) with the needle-like sample ($\sim$0.2 mg). For all measurements, no specific orientation was chosen since the poly-crystalline nature could guarantee isotropic transport properties.

The electronic structure of Fe_3Si was obtained by using the OpenMX code [17], where the exchange-correlation functional within the generalized gradient approximation and norm-conserving pseudopotentials were employed [18]. The spin-orbit coupling was induced by using total angular momentum-dependent pseudopotentials. The wave functions were expanded by a linear combination of multiple pseudoatomic orbitals [19]. A set of pseudoatomic orbital basis was specified as Fe5.5-$s3p2d2f1$, Si7.0-$s3p3d2$, where the number after each element stands for the radial cutoff in the unit of Bohr and the integer after s, p, d, and f indicates the radial multiplicity of each angular momentum component. The lattice constant was set to the experimental lattice constant of 5.65 Å. The cutoff energies for a charge density of 500 Ry and a k-point mesh of $36 \times 36 \times 36$ were used.

Figure 1(b) shows the magnetic field dependence of magnetization M at 300 K for Fe_3Si ferromagnet. The saturated magnetization M_s is around 4.6 μ_B/f.u at 300 K, which is comparable to the previous research ($M \sim 4.5\ \mu_B$/f.u) [20]. As predicted by the Slater-Pauling rule, M_s for Fe_3Si is smaller than Fe_3Ga and Fe_3Al owing to the smaller number of valence electrons [7]. Figure 1 (c) and (d) show the B dependence of the Nernst coefficient S_{yx} and the Hall resistivity ρ_{yx}, respectively. Both S_{yx} and ρ_{yx} saturate at $B \sim 0.9$ T. The difference between the saturated magnetic field in magnetization (M) and transport properties (S_{yx} and ρ_{yx}) originates from the demagnetization effect owing to the shape anisotropy. S_{yx} reaches $\sim 2\ \mu$V/K, which is nearly 10 times larger than that of pure iron [21]. This value is also comparable to the recent topological magnets such as kagome metal Fe_3Sn, Fe_3Sn_2 and $TbMn_6Sn_6$ [22–24], suggesting some topological feature in the band structure may also be important in Fe_3Si.

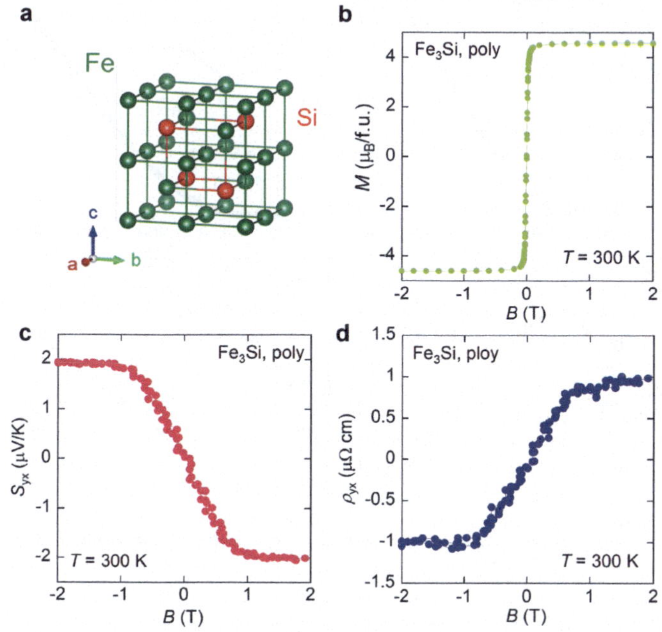

Fig. 1. (Color online) (a) The ordered cubic $D0_3$ structure for Fe_3Si (Space group: $Fm\bar{3}m$) (b)-(d) The magnetic field dependence of the magnetization M (b), the Nernst coefficient S_{yx} (c) and the Hall resistivity ρ_{yx} (d) for polycrystal Fe_3Si at 300 K.

To understand the mechanism of the large S_{yx} in Fe_3Si, we also measure T dependence of both transverse and longitudinal electric (ρ_{yx} and ρ_{xx}) and thermoelectric properties (S_{yx} and S_{xx}) as shown in Fig. 2. On cooling, $-S_{yx}$ peaks at $T \sim 340$ K and then monotonically decreases down to the lowest temperature. $-S_{yx}$ becomes slightly negative below ~ 70 K due to the carrier type change. Similarly, Seebeck coefficient S_{xx} at zero fields also shows a peak around 340 K and monotonically decreases down to ~ 70 K accompanied by the sign change at ~ 160 K as shown in Fig. 2 (c). The sign change of S_{xx} indicates the change of the dominant carrier type from the electron at high T to the hole low T.

On the other hand, ρ_{yx} monotonically decreases down to the lowest temperature (Fig. 2 (b)). Although ρ_{yx} is governed by the AHE at high temperatures, clear B-linear ordinary Hall effect (OHE) contribution appears at low temperatures $T <\sim 100$ K and finally dominates at $T \sim 2$ K as shown in the inset of Fig. 2 (b). As shown in Fig. 2 (d), T dependence of ρ_{xx} exhibits a typical metallic behavior, monotonically decreasing on cooling to the lowest temperature. Although the residual resistivity ratio (RRR) is similar to Fe_3Ga single crystals, the absolute value of ρ_{xx} for Fe_3Si is only about 70% of that of Fe_3Ga [7].

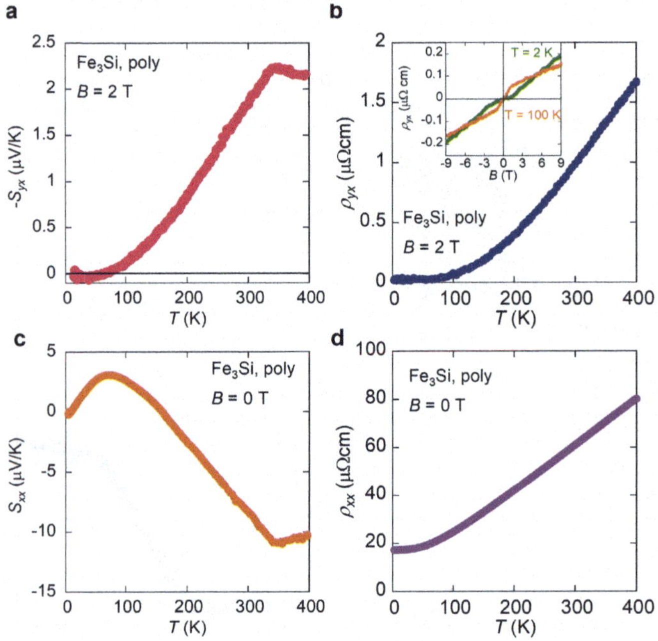

Fig. 2. (Color online) (a, b) The temperature T dependence of the Nernst coefficient -S_{yx} (a) and Hall resistivity ρ_{yx} (b) under the magnetic field of 2 T. The inset of Fig. 2 (b) is the field dependence of ρ_{yx} at 2 K and 100 K. (c, d) The temperature T dependence of the Seebeck coefficient S_{xx} (c) and longitudinal resistivity ρ_{xx} (d) under zero magnetic field.

To check the intrinsic contribution for ANE and AHE, we experimentally estimate the Hall conductivity σ_{yx} and transverse thermoelectric conductivity α_{yx} based on the following formulas,

$$\alpha_{yx} = \frac{1}{\rho_{xx}}[S_{yx} - \frac{\rho_{yx}}{\rho_{xx}}S_{xx}], \tag{1a}$$

$$\sigma_{yx} = -\rho_{yx}/\rho_{xx}^2. \tag{1b}$$

The obtained $-\sigma_{yx}$ and $-\alpha_{yx}$ are shown in Figs. 3 (a) and 3(b), respectively. Here, we only show the data above $\sim$ 100 K since we cannot easily separate the ordinal

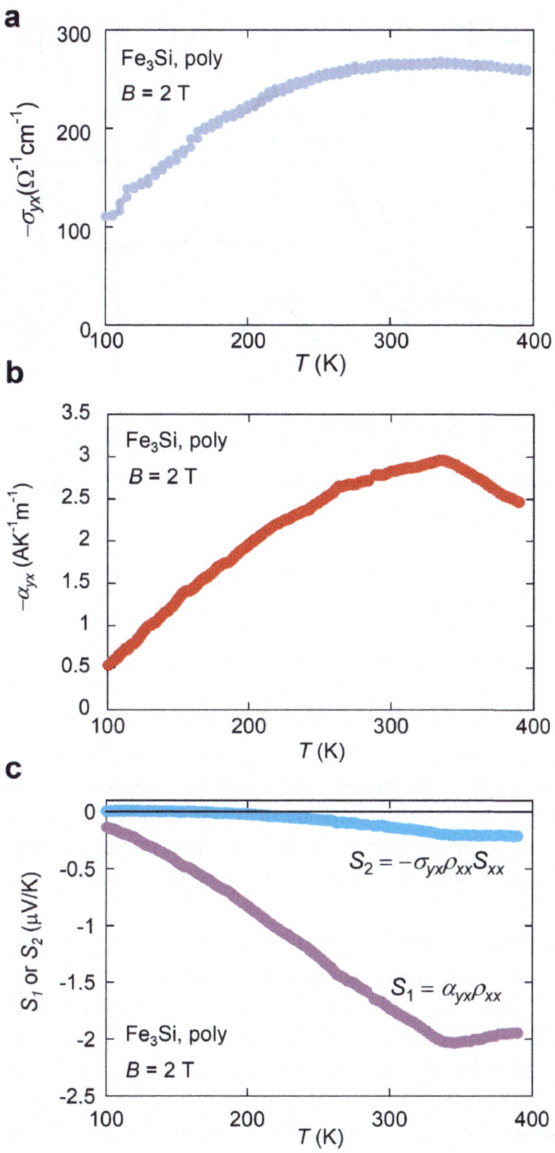

Fig. 3. (Color online) The temperature T dependence of the Hall conductivity $-\sigma_{yx}$ (a), the transverse thermoelectric conductivity $-\alpha_{yx}$ (b), and two contributions S_1 and S_2 to anomalous Nernst effect (c) under the magnetic field of 2 T.

Hall/Nernst contribution at low temperatures as discussed above. As shown in Fig. 3 (a), $-\sigma_{yx}$ shows a broad peak around 300 K and gradually decreases down to 100 K. On the other hand, $-\alpha_{yx}$ increases to 3 $\mathrm{Am^{-1}K^{-1}}$ with a kink at $T \sim 340$ K and then monotonically decreases on cooling.

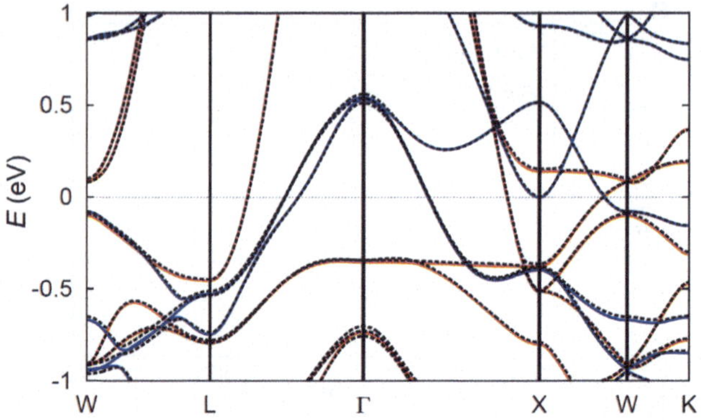

Fig. 4. (Color online) Band structure around the Fermi energy for Fe_3Si obtained from first-principles calculations for the case of magnetization M = 5.1 μ_B/f.u. along [001]. The red and blue lines represent the majority and minority bands without SOC, respectively. The black dotted line corresponds to the bands with SOC.

The Eq. (1a) can be rewritten as,

$$S_{yx} = \alpha_{yx}\rho_{xx} - \sigma_{yx}\rho_{xx}S_{xx} \equiv S_1 + S_2, \tag{2}$$

The second term could be also expressed as S_2 =-tan$(\theta_{AHE})S_{xx}$, where tan (θ_{AHE}) is anomalous Hall angle. The first term S_1 represents the transverse voltage directly driven by the transverse thermoelectric coefficient α_{yx} while the second term S_2 can be regarded as the Hall effect of the current flow generated by the Seebeck effect. As shown in Fig. 3 (c), the contribution from S_2 is almost negligible around room temperature ($\sim$10% of S_1). This indicates that the origin of the large ANE at room T is indeed the large α_{yx}.

In fact, the room temperature $|\alpha_{yx}|$ for Fe_3Si is even larger than that for some topological materials with large ANE, such as Co_2MnGa [$|\alpha_{yx}| \sim$ 2.7 $Am^{-1}K^{-1}$] and Fe_3Sn [$|\alpha_{yx}| \sim$ 2.3 $Am^{-1}K^{-1}$] [9,23]. However, the experimental S_{yx} for Fe_3Si at 300 K is much smaller than Co_2MnGa [$S_{yx} \sim$ 6.5 μV/K]. According to Eq. (2), this difference can be explained by two reasons. Firstly, the ρ_{xx} for Fe_3Si is smaller than those topological semimetals [7,9,23]. The Nernst voltage is the combination of the current flow induced by spin-orbital coupling (SOC) and the material resistance. Thus, the larger the longitudinal resistivity is, the larger ANE becomes if $|\alpha_{yx}|$ is the same. However, We note that the large resistivity suppresses the magnitude of the figure of merit $ZT = \sigma S^2 T/\kappa$, power factor $PF = \sigma S^2$ and the specific power generation capacity $\Gamma_P = P_{max}/\left(A\left(\Delta T\right)^2\right)$ in thermoelectric devices. Therefore, semimetals with a resistivity of 100 $\sim$ 200 $\mu\Omega$ cm at room temperature are more suitable for practical applications. Secondly, the second term S_2 doesn't contribute too much to the total S_{yx} in Fe_3Si. In Co_2MnGa, this contribution accounts for roughly 50 % of the total S_{yx} because of the large Hall angle tan (θ_{AHE}) [9]. This phenomenon is also found in $Co_3Sn_2S_2$ [10,25]. In order to utilize S_2 to improve the total S_{yx}, the tan(θ_{AHE}) should be large ~ 0.1, which is often discovered in topological ferromagnets with Weyl points [26,27].

Figure 4 shows the band structure around the Fermi energy for Fe_3Si obtained from first-principles calculation. Around the L point, nodal web structures similar to Fe_3Ga, composed of minority bands [28], are found around $E = E_F - 0.6$ eV. Unlike Fe_3Ga, the distance of the nodal web from E_F is far to affect the transport properties. It indicates the origin of the large ANE for Fe_3Si might be different. In fact, we also found a flat band-like structure on the Γ-X line. Since topological flat band structures could also be the source of giant ANE and AHE [29,30], a further theoretical investigation is required to reveal the origin of the large transverse thermoelectric coefficient.

3 Conclusions

We have discovered a large ANE in the polycrystalline Fe_3Si, the sister compound of Fe_3Ga and Fe_3Al. The one-step synthesis method and the low material cost make it the most promising material in the Fe_3X system for future applications. In addition, for thin film applications, Fe_3Si could have better compatibility with other silicon-based electronic devices [31]. The electronic band structure for Fe_3Si indicates that the dispersionless band on the Γ-X line might induce a large Berry curvature instead of the nodal web structure like Fe_3Ga. However, the topological electronic structure as an origin of large ANE in Fe_3Si is still an open question. For this purpose, systematic research based on single-crystal Fe_3Si is also expected in the future.

Acknowledgments. This work was partially supported by JST-Mirai Program (JPMJMI20A1), JST-ASPIRE(JPMJAP2317), JST-CREST (JPMJCR18T3), New Energy and Industrial Technology Development Organization (NEDO), and JSPS-KAKENHI (JP19H00650, JP20K22479, JP21J22318, JP22K14587). The work at the Institute for Quantum Matter, an Energy Frontier Research Center was funded by DOE, Office of Science, Basic Energy Sciences under Award # DE-SC0019331. The computations in this research were partially carried out using the Fujitsu PRIMERGY CX400M1/CX2550M5 (Oakbridge-CX) in the Information Technology Center, The University of Tokyo. And the use of the facilities of the Materials Design and Characterization Laboratory at the Institute for Solid State Physics, The University of Tokyo, is gratefully acknowledged.

References

1. Zhang, Q., Deng, K., Wilkens, L., Reith, H., Nielsch, K.: Nat. Electron. **5**, 333 (2022)
2. Jaziri, N., Boughamoura, A., Müller, J., Mezghani, B., Tounsi, F., Ismail, M.: Energy Rep. **6**, 264 (2020)
3. Bell, L.E.: Science **321**, 1457 (2008)
4. Nakatsuji, S., Arita, R.: Ann. Rev. Condens. Matter Phys. **13**, 119 (2022)
5. Uchida, K.-I., Zhou, W., Sakuraba, Y.: Appl. Phys. Lett. **118**, 140504 (2021)
6. Ikhlas, M., et al.: Nat. Phys. **13**, 1085 (2017)
7. Sakai, A., et al.: Nature **581**, 53 (2020)
8. Wang, Y., et al.: Robust giant anomalous Nernst effect in polycrystalline nodal web ferromagnets. Appl. Phys. Lett. **125**, 081901 (2024). https://pubs.aip.org/aip/apl/article/125/8/081901/3309483/Robust-giant-anomalous-Nernst-effect-in

9. Sakai, A., et al.: Nat. Phys. **14**, 1119 (2018)
10. Guin, S.N., et al.: Adv. Mater. **31**, 1806622 (2019)
11. Pan, Y., et al.: Nat. Mater. **21**, 203 (2022)
12. Ma, R., Xie, Q., Huang, J., Yan, W., Guo, X.: J. Alloy. Compd. **552**, 324 (2013)
13. Niculescu, V., Raj, K., Budnick, J., Burch, T., Hines, W., Menotti, A.: Phys. Rev. B **14**, 4160 (1976)
14. Kawamiya, N., Adachi, K., Nakamura, Y.: J. Phys. Soc. Jpn. **33**, 1318 (1972)
15. Shinohara, T.: J. Phys. Soc. Jpn. **19**, 51 (1964)
16. Hamada, Y., Kurokawa, Y., Yamauchi, T., Hanamoto, H., Yuasa, H.: Appl. Phys. Lett. **119**, 152404 (2021)
17. Openmx: Open source package for material explorer. http://www.openmx-square.org/
18. Morrison, I., Bylander, D.M., Kleinman, L.: Phys. Rev. B **47**, 6728 (1993)
19. Theurich, G., Hill, N.A.: Phys. Rev. B **64**, 073106 (2001)
20. Shinjo, T., Nakamura, Y., Shikazono, N.: J. Phys. Soc. Jpn. **18**, 797 (1963)
21. Watzman, S.J., et al.: Phys. Rev. B **94**, 144407 (2016)
22. Zhang, H., Xu, C., Ke, X.: Phys. Rev. B **103**, L201101 (2021)
23. Chen, T., et al.: Science advances 8, eabk1480 (2022)
24. Zhang, H., Koo, J., Xu, C., Sretenovic, M., Yan, B., Ke, X.: Nat. Commun. **13**, 1 (2022)
25. Liu, E., et al.: Nat. Phys. **14**, 1125 (2018)
26. Li, P., et al.: Nat. Commun. **11**, 1 (2020)
27. Armitage, N., Mele, E., Vishwanath, A.: Rev. Mod. Phys. **90**, 015001 (2018)
28. Minami, S., Ishii, F., Hirayama, M., Nomoto, T., Koretsune, T., Arita, R.: Phys. Rev. B **102**, 205128 (2020)
29. Kato, Y.D., et al.: NPJ Quant. Mat. **7**, 1 (2022)
30. Roychowdhury, S., et al.: Adv. Mater. **34**, 2201350 (2022)
31. Hines, W., et al.: Phys. Rev. B **13**, 4060 (1976)

Astrophysics, Cosmology, and Gravitation

Solving Information Loss Paradox via Euclidean Path Integral

Pisin Chen[1,2,3,4], Misao Sasaki[1,5,6], Dong-han Yeom[1,7,8(✉)], and Junggi Yoon[9]

[1] Leung Center for Cosmology and Particle Astrophysics, National Taiwan University, Taipei 10617, Taiwan
pisinchen@phys.ntu.edu.tw, misao.sasaki@ipmu.jp
[2] Department of Physics, National Taiwan University, Taipei 10617, Taiwan
[3] Graduate Institute of Astrophysics, National Taiwan University, Taipei 10617, Taiwan
[4] Kavli Institute for Particle Astrophysics and Cosmology, SLAC National Accelerator Laboratory, Stanford University, Stanford, CA 94305, USA
[5] Kavli Institute for the Physics and Mathematics of the Universe (WPI), University of Tokyo, Chiba 277-8583, Japan
[6] Center for Gravitational Physics, Yukawa Institute for Theoretical Physics, Kyoto University, Kyoto 606-8502, Japan
[7] Department of Physics Education, Pusan National University, Busan 46241, Republic of Korea
[8] Research Center for Dielectric and Advanced Matter Physics, Pusan National University, Busan 46241, Republic of Korea
innocent.yeom@gmail.com
[9] Asia Pacific Center for Theoretical Physics, Pohang 37673, Republic of Korea

Abstract. The information loss paradox associated with black hole Hawking evaporation is an unresolved problem in modern theoretical physics. In this paper, we revisit the entanglement entropy via the Euclidean path integral (EPI) of the quantum state and allow for the branching of semi-classical histories along the Lorentzian evolution. We posit that there exist at least two histories that contribute to EPI, where one is an information-losing history while the other is information-preserving. At early times, the former dominates EPI, while at late times the latter becomes dominant. By so doing we recover the essence of the Page curve and thus the unitarity, albeit with the turning point, i.e., the Page time, much shifted toward the late time. One implication of this modified Page curve is that the entropy bound may thus be violated. We comment on the similarity and difference between our approach and that of the replica wormholes and the island conjecture (Proceedings of the 15th Asia-Pacific Physics Conference, talk on August 23, 2022, Republic of Korea. Based on [1].).

Keywords: information loss paradox · entanglement entropy · Euclidean path integral

1 Introduction

The information loss paradox of black holes [2] is an unresolved problem in modern theoretical physics. This paradox implies a contradiction between general relativity (GR)

H. J. Choi et al. (Eds.): APPC 2022, *Proceedings of the 15th Asia Pacific Physics Conference*, pp. 25–34, 2024.
https://doi.org/10.1007/978-981-96-0191-2_3

and local quantum field theory (QFT) [3,4]. There have been attempts to solve the paradox within GR and QFT. For example, the 'soft hair' proposed by Hawking, Perry and Strominger [5] invokes the BMS symmetry within GR, but it was soon argued that the soft hair cannot carry information [6,7]. The 'firewall' conjecture [4], on the other hand, attempts to solve the paradox by violating GR equivalence principle near the black hole horizon, but was argued to be problematic [8,9]. In order to render the black hole evaporation process unitary, it may be necessary to invoke some unknown new mechanism or a hidden sector [10–12] that lies outside proper GR and QFT. A natural consequence is that such new element must be derived from *quantum gravity*.

It is important to stress that *entanglement entropy* is the physical quantity that measures the information flow from a black hole to its radiation [13]. In this bipartite system of a black hole and its Hawking radiation, the radiation entropy will increase as the evaporation proceeds. On the other hand, the black hole entropy, known as the Bekenstein entropy, which is supposed to describe the black hole's microscopic degrees of freedom, will decrease as the black hole mass decreases. Then in the middle of the process, the two versions of the entropy coincide with each other. Page asserted that the entanglement entropy of the system is approximately the minimum of the two [14]. This curve of the evolution of the entanglement entropy, known as the Page curve, with the turning point, the Page time, occurring at a time when the black hole mass reduces to roughly half, plays an essential role in the investigation of the information loss problem.

In spite of Page's demonstration in a quantum mechanical system, the attempt to derive the Page curve in GR has failed [15], which is often regarded as the deficiency of the semi-classical perturbative calculations in GR. In particular, the decrease of the entanglement entropy after Page time would require non-perturbative effects in quantum gravity beyond our current understanding of GR. One possible circumvention of a full-blown quantum gravity calculation, which does not yet exist, would be via a new classical saddle point, e.g., a soliton, deduced from a valid approximation of quantum gravity, where a semi-classical tunneling around such a new saddle-point, e.g., via instantons, might be able to capture the essence of non-perturbative quantum effects evaluated around the original saddle-point. Under this light, the Page curve would result from nothing but a transition between those two saddles.

Under this philosophy, that there exist two stages in black hole evaporation. First, before a modified Page time, GR and QFT work well and any hidden contributions are negligible. After the modified Page time, however, the hidden contributions are no more negligible. In fact, it must be dominant at late times. Otherwise, the contributions via proper GR and QFT may erase the information.

Inspired by the above thinking, we investigate the information loss paradox via the Euclidean path integral (EPI) approach [16]. The EPI formalism is widely regarded as one of the most promising candidates of quantum gravity that can describe a non-purturbative processes [17]. Though not the final theory, EPI manages to capture the essence of a full-blown quantum gravity theory by dealing with the *entire* wave-function, which includes not only perturbative but also non-perturbative gravity effects [18] via Wheeler-DeWitt equation [19].

2 Essential Conditions

We posit that a successful solution to the black hole information loss paradox must satisfy the following two conditions during the black hole evaporation.
1. Multi-history condition: There exist at least two histories, say h_1 and h_2, that contribute to EPI, where h_1 is an information-losing history while h_2 is an information-preserving history [20–22].
2. Late-time dominance condition: At early times, the probability p_1 of h_1 dominates EPI. At late times, the probability p_2 of h_2 becomes dominant.

Here by "information-losing history" we mean the semi-classical history of an evaporating black hole in which the unitary evolution would be lost when the black hole has completely evaporated. Based on these two conditions, the entanglement entropy S_{ent} can be approximately expressed as (we will prove this equation later)

$$S_{\mathrm{ent}} \simeq p_1 S_1 + p_2 S_2, \tag{1}$$

where $S_{1,2}$ is the entanglement entropy of $h_{1,2}$, respectively. Initially, $p_2 \ll p_1 \simeq 1$ and eventually, $p_1 \ll p_2 \simeq 1$. S_1 increases monotonically or never approaches zero due to the information-losing nature of the history [15], whereas S_2 eventually decreases to zero. Therefore, if these two conditions are satisfied, then the unitary Page curve would be explained, because the entanglement entropy eventually becomes zero again at the end. We will demonstrate that EPI can indeed deliver such a conclusion.

3 Euclidean Path Integral

Let us focus on the following EPI [16]:

$$\langle \Psi_j^{\mathrm{cl}} | \mathrm{in} \rangle = \int \mathcal{D}g \mathcal{D}\phi \; e^{-S_{\mathrm{E}}[g,\phi]}, \tag{2}$$

where $|\mathrm{in}\rangle$ is the initial state, g is the metric, ϕ is a matter field, S_{E} is the Euclidean action, and we sum over all histories that connect $|\mathrm{in}\rangle$ with a classical boundary $|\Psi_j^{\mathrm{cl}}\rangle$. The final out-state can therefore be expressed as a superposition of many classical boundaries [23, 24]:

$$|\mathrm{out}\rangle = \sum_j a_j |\Psi_j^{\mathrm{cl}}\rangle, \tag{3}$$

where $a_j = \langle \Psi_j^{\mathrm{cl}} | \mathrm{in} \rangle$ is the coefficient of a specific classical future boundary.

Let h_1 be the usual history with a semi-classical black hole geometry (left of Fig. 1), while h_2 is an information-preserving geometry due to the non-perturbative tunneling (right of Fig. 1). Under normal situations the information-preserving geometry is exponentially suppressed. We will show that such a history becomes dominant at late times. Since tunneling can happen at any time, the histories will continue to branch out. Eventually, infinitely many histories appear in the out-state, while the in-state is fixed (Fig. 2) [23, 24].

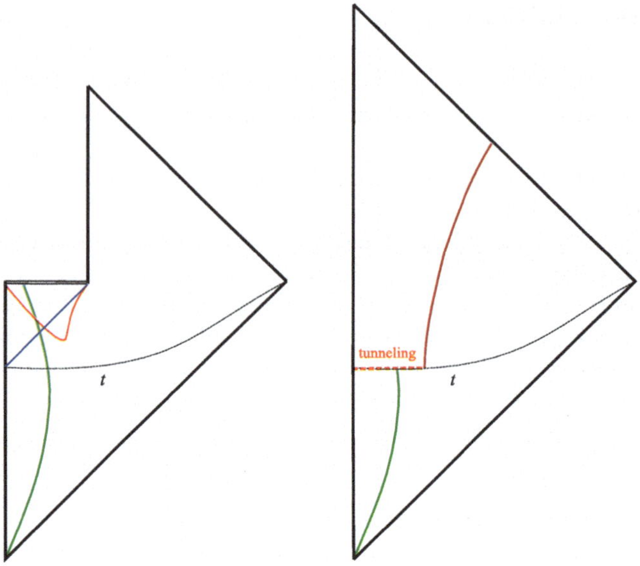

Fig. 1. Left: the causal structure of the usual semi-classical black hole, where the green curve is the trajectory of the collapsing matter, the red curve is the apparent horizon, and the blue line is the event horizon. Right: the causal structure after a quantum tunneling at the time slice t. After the tunneling, matter or information (red curve) is emitted and the black hole structure disappears.

4 Thermal Thin-Shell Model

In order to realize the information-preserving history in the EPI approach, we introduce the thin-shell model [25] without black hole interiors. We consider a spacetime with spherically symmetric metric: $ds_{\pm}^2 = -f_{\pm}(R)dT^2 + f_{\pm}^{-1}(R)dR^2 + R^2 d\Omega^2$. Let a thin-shell be located at r, which follows the metric $ds^2 = -dt^2 + r^2(t)d\Omega^2$. The region outside the shell corresponds to $r < R$ (denoted by $+$), and that inside the shell is $R < r$ (denoted by $-$). We impose the metric ansatz for outside and inside the shell, $f_{\pm}(R) = 1 - 2M_{\pm}/R$, where M_+ and M_- are the mass parameters of each region. We assume $M_+ > M_- = 0$, and hence the inside has no black hole.

The equation of motion of the thin-shell is determined by the junction equation [25]:

$$\epsilon_- \sqrt{\dot{r}^2 + f_-(r)} - \epsilon_+ \sqrt{\dot{r}^2 + f_+(r)} = 4\pi r \sigma(r). \tag{4}$$

Here, we impose $\epsilon_{\pm} = +1$. $\sigma(r)$ is the tension parameter, where it satisfies the energy conservation relation

$$\sigma' = -\frac{2\sigma\,(1+w)}{r}, \tag{5}$$

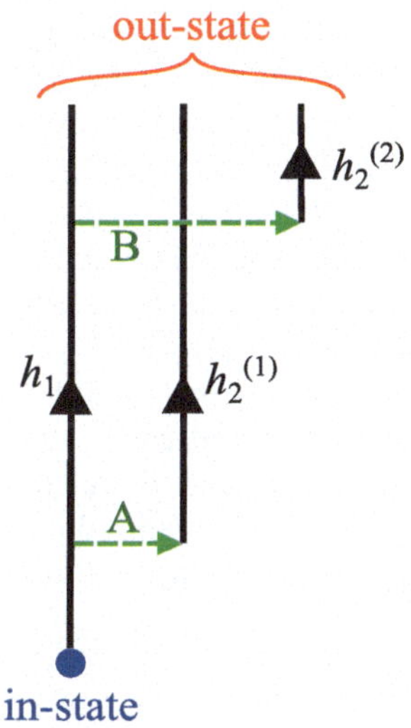

Fig. 2. Conceptual figure for interpretations. h_1 is the information-losing history, while $h_2^{(1,2)}$ are the information-preserving history. For any time, histories can be branched; the tunneling probability must be dominated at the late time.

where w is the equation of state of the shell. $w \geq -1$ is necessary for the null energy condition. In general, we assume the following form of the tension [24]

$$\sigma(r) = \sum_{i=1}^{n} \frac{\sigma_i}{r^{2(1+w_i)}}, \tag{6}$$

where σ_i and w_i are constants.

The conditions for the existence of thermal thin-shell solutions are as follows: there exists a value r_0 such that $\dot{r}|_{r=r_0} = 0$. This is possible, for example, if one chooses $w_1 = 0.5$, $w_2 = 1$, and tune $\sigma_{1,2}$. This ensures the asymptotic flatness of the spacetime, and can thus mimic causal structures of both perturbative and non-perturbative processes due to Hawking radiation.

5 Late-Time Dominance Condition

For each on-shell history of EPI, the nucleation rate is $\Gamma \propto e^{-2B}$, where

$$B = S_{\mathrm{E}}\,(\text{solution}) - S_{\mathrm{E}}\,(\text{background})\,. \tag{7}$$

Due to the thermal condition, the shell is stationary in Euclidean signatures. Hence the Euclidean action is simply

$$S_{\mathrm{E}} = -\int \sqrt{+g}d^4x \left(\frac{\mathcal{R}}{16\pi} - V\right) + \sigma \int_{\Sigma} \sqrt{+h}d^3x - \int_{\partial\mathcal{M}} \frac{\mathcal{K} - \mathcal{K}_o}{8\pi}\sqrt{+h}d^3x, \tag{8}$$

where $\mathcal{R}$ is the Ricci scalar, V is the vacuum energy of the matter fields, Σ is the hypersurface of the thin-shell, and $\mathcal{K}$ and $\mathcal{K}_o$ are the Gibbons-Hawking boundary terms at infinity for the solution and the Minkowski background, respectively [26]. After straightforward computations, we obtain $2B = 4\pi M_+^2$ [27].

If the shell is static, then the solution has two parts, one from the bulk integration, $4\pi M_+^2$, and the other from the boundary term at infinity, $4\pi M_+^2$. The last term in Eq. (8) must be cancelled between the solution of the boundary term at infinity and that of the background. Eventually, one obtains the result $2B = 4\pi M_+^2$.

There is an important point in this computation that we like to emphasize. In general, the Euclidean time of the background cannot be chosen freely. The Euclidean periodicity of the solution, on the other hand, is free to adjust. In principle, as long as the final boundary remains the same, one can introduce an arbitrary periodicity for the Euclidean time. One then obtains the following factorization:

$$\begin{aligned} 2B = n\,(&(\text{bulk term of solution}) \\ &+ (\text{boundary term of solution})) \\ &-(\text{boundary term of background}), \end{aligned} \tag{9}$$

where $n \geq 1$. Interestingly, $n = 1$ is the most dominant contribution (one can thus ignore higher terms in general), but the subdominant contributions become important as M_+ decreases. The tunneling probability is therefore

$$\sum_{n=1}^{\infty} e^{-S(2n-1)} = \frac{1}{e^S - e^{-S}}, \tag{10}$$

where $S = 4\pi M_+^2$.

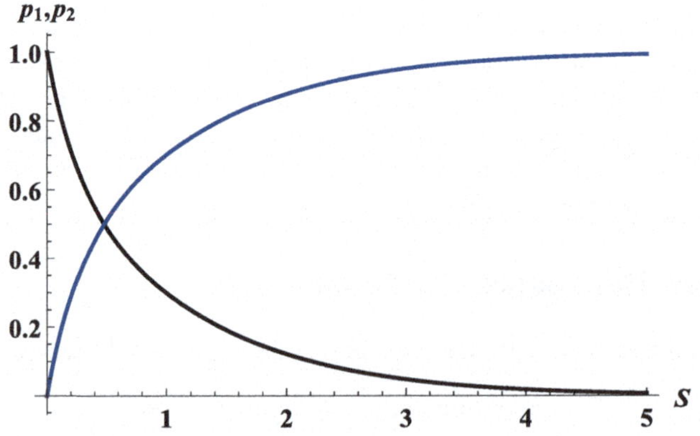

Fig. 3. Probabilities of the semi-classical history (p_1, blue curve) and histories with thermal-shell emission (p_2, black curve) as a function of the entropy S. For large S, p_1 is dominated; however, there exists a golden cross between two probabilities, and eventually, p_2 is dominated.

If one considers only two histories, where one is the semi-classical black hole and the other is the black hole with a thermal-shell emission, then the probability of the

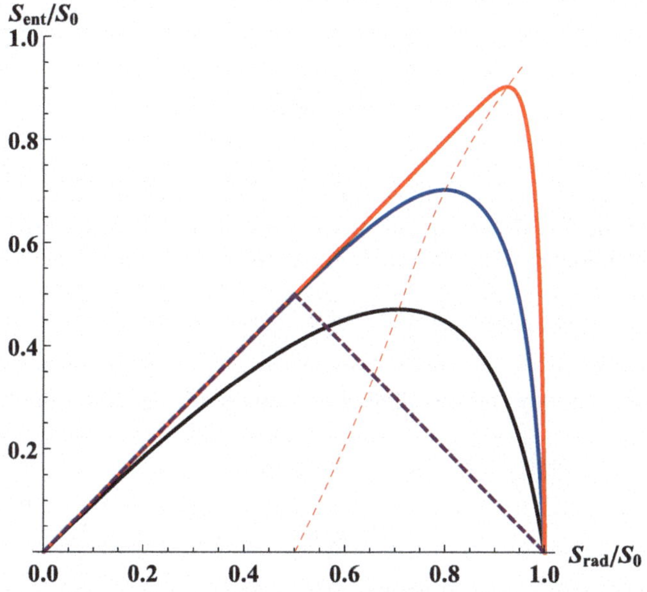

Fig. 4. Entropy of emitted radiation $S_{\text{rad}}/S_0 \equiv 1 - S/S_0$ vs. entanglement entropy S_{ent}/S_0, where $S_0 = 3$ (black), 10 (blue), 50 (red) are initial black hole entropies, respectively. The purple dashed curve is the usually expected Page curve, while there exists a region where the curve is outside the purple dashed triangle region. The thin red dashed curve is the location of the Page time, i.e., $dS_{\text{ent}}/dS_{\text{rad}} = 0$.

history, p_1 and p_2, respectively, are as follows:

$$p_1 = \frac{e^S - e^{-S}}{1 + e^S - e^{-S}}, \quad p_2 = \frac{1}{1 + e^S - e^{-S}}. \tag{11}$$

Interestingly, there is a golden-cross point near $S \simeq 1$ (Fig. 3). Such a golden-cross is actually generic. That is, even if one does not invoke multiple Euclidean time-periods, one can still obtain the same qualitative result of such a crossing, e.g., by assuming the time-accumulation, etc.

6 Modification of the Page Curve

A generic quantum state with two classical histories can be described as follows: $|\psi\rangle = c_1|\psi_1\rangle + c_2|\psi_2\rangle$, where 1 and 2 denote two different histories and $c_{1,2}$ are complex coefficients. The total density matrix ρ is

$$\rho = \begin{pmatrix} |c_1|^2 & c_1^* c_2 \\ c_1 c_2^* & |c_2|^2 \end{pmatrix}. \tag{12}$$

We assume that after averaging over a long time, the off-diagonal terms will become less dominated; this is in accordance with the decoherence condition. With this assumption, one can write

$$\rho \simeq \begin{pmatrix} p_1 & 0 \\ 0 & p_2 \end{pmatrix}, \tag{13}$$

where $p_1 = |c_1|^2$ and $p_2 = |c_2|^2$.

Now let us assume that for each $|\psi_1\rangle$ and $|\psi_2\rangle$, one can define subsystems A and B, respectively. Physically, A corresponds to the subsystem outside the horizon, while B is that inside the horizon. (If there is no horizon, then B is empty.) One can then write

$$|\psi_\nu\rangle = \sum_i a_{\nu,i} |A_i^{(\nu)}\rangle |B_i^{(\nu)}\rangle \qquad (\nu = 1, 2)\,. \tag{14}$$

The entanglement entropy should follow the Shannon information formula:

$$S\,(A|B) = \sum_{\nu,i} p_\nu |a_{\nu,i}|^2 \log p_\nu |a_{\nu,i}|^2 \tag{15}$$

$$= p_1 S_1 \left(A^{(1)}|B^{(1)}\right) + p_2 S_2 \left(A^{(2)}|B^{(2)}\right) + p_1 \log p_1 + p_2 \log p_2, \tag{16}$$

where the last two terms are negligible if the number of degrees of freedom of the system is much greater than 2. This proves $S_{\text{ent}} \simeq p_1 S_1 + p_2 S_2$.

We can finally coin the entanglement entropy as a function of S as

$$S_{\text{ent}} = (S_0 - S) \times p_1 + 0 \times p_2 \tag{17}$$

$$= (S_0 - S) \left(\frac{e^S - e^{-S}}{1 + e^S - e^{-S}} \right), \tag{18}$$

where S_0 is the initial entropy of the black hole. Here, for simplicity we assume that the entanglement entropy p_1 monotonically increases, while the entanglement entropy p_2 goes to zero (because there is no black hole at the end, Fig. 4). This preserves the unitary of the black hole evolution.

Two salient feature of our new result is at hand. First, the Page time (thin red dashed curve in Fig. 4) deviates from Page's original value. In our theory it is located much later than the half-way point of the radiation entropy. Second, at late times, the entanglement entropy can become much greater than what expected in the conventional Page curve (purple dashed curve in Fig. 4). This implies that there exists a situation where the entanglement entropy is greater than that of the areal entropy. Thus the proposed monster state or the remnant-like state [11,28] seems to be realized. However, this is not so surprising if the total number of states varies as time goes on.

7 Conclusion

In this paper, we demonstrate the essence of the Page curve through the EPI approach. This approach can explain the existence of the information-preserving history, as well as the late time dominance condition. The original Page curve, however, is modified. One implication is that the entropy bound may be violated and at the end of the evaporation a monster or remnant state [28] would seem permissible.

At this point, it is worthwhile to compare our result with the recent remarkable developments in string theory about the information loss paradox. The basic idea stems from the gravitational fine-grained entropy that can be computed by quantum extremal surfaces (QES) [29], i.e., the global minimum among saddles upon extremizing the generalized entropy. It was shown [30–33] that in addition to the conventional saddle, which reproduces Hawking's result, there exists another saddle that induces an *island* region inside the black hole horizon. This second saddle with an island dominates the black hole evolution after the Page time and leads to the Page curve. It has been shown that such a contribution from the new saddle can be reproduced by the replica wormhole [34,35].

Although our interpretation of the thin-shell tunneling shares common features with the island conjecture and the replica wormhole, there exists a clear difference. The replica wormhole is based on Euclidean path integral whereas our approach deals with branching of semi-classical histories along the Lorentzian evolution. In this sense, our interpretation of thin-shell tunneling may be more closely connected with the recent studies in the context of real-time gravitational replica wormholes [36,37] for the generalization of the island conjecture and the replica wormholes with baby universes [38–40]. More detailed comparisons of our approach with the replica wormhole approach, either Euclidean or Lorentzian, are left for future investigations.

Acknowledgments. PC is supported by Taiwan's Ministry of Science and Technology (MOST) and the Leung Center for Cosmology and Particle Astrophysics (LeCosPA), National Taiwan University. MS is supported in part by JSPS KAKENHI grant Nos. 19H01895, 20H04727, and 20H05853. DY is supported by the National Research Foundation of Korea (Grant no.: 2021R1C1C1008622, 2021R1A4A5031460). JY is supported by the National Research Foundation of Korea (Grant no.: 2019R1F1A1045971). JY is also supported by an appointment to the JRG Program at APCTP through the Science and Technology Promotion Fund, the Lottery Fund of the Korean government, and by Gyeongsangbuk-do and Pohang-si.

References

1. Chen, P., Sasaki, M., Yeom, D., Yoon, J.: arXiv:2206.10251 [gr-qc]
2. Hawking, S.W.: Commun. Math. Phys. **43**, 199 (1975). Erratum: [Commun. Math. Phys. **46**, 206 (1976)]
3. Yeom, D., Zoe, H.: Int. J. Mod. Phys. A **26**, 3287–3314 (2011). arXiv:0907.0677 [hep-th]
4. Almheiri, A., Marolf, D., Polchinski, J., Sully, J.: JHEP **02**, 062 (2013). arXiv:1207.3123 [hep-th]
5. Hawking, S.W., Perry, M.J., Strominger, A.: Phys. Rev. Lett. **116**(23), 231301 (2016). arXiv:1601.00921 [hep-th]
6. Bousso, R., Porrati, M.: Class. Quant. Grav. **34**(20), 204001 (2017). arXiv:1706.00436 [hep-th]
7. Giddings, S.B.: Phys. Rev. D **100**(12), 126001 (2019). arXiv:1903.06160 [hep-th]
8. Chen, P., Ong, Y.C., Page, D.N., Sasaki, M., Yeom, D.: Phys. Rev. Lett. **116**(16), 161304 (2016). arXiv:1511.05695 [hep-th]
9. Hwang, D., Lee, B.H., Yeom, D.: JCAP **01**, 005 (2013). [arXiv:1210.6733 [gr-qc]]

10. Ong, Y.C., Yeom, D.: Proceedings of the Second LeCosPA Symposium. arXiv:1602.06600 [hep-th]
11. Chen, K.Y., Chen, P., Chiang, H.W., Yeom, D.: arXiv:2108.03593 [hep-th]
12. Hotta, M., Nambu, Y., Yamaguchi, K.: Phys. Rev. Lett. **120**(18), 181301 (2018). arXiv:1706.07520 [gr-qc]
13. D. N. Page, Phys. Rev. Lett. **71**, 3743–3746 (1993). arXiv:hep-th/9306083 [hep-th]
14. Page, D.N.: Phys. Rev. Lett. **71**, 1291–1294 (1993). arXiv:gr-qc/9305007 [gr-qc]
15. Hwang, J., Park, H., Yeom, D., Zoe, H.: J. Korean Phys. Soc. **73**(10), 1420–1430 (2018). arXiv:1712.00347 [hep-th]
16. Hartle, J.B., Hawking, S.W.: Phys. Rev. D **28**, 2960 (1983). [Adv. Ser. Astrophys. Cosmol. **3**, 174 (1987)]
17. Gibbons, G.W., Hawking, S.W.: Euclidean quantum gravity. World Scientific (1993)
18. Chen, P., Sasaki, M., Yeom, D.: Eur. Phys. J. C. **79**(7), 627 (2019). arXiv:1806.03766 [hep-th]
19. DeWitt, B.S.: Phys. Rev. **160**, 1113–1148 (1967)
20. Maldacena, J.M.: JHEP **0304**, 021 (2003). arXiv:hep-th/0106112
21. Hawking, S.W.: Phys. Rev. D **72**, 084013 (2005). [arXiv:hep-th/0507171]
22. Sasaki, M., Yeom, D.: JHEP **1412**, 155 (2014). arXiv:1404.1565 [hep-th]
23. Hartle, J., Hertog, T.: Phys. Rev. D **92**(6), 063509 (2015). arXiv:1502.06770 [hep-th]
24. Chen, P., Domènech, G., Sasaki, M., Yeom, D.: JCAP **04**, 013 (2016). [arXiv:1512.00565 [hep-th]]
25. Israel,W.: Nuovo Cim. B **44S10** (1966). 1 [erratum: Nuovo Cim. B **48** (1967), 463]
26. Garriga, J., Megevand, A.: Int. J. Theor. Phys. **43**, 883 (2004). [hep-th/0404097]
27. Chen, P., Domènech, G., Sasaki, M., Yeom, D.: JHEP **1707**, 134 (2017). arXiv:1704.04020 [gr-qc]
28. Chen, P., Ong, Y.C., Yeom, D.: Phys. Rept. **603**, 1–45 (2015). arXiv:1412.8366 [gr-qc]
29. Engelhardt, N., Wall, A.C.: JHEP **01**, 073 (2015). arXiv:1408.3203 [hep-th]
30. Penington, G.: JHEP **09**, 002 (2020). arXiv:1905.08255 [hep-th]
31. Almheiri, A., Engelhardt, N., Marolf, D., Maxfield, H.: JHEP **12**, 063 (2019). arXiv:1905.08762 [hep-th]
32. Almheiri, A., Mahajan, R., Maldacena, J., Zhao, Y.: JHEP **03**, 149 (2020). arXiv:1908.10996 [hep-th]
33. Bak, D., Kim, C., Yi, S.H., Yoon, J.: JHEP **01**, 155 (2021). arXiv:2006.11717 [hep-th]
34. Penington, G., Shenker, S.H., Stanford, D., Yang, Z. arXiv:1911.11977 [hep-th]
35. Almheiri, A., Hartman, T., Maldacena, J., Shaghoulian, E., Tajdini, A.: JHEP **05**, 013 (2020). arXiv:1911.12333 [hep-th]
36. Colin-Ellerin, S., Dong, X., Marolf, D., Rangamani, M., Wang, Z.: JHEP **05**, 117 (2021). arXiv:2012.00828 [hep-th]
37. Colin-Ellerin, S., Dong, X., Marolf, D., Rangamani, M., Wang, Z.: arXiv:2105.07002 [hep-th]
38. Marolf, D., Maxfield, H.: JHEP **08**, 044 (2020). arXiv:2002.08950 [hep-th]
39. Giddings, S.B., Turiaci, G.J.: JHEP **09**, 194 (2020). arXiv:2004.02900 [hep-th]
40. Marolf, D., Maxfield, H.: JHEP **04**, 272 (2021). arXiv:2010.06602 [hep-th]

Gravitational Wave Signatures of Hybrid Defects

David I. Dunsky[1,2], Anish Ghoshal[3], Hitoshi Murayama[1,4,5], Yuki Sakakihara[6(✉)], and Graham White[5]

[1] Berkeley Center for Theoretical Physics, University of California, Berkeley, CA 94720, USA
[2] Theoretical Physics Group, Lawrence Berkeley National Laboratory, Berkeley, CA 94720, USA
[3] Institute of Theoretical Physics, Faculty of Physics, University of Warsaw, ul. Pasteura 5, 02-093 Warsaw, Poland
[4] Theory Group, Lawrence Berkeley National Laboratory, Berkeley, CA 94720, USA
[5] Kavli IPMU (WPI), UTIAS, The University of Tokyo, Kashiwa, Chiba 277-8583, Japan
[6] Graduate School of Science and Engineering, Yamagata University, Yonezawa, Yamagata 992-8510, Japan

yuki.sakakihara@yz.yamagata-u.ac.jp

Abstract. The breakings of a large gauge symmetry into a smaller gauge symmetry through an intermediate stage with discrete symmetry can lead to the formation of string-domain wall hybrid defects in the early universe. The gravitational wave spectrum from string-domain wall system has significant differences from the normal string spectrum, a steep decay in low frequencies by the destruction of the string network by the domain walls and furthermore an enhancement in the amplitude before the decay if domain wall energy is saturated at the domain wall formation. We show the consistency between the condition for the domain wall saturation and that for the gravitational wave enhancement. We evaluate the degree of the enhancement of the gravitational wave amplitude and the peak frequency in a large parameter space. The saturation does not require fine-tuning in the energy scales of the symmetry breakings.

Keywords: gravitational waves · hybrid defects · cosmic strings · domain walls · grand unified models

1 Introduction

Many spontaneous symmetry breaking paths in grand unified theory including right handed neutrinos give hybrid-type topological defects, i.e., hybrid defects. There are two kinds of hybrid defects, one is cosmic strings bounded by monopoles and the other is domain walls bounded by cosmic strings [1–7]. There are two processes giving each configuration depending on when inflation occurs [8]. For domain walls bounded by strings, if inflation occurs earlier than the string formation, a classical process becomes important, where the strings become the boundary of the domain walls at the domain wall formation [3–5]. If inflation occurs after the string formation, the strings go beyond

H. J. Choi et al. (Eds.): APPC 2022, *Proceedings of the 15th Asia Pacific Physics Conference*, pp. 35–43, 2024.
https://doi.org/10.1007/978-981-96-0191-2_4

the horizon [9, 10] and the domain walls form and evolve as normal domain walls without string boundaries. However, the domain walls can be broken by the formation of string holes on them through the tunneling process [1–4, 11–15]. The authors recently found these four types of hybrid defects are distinguishable from each other by combining gravitational wave observations in high frequencies and in low frequencies [16]. It should be noted that these gravitational waves [16–25] can be distinguished from other gravitational wave sources, e.g., first-order phase transition [26, 27]. These gravitational wave signals can be used to probe high energy physics, e.g., right-handed neutrinos heavier than 10^9 GeV, which are favourable for the explanation of the smallness of the neutrino masses and of the baryon anti-baryon asymmetry [28–31].

In this paper, we focus on domain walls bounded by cosmic strings forming through the classical process, led by a symmetry breaking path

$$G \to H \times Z_2 \to H \,, \tag{1}$$

where G and H are gauge groups and $\pi_0(G/H) = I$ [8]. Z_2 can be D-parity in grand unification [2–4] or R-parity as a remnant of gauged $U(1)_{B-L}$ [32–37]. The advantage of this scenario is that it predicts a distinguishable gravitational wave signal from the standard string and domain wall spectra in a large parameter space, while hybrid defects forming through the quantum process requires the coincidence between the energy scales of the string formation and the domain wall formation because of the exponential suppression of the quantum process. We confirm the consistency of the gravitational wave enhancement, leading to a "bump" in gravitational wave spectrum, and the domain wall domination for string loops. We evaluate how much the gravitational wave signal are enhanced at the certain frequency in a large parameter space of the symmetry breaking energy scales. We also discuss how much the frequency where the bump appears can be lowered depending on the temperatures of two phase transitions.

2 Features of Gravitational Waves from String-Domain Wall System

We examine the scenario that strings form first and later they become the boundary of domain walls. We discuss the relation between the formation and the dynamics of the string-domain wall system and the features of the gravitational wave spectrum from the system. We are based on the Nambu-Goto description.

After the string formation at $T \sim v_1$, $t_1 \sim M_{\rm pl}/v_1^2$, the domain wall formation happens at $T \sim v_2$,

$$t_2 \sim \frac{3\sqrt{5}M_{\rm pl}}{4\pi^{3/2}\sqrt{g_*}v_2^2} \,, \tag{2}$$

where we assume $g_* = 100$. We have another time scale where the domain wall energy dominate over the string energy of string-domain wall system for the string network, i.e.,

$$t_{\rm D} \sim \frac{R_c}{\xi} = \frac{2v_1^2}{\xi v_2^3} \,, \tag{3}$$

where

$$R_c = \frac{2v_1^2}{v_2^3}\,, \tag{4}$$

[4,5] and $\xi = 0.27$ is the correlation length divided by the cosmic time [38]. We also have a similar but slightly different time scale defined by the domain wall energy domination over the string energy at the size of loops leaving from the string network,

$$t_{\rm L} \sim \frac{R_c}{\alpha} = \frac{2v_1^2}{\alpha v_2^3}\,, \tag{5}$$

where $\alpha = 0.1$ is the initial loop size divided by the cosmic time [39,40]. Note that $t_{\rm L} \gtrsim t_{\rm D}$. Depending on which time scales of t_2 and t_D is larger than the other, there are two regimes in the formation and the evolution of string-domain wall system. In the first case where $t_2 < t_D$, the domain wall forms after the formation of the strings but the domain wall is less energetic than the string bounding the domain wall until $t = t_D$. We expect that the behaviour of the string-domain wall system is similar to the normal string system for $t < t_D$, leaving a flat gravitational wave spectrum in high frequencies as is in the normal string system. After $t = t_D$, the string-domain wall network quickly decays because the domain wall tension dominates over the string tension. Because of the quick decay of the string-domain wall network, the gravitational wave spectrum decays as $\Omega_{\rm GW} \propto f^3$, derived from the causality [41,42], in the low frequencies much steeper than that of the normal string system. The signal can be further different in the second case where $t_2 > t_D$. In this case, the behaviour of the string network is the same as that of the normal string until the formation of the domain wall at $t = t_2$. However, the domain wall tension is "already" greater than the string tension at the domain wall formation and the string network quickly decays just after $t = t_2$. Furthermore, if the domain wall tension dominates over the string tension for the loops forming at $t = t_2$, which happens if and only if $t_{\rm L} < t_2$, the string-domain wall segments becomes more massive than that of the normal string loops. This results in the enhancement of the gravitational wave signal by

$$\Omega_{\rm GW} \sim \frac{1}{2\pi}\sqrt{\frac{\Gamma\alpha t_2}{2R_c}}\,\Omega_{\rm GW}^{\rm (n.s.)}\,, \tag{6}$$

where $\Gamma = 50$ is the gravitational radiation power from a normal string loop in $G\mu \sim Gv_1^2$, $\Omega_{\rm GW}^{\rm (n.s.)}$ is the gravitational wave amplitude from normal string loops, for

$$f \sim \frac{1}{\alpha t_2}\left(\frac{g_{*S0}^{1/3} T_0}{g_{*S}^{1/3} T_{\rm dec}}\right) \tag{7}$$

where g_{*S} and $T_{\rm dec}$ are the effective number of relativistic degree of freedom for entropy and the temperature at

$$t_{\rm dec} \sim \frac{1}{G\sigma}\,, \tag{8}$$

which is the time scale of the decay of the string-domain wall segment [16]. Similarly, $g_{*S0} = 3.90$ and $T_0 = 2.73\text{K}$ are these values of the present time. The time scale is evaluated by the gravitational wave emission power from the string-domain wall segment,

$$P_{\text{GW}} \sim G\omega^2 \times (\sigma l^2)^2 = G\sigma^2 l^2 \,, \tag{9}$$

where we assume $\omega \sim 1/l$, which is a consequence of relativity and confirmed by numerically solving the equations of motion for a string-domain wall segment in Nambu-Goto description [8, 16, 43], and thus,

$$t_{\text{dec}} \sim \frac{E_{\text{seg.}}}{P_{\text{GW}}} \sim \frac{\sigma l^2}{G\sigma^2 l^2} = \frac{1}{G\sigma} \,. \tag{10}$$

Since the string network vanishes and stops producing new string loops just after $t = t_2$, $\Omega_{\text{GW}} \propto f^3$ in the low frequencies irrespective of whether $t_2 < t_{\text{L}}$ or $t_{\text{L}} < t_2$. The point is that the f^3 decay in the low frequency band happens because of the sudden vanishment of the string-domain wall network, but the bump-like feature appears because the domain wall energy dominates over the string energy for the string loops leaving from the network at the formation of the domain wall.

3 Bump Appearing Condition

In Fig. 1, we show the energy dependence of the time scales determining the shape of the gravitational wave spectrum from the string-domain wall system. We in principle have an enhancement of the gravitational wave signature if the domain wall formation occurs while the domain wall energy is dominant for the string loops, i.e., string-domain wall segments. For the same value for the first phase transition temperature, where the strings form, the domain wall formation is late enough if the second phase transition temperature is high enough. If we vary the first phase transition temperature, the lower the temperature is, the broader the parameter region accommodating the bump becomes.

In Fig. 2, we show the parameter region where the bump-like feature appears in the gravitational wave spectrum on v_1-v_2 parameter space. We also show the regions where the domain wall energy dominates over the string energy for loops and for the network at the domain wall formation. The first two regions almost coincide with each other, confirming our scenario that the formation of massive domain wall for the loops injects additional gravitational wave energy compared to the normal string case, leaving a bump-like feature of the gravitational wave spectrum.

4 Peak Height and Frequency

In Fig. 3, we show how much the gravitational wave amplitude is enhanced compared to the normal string spectrum amplitude,

$$\frac{\Omega_{\text{GW}}}{\Omega_{\text{GW}}^{(\text{n.s.})}} \sim 3.4 \times 10^7 \sqrt{\frac{v_2}{10^1\text{GeV}}} \left(\frac{10^1\text{GeV}}{v_1}\right) , \tag{11}$$

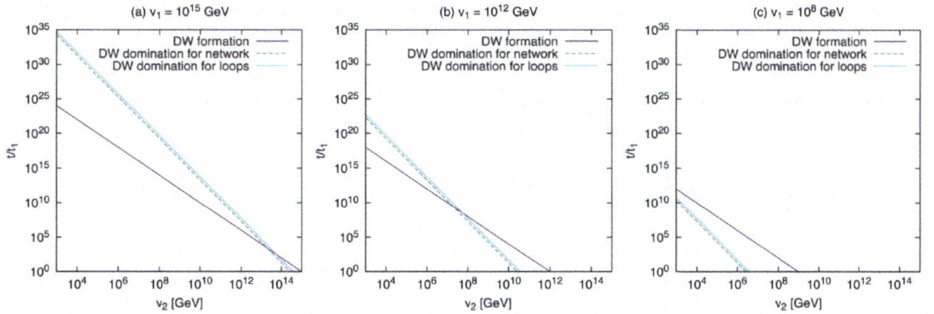

Fig. 1. Time scales determining the gravitational wave spectrum features. The purple line indicates the time of the domain wall formation. Green broken line indicates that the domain wall energy dominates over the string energy for the string(-domain wall) network later than the time. Blue line indicates that the domain wall energy dominates over the string energy for string(-domain wall) segments leaving from the network later than the time. The times scales are normalized by the time when the strings form.

at the frequency (7) for each mass parameter. The lower the temperatures are, the more significant the enhancement is. The amplitude is independent of the temperature of the first phase transition,

$$\Omega_{\rm GW} \sim 6.4 \times 10^{-16} \sqrt{\frac{v_2}{10^1 {\rm GeV}}} \, . \tag{12}$$

In Fig. 4, we show the peak frequency for $t_{\rm L} < t_2$ and the frequency of the turning point in the spectrum for $t_2 < t_{\rm L}$ for different energy scales of the first and the second phase transitions. For $v_1 < 7.04 \times 10^8$ GeV, the bump appears in the spectrum for any $v_2 > 10$ GeV as indicated by the solid line. Even for higher first phase transition temperatures, the bump appears for higher v_2 along the solid line until the intersection of the solid line and the dashed line, while the bump does not appear for lower v_2 and we indicate the frequency of the turning point from the flat string spectrum to f^3 in low frequencies by the dashed line. The intersection point is for instance $v_2 = 2.0 \times 10^7$ GeV for $v_1 = 10^{12}$ GeV. The frequency of the peak is evaluated as

$$f \sim 2.1 \times 10^5 \sqrt{\frac{v_2}{10^1 {\rm GeV}}} \, [{\rm Hz}] \, . \tag{13}$$

5 Discussion and Conclusion

In this paper, we discuss the string-domain wall system emanating from the process that the cosmic string formation is followed by the domain wall formation and the strings become the boundary of the domain walls. Such a configuration can be formed in sequences of symmetry breakings from a larger symmetry group to a smaller symmetry group if they include an intermediate stage with Z_2 discrete symmetry and the discrete

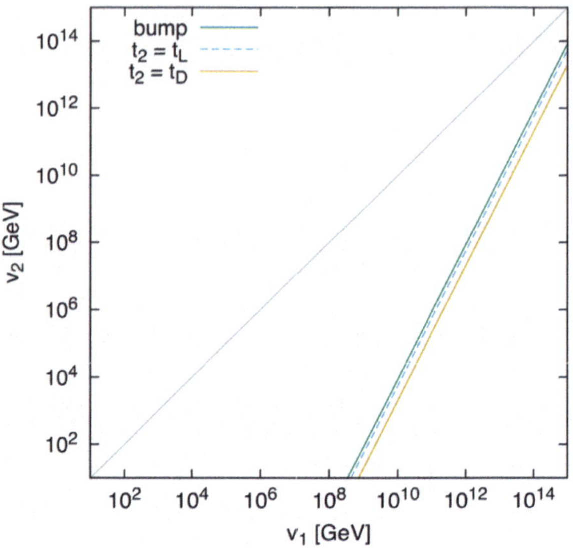

Fig. 2. The parameter region accommodating a bump-like feature in gravitational wave spectrum. The gravitational wave spectrum has a bump-like feature above the green line. The domain wall energy is dominant at the domain wall formation for the loops (above the blue broken line) and for the network (above the orange line). Upper triangular above the gray line is out of our interest of the present paper because of the inverse hierarchy in the mass scales.

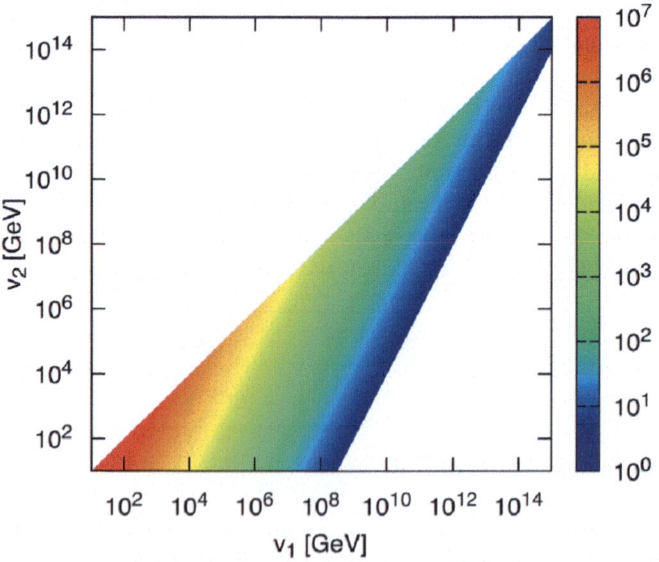

Fig. 3. Degree of the enhancement of the gravitational wave amplitude at the bump compared to the amplitude for the normal string loops.

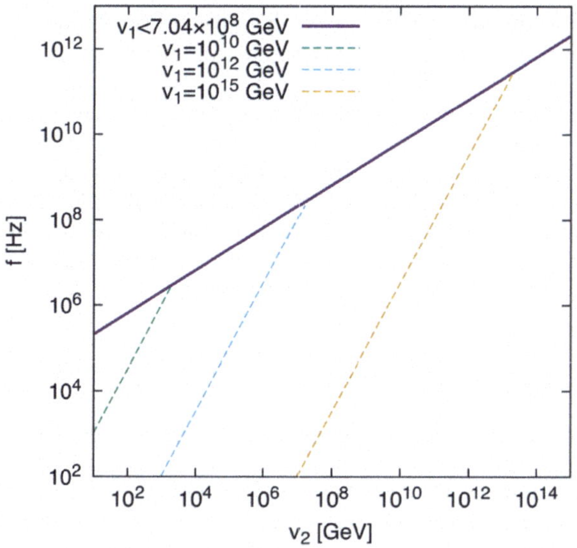

Fig. 4. Frequencies of the peak and the turning point of the gravitational wave spectrum. Purple line corresponds to the peak frequency. Dashed lines are for the frequency of the turning point for each first phase transition temperature.

symmetry is broken by the last stage of the symmetry breakings. The gaugedness bans the inverse hierarchy between the symmetry breaking producing cosmic strings and that producing domain walls.

We compare time scales determining the formation and the dynamics of string-domain wall system in the parameter space of the string and domain wall energy scales. We confirm that an enhancement of the gravitational wave signal found in our previous paper is consistent with the condition that the domain wall energy dominates over the string energy for the string-domain wall segments leaving from the string-domain wall network at the domain wall formation. We evaluate the degree of the enhancement of the gravitational wave signal compared to the gravitational wave amplitude from normal string loops without domain walls. The enhancement can be $\mathcal{O}(10^7)$ at maximum in the range between 10^1 GeV and 10^{15} GeV for the temperatures of the first and second phase transitions. We find that the peak amplitude is irrespective of the first phase transition temperature and is above $\mathcal{O}(10^{-16})$ in the same range. The peak frequency of the enhancement can be low as $\mathcal{O}(10^5)$ Hz as well. It is remarkable that the enhancement, i.e., the difference from the normal string spectrum, is more significant and the peak frequency is lowered especially for lower energy scales. We also note that the dependence of the peak frequency on the energy scale of the domain wall formation is $v_2^{1/2}$ and not strong, keeping the frequency relatively low even for high scale second phase transition. Detecting the enhancement in the gravitational wave amplitude at the peak frequency in the parameter range requires gravitational wave detectors incorporating new techniques and covering the high frequency region above 10^5 Hz [44]. Also, it will be important to examine the property of the dynamics of the string-domain wall system not in Nambu-Goto description but in a concrete field theory [45].

Acknowledgments. HM is supported by the Director, Office of Science, Office of High Energy Physics of the U.S. Department of Energy under the Contract No. DE-AC02-05CH11231, by the NSF grant PHY-1915314, by the JSPS Grant-in-Aid for Scientific Research JP20K03942, MEXT Grant-in-Aid for Transformative Research Areas (A) JP20H05850, JP20A203, by World Premier International Research Center Initiative, MEXT, Japan, and Hamamatsu Photonics, K.K. and received the title of Hamamatsu Professor. YS is supported by JSPS KAKENHI Grant Number JP22K20365.

References

1. Lazarides, G., Shafi, Q., Walsh, T.F.: Cosmic strings and domains in unified theories. Nucl. Phys. B **195**, 157 (1982)
2. Kibble, T.W.B., Lazarides, G., Shafi, Q.: Strings in SO(10). Phys. Lett. B **113**, 237 (1982)
3. Kibble, T., Lazarides, G., Shafi, Q.: Walls bounded by strings. Phys. Rev. D **26**, 435 (1982)
4. Everett, A.E., Vilenkin, A.: Left-right symmetric theories and vacuum domain walls and strings. Nucl. Phys. B **207**, 43 (1982)
5. Vilenkin, A.: Cosmic strings and domain walls. Phys. Rept. **121**, 263 (1985)
6. Preskill, J.: Vortices and monopoles. In: Les Houches School of Theoretical Physics: Architecture of Fundamental Interactions at Short Distances, pp. 235–338 (1987)
7. Martin, X., Vilenkin, A.: Gravitational wave background from hybrid topological defects. Phys. Rev. Lett. **77**, 2879 (1996). arXiv:astro-ph/9606022
8. Vilenkin, A., Shellard, E.P.S.: Cosmic strings and other topological defects. Cambridge University Press, Cambridge (2000)
9. Guth, A.H.: The inflationary universe: a possible solution to the horizon and flatness problems. Phys. Rev. D **23**, 347 (1981)
10. Linde, A.D.: A new inflationary universe scenario: a possible solution of the horizon, flatness, homogeneity, isotropy and primordial monopole problems. Phys. Lett. B **108**, 389 (1982)
11. Langacker, P., Pi, S.-Y.: Magnetic monopoles in grand unified theories. Phys. Rev. Lett. **45**, 1 (1980)
12. Vilenkin, A.: Cosmological evolution of monopoles connected by strings. Nucl. Phys. B **196**, 240 (1982)
13. Vilenkin, A.: Cosmological evolution of monopoles connected by strings, Nuclear Phys. B **196**, 240 (1982)
14. Preskill, J., Vilenkin, A.: Decay of metastable topological defects. Phys. Rev. D **47**, 2324 (1993). arXiv:hep-ph/9209210
15. Monin, A., Voloshin, M.B.: The spontaneous breaking of a metastable string. Phys. Rev. D **78**, 065048 (2008). arXiv:0808.1693 [hep-th]
16. Dunsky, D.I., Ghoshal, A., Murayama, H., Sakakihara, Y., White, G.: Gravitational Wave Gastronomy (2021). arXiv:2111.08750 [hep-ph]
17. Leblond, L., Shlaer, B., Siemens, X.: Gravitational waves from broken cosmic strings: the bursts and the beads. Phys. Rev. D **79**, 123519 (2009). arXiv:0903.4686 [astro-ph.CO]
18. Dror, J.A., Hiramatsu, T., Kohri, K., Murayama, H., White, G.: Testing the seesaw mechanism and leptogenesis with gravitational waves. Phys. Rev. Lett. **124**, 041804 (2020). arXiv:1908.03227 [hep-ph]
19. Buchmuller, W., Domcke, V., Murayama, H., Schmitz, K.: Probing the scale of grand unification with gravitational waves. Phys. Lett. B **809**, 135764 (2020). arXiv:1912.03695 [hep-ph]

20. King, S.F., Pascoli, S., Turner, J., Zhou, Y.-L.: Gravitational waves and proton decay: complementary windows into grand unified theories. Phys. Rev. Lett. **126**, 021802 (2021). arXiv:2005.13549 [hep-ph]
21. Buchmuller, W., Domcke, V., Schmitz, K.: From NANOGrav to LIGO with metastable cosmic strings. Phys. Lett. B **811**, 135914 (2020). arXiv:2009.10649 [astro-ph.CO]
22. Samanta, R., Datta, S.: Gravitational wave complementarity and impact of NANOGrav data on gravitational leptogenesis, JHEP **05**, 211. arXiv:2009.13452 [hep-ph]
23. Chakrabortty, J., Lazarides, G., Maji, R., Shafi, Q.: Primordial Monopoles and Strings, Inflation, and Gravity Waves. JHEP **02**, 114. arXiv:2011.01838 [hep-ph]
24. Chigusa, S., Nakai, Y., Zheng, J.: Implications of gravitational waves for supersymmetric grand unification. Phys. Rev. D **104**, 035031 (2021). arXiv:2011.04090 [hep-ph]
25. Buchmuller, W., Domcke, V., Schmitz, K.: Stochastic gravitational-wave background from metastable cosmic strings. arXiv:2107.04578 [hep-ph] (2021)
26. Mazumdar, A., White, G.: Review of cosmic phase transitions: their significance and experimental signatures. Rept. Prog. Phys. **82**, 076901 (2019). arXiv:1811.01948 [hep-ph]
27. Hindmarsh, M.B., Lüben, M., Lumma, J., Pauly, M.: Phase transitions in the early universe. SciPost Phys. Lect. Notes **24**, 1 (2021). arXiv:2008.09136 [astro-ph.CO]
28. Fukugita, M., Yanagida, T.: Baryogenesis Without Grand Unification. Phys. Lett. B **174**, 45 (1986)
29. Davidson, S., Ibarra, A.: A Lower bound on the right-handed neutrino mass from leptogenesis. Phys. Lett. B **535**, 25 (2002). arXiv:hep-ph/0202239
30. Buchmuller, W., Di Bari, P., Plumacher, M.: Leptogenesis for pedestrians. Annals Phys. **315**, 305 (2005). arXiv:hep-ph/0401240
31. Drewes, M.: The phenomenology of right handed neutrinos. Int. J. Mod. Phys. E **22**, 1330019 (2013). arXiv:1303.6912 [hep-ph]
32. Krauss, L.M., Wilczek, F.: Discrete gauge symmetry in continuum theories. Phys. Rev. Lett. **62**, 1221 (1989)
33. Ibanez, L.E., Ross, G.G.: Discrete gauge symmetry anomalies. Phys. Lett. B **260**, 291 (1991)
34. Martin, S.P.: Some simple criteria for gauged R-parity. Phys. Rev. D **46**, R2769 (1992). arXiv:hep-ph/9207218
35. Buchmuller, W., Peccei, R.D., Yanagida, T.: Leptogenesis as the origin of matter. Ann. Rev. Nucl. Part. Sci. **55**, 311 (2005). arXiv:hep-ph/0502169
36. Barbier, R., et al.: R-parity violating supersymmetry. Phys. Rept. **420**, 1 (2005). arXiv:hep-ph/0406039
37. Buchmuller, W., Covi, L., Hamaguchi, K., Ibarra, A., Yanagida, T.: Gravitino dark matter in R-parity breaking vacua. JHEP **03**, 037. arXiv:hep-ph/0702184
38. Gouttenoire, Y., Servant, G., Simakachorn, P.: Beyond the standard models with cosmic strings. JCAP **07**, 032. arXiv:1912.02569 [hep-ph]
39. Blanco-Pillado, J.J., Olum, K.D., Shlaer, B.: The number of cosmic string loops. Phys. Rev. D **89**, 023512 (2014). arXiv:1309.6637 [astro-ph.CO]
40. Blanco-Pillado, J.J., Olum, K.D.: Stochastic gravitational wave background from smoothed cosmic string loops. Phys. Rev. D **96**, 104046 (2017). arXiv:1709.02693 [astro-ph.CO]
41. Caprini, C., Durrer, R., Konstandin, T., Servant, G.: General properties of the gravitational wave spectrum from phase transitions. Phys. Rev. D **79**, 083519 (2009). arXiv:0901.1661 [astro-ph.CO]
42. Hiramatsu, T., Kawasaki, M., Saikawa, K.: On the estimation of gravitational wave spectrum from cosmic domain walls. JCAP **02**, 031. arXiv:1309.5001 [astro-ph.CO]
43. Garriga, J., Sakellariadou, M.: Effects of friction on cosmic strings. Phys. Rev. D **48**, 2502 (1993). arXiv:hep-th/9303024
44. Howl, R., Fuentes, I.: Quantum frequency interferometry: with applications ranging from gravitational wave detection to dark matter searches. arXiv:2103.02618 [quant-ph] (2021)
45. Eto, M., Hiramatsu, T., Saito, I., Sakakihara, Y.: (forthcoming)

Dynamical Analysis in Regularized 4D Einstein-Gauss-Bonnet Gravity

Bilguun Bayarsaikhan[1,2], Sunly Khimphun[3], Phearun Rithy[3](✉), and Gansukh Tumurtushaa[4]

[1] Institute for Theoretical Physics, ELTE Eotvos Lorand University, Pazmany Peter setany 1/A, Budapest 1117, Hungary

[2] Institute of Physics and Technology, Mongolian Academy of Sciences, Ulaanbaatar 13330, Mongolia

[3] Graduate School of Science, Royal University of Phnom Penh, Phnom Penh 12150, Cambodia
khimphun.sunly@rupp.edu.kh, phearunrain0@gmail.com

[4] Department of Science Education, Jeju National University, Jeju 63243, Korea
gansukh@jejunu.ac.kr

Abstract. We consider an extension of the 4D Einstein-Gauss-Bonnet (EGB) gravity by proposing a coupling between the scalar field and the Gauss-Bonnet term and demonstrate that the additional contributions to the equations of motion (EoM) come from both the scaling of a coupling constant and the non-minimal coupling between the scalar field and gravity. By choosing the exponential functions for the potential and the coupling function to the Gauss-Bonnet term, we investigate the system's stability based on the fixed points and the dynamical system approach (DSA). Our result shows that the system is not only stable but also can explain the late-time acceleration of the universe. However, stable accelerating solutions exist under specific conditions that the model parameters, including the potential parameter, must satisfy.

Keywords: 4D Gauss-Bonnet gravity · Non-minimal coupling · Dynamical Analysis · fixed points · accelerating universe

1 Introduction

The accelerating phase of the present universe has been confirmed by the observations of Type Ia Supernovae (SNe Ia) [1,2] along with the measurements of Cosmic Microwave Background (CMB) and Baryon Acoustic Oscillation (BAO) from Planck 2015 and 2018 [3,4]. The accelerated expansion influenced by unknown energy called dark energy is well explained by introducing the Cosmological Constant (CC) or ΛCDM model. Even though the model is the most favored by observational data in the low red-shift which is reasonable due to its nature as an effective theory, its shortcoming in terms of its fundamental interpretation and inconsistency remains unsolved. This motivated people to work on the modified gravity theory. There are many modified theories having been recently proposed and resurrected which can be clustered in terms of geometry, extra-degree of freedom, and higher dimension points of view. In this work,

H. J. Choi et al. (Eds.): APPC 2022, *Proceedings of the 15th Asia Pacific Physics Conference*, pp. 44–54, 2024.
https://doi.org/10.1007/978-981-96-0191-2_5

we will consider all these views altogether. As the primary purpose of these proposals is to solve CC problems, the dark energy models which based on these modified theories will accommodate. There are many dark energy models, but we list down some such as the Chevalliear Polarski Linder (CPL) [7,8], Holographic Dark Energy and its constraining with the observational data [9–11], generalized holographic cosmology and its constraint [12], Dvali Gabadadze Porrati (DGP) braneworld [13,14] and Chaplygin gas model [15,16]. For the comparison of different dark energy models by using the observational constraints where ΛCDM has been used as a reference model, one is referred to [17]. One of the possible modifications of gravity is the higher order of curvature ($R, R_{\mu\nu}R^{\mu\nu}, R_{\mu\nu\rho\sigma}R^{\mu\nu\rho\sigma}$) so-called Gauss-Bonnet (GB) gravity. It is well known that the GB gravity vanishes identically in $D = 4$, so it does not modify the General Relativity (GR) and does not contribute to gravitational dynamics.

Recently, the model of 4D Einstein-Gauss-Bonnet (EGB) gravity has been proposed by Glavan and Lin [18], in which the constant is scaled as $\alpha \to \alpha/(D-4)$ and then take the limit $D \to 4$, where D is the space-time dimensions that assumed to be continuous. In 4D the GB term vanishes due to the overall factor $(D-4)$, where the rescaling constant gives rise to nontrivial contribution to modified Einstein's gravity in $D = 4$ with the additional quartic term in the Hubble parameter to the Friedmann equations that would be interested in the cosmological frameworks. However, the original 4D EGB theory only deals up to EoM level so that some debates have been discussed [19–25]. Furthermore, its regularization up to action level has been investigated in [26–30]. However, consistent theory of $D \to 4$ EGB theory without extra degree of freedom up to spatially diffeomophism invariance has been considered in [31]. From this development, it is clear that taking the limit provides a scheme of dimensional regularization to the vanishing terms in a particular dimension but may not bypass Lovelock's theorem without restriction. In this regard, we exploit the regularized 4D EGB scheme and introduce the non-minimal coupling function to the GB term and a scaling constant. In this sense, the role of the non-minimal coupling function may affect the EoM that would explain other dynamics of the universe.

In this work, we study the extension of the 4D EGB gravity with a scaling constant $\alpha \to \alpha/(D-4)$ and the coupling function of the scalar field $\xi(\phi)$. When we introduce $\xi(\phi)$ and the scaling constant to GB term in D-dimensions, there appears a divergent term in the EoM due to the presence of the GB coupling function which leads to the overall scaling factor $1/(D-4)$. Hence we introduce

$$\xi(\phi) \to \xi^{(D-4)}(\phi),$$

with flat Friedmann-Lemaître-Robertson-Walker (FLRW) metric in D-dimensions and take the limit $D \to 4$. The motivation is that we derive well-defined EoM with the extension of the equations to the 4D EGB gravity that this investigation might lead to considering the other dynamics of the universe. From the model, we analyze the cosmological dynamics by using DSA and consider the exponential functions of potential and $\xi(\phi)$. From this approach, the fixed points can be obtained so that stability can be realized [32–37]. Beyond the stability considerations of the model, further study on the cosmic evolution of cosmological parameters, the evolution of the parameter spaces, and the constraint of the potential parameters with the observational bounds from SNIa,

BAO, CMB, and H_0 measurement have been further investigated in a recently published paper [38].

We organized this work as the following. In Sect. 2, we introduce the action and equations of motion. In Sect. 3, we consider the setup of the autonomous equations. In Sect. 4, we discuss the fixed points and stability. In the last section, we refer to the conclusions.

2 The Action and Equations of Motion

We study the combinations of Einstein-Hilbert action with the canonical scalar field, the GB term with a scaling coupling constant $\alpha \rightarrow \alpha/(D-4)$ and the coupling function $\xi(\phi)$ in D-dimensions, thus action equation is given by,

$$S = \int \mathrm{d}^D x \sqrt{-g} \left[\frac{1}{2\kappa^2} R - \frac{1}{2} g^{\mu\nu} \partial_\mu \phi \partial_\nu \phi - V(\phi) \right. \\ \left. - \frac{\alpha}{2(D-4)} \xi(\phi) \mathcal{G} \right] + S_{\text{m,r}}, \tag{1}$$

where $\kappa^2 = 8\pi G$. R and G are Ricci scalar and Newton's constant respectively. $S_{\text{m,r}}$ represent the standard matter and radiation components, and

$$\mathcal{G} = R^2 - 4R_{\mu\nu}R^{\mu\nu} + R_{\mu\nu\rho\sigma}R^{\mu\nu\rho\sigma},$$

is the Gauss-Bonnet term. When we introduce $\xi(\phi)$ and a scaling constant to GB term in D-dimensions, there appears a divergent term in the EoM due to the presence of the GB coupling function because of the overall scaling factor of $1/(D-4)$. Hence, in order to derive well-defined EoM we regularize the coupling function

$$\xi(\phi) \rightarrow \xi^{(D-4)}(\phi),$$

and take the limit $D \rightarrow 4$. Interestingly, the re-scaling factor $1/(D-4)$ can cancel with the overall factor $(D-4)$ in the EoM level, which comes from the derivative of regularized coupling function $\xi(\phi)$. Varying the action with respect to the inverse metric $g^{\mu\nu}$ and scalar field ϕ yields the Einstein field equation and scalar field equation as the following

$$R_{\mu\nu} - \frac{1}{2} g_{\mu\nu} R = \kappa^2 \left(T^{(\phi)}_{\mu\nu} + T^{(\text{GB})}_{\mu\nu} + T^{(\text{matter})}_{\mu\nu} \right), \tag{2}$$

$$\Box\phi - V_\phi - \frac{\alpha}{2(D-4)} (D-4) \xi^{(D-5)} \xi_\phi \mathcal{G} = 0, \tag{3}$$

where $\Box \equiv \nabla_\mu \nabla^\mu$, $V_\phi = dV/d\phi$, $\xi_\phi = d\xi/d\phi$, and $T^{(\phi)}_{\mu\nu}$, $T^{(\text{GB})}_{\mu\nu}$, $T^{(\text{matter})}_{\mu\nu}$ are stress-energy-momentum tensor of scalar field, GB and matter term respectively; the explicit expressions are given in [38]. The FLRW metric in D-dimensions of spatially flat $k = 0$ with a scale factor $a(t)$ is given by,

$$\mathrm{d}s^2 = -\mathrm{d}t^2 + a^2(t) \left(\mathrm{d}\chi_1^2 + \mathrm{d}\chi_2^2 + \mathrm{d}\chi_3^2 + \ldots + \mathrm{d}\chi_{D-1}^2 \right). \tag{4}$$

Substituting the metric into the field equations with arbitrary dimensions. Hence we can derive well-defined EoM in D-dimensions for the temporal and spatial components and scalar field equation

$$\begin{aligned}\frac{(D-2)(D-1)H^2}{2\kappa^2} &= \rho_{\rm m} + \rho_{\rm r} + \frac{1}{2}\dot{\phi}^2 + V \\ &+ 2\alpha(D-3)(D-2)(D-1)\dot{\xi}\xi^{(D-5)}H^3 \\ &+ \frac{1}{2}\alpha(D-3)(D-2)(D-1)\xi^{(D-4)}H^4,\end{aligned} \tag{5}$$

$$\begin{aligned}&-\frac{(D-2)}{\kappa^2}\left(\frac{(D-3)H^2}{2}+\frac{\ddot{a}}{a}\right) = P_{\rm r} + \frac{1}{2}\dot{\phi}^2 - V \\ &- 4\alpha(D-3)(D-2)\dot{\xi}\xi^{(D-5)}H\frac{\ddot{a}}{a} \\ &- \frac{1}{2}\alpha(D-3)(D-2)(D-5)\xi^{(D-4)}H^4 \\ &- 2\alpha(D-4)(D-3)(D-2)\dot{\xi}\xi^{(D-5)}H^3 \\ &- 2\alpha(D-3)(D-2)\xi^{(D-4)}H^2\frac{\ddot{a}}{a} \\ &- 2\alpha(D-3)(D-2)\left[(D-5)\dot{\xi}^2\xi^{(D-6)} + \ddot{\xi}\xi^{(D-5)}\right]H^2,\end{aligned} \tag{6}$$

$$\begin{aligned}&2\alpha(D-3)(D-2)(D-1)\xi_\phi\xi^{(D-5)}H^2\frac{\ddot{a}}{a} \\ &+ \frac{\alpha}{2}(D-4)(D-3)(D-2)(D-1)\xi_\phi\xi^{(D-5)}H^4 \\ &+ (D-1)\dot{\phi}H + V_\phi + \ddot{\phi} = 0,\end{aligned} \tag{7}$$

respectively, where $\xi_{\phi\phi} = d^2\xi/d\phi^2$ and $\dot{\xi} = \xi_\phi\dot{\phi}$, dot means the derivative with respect to the cosmic time. Also, $H \equiv \dot{a}/a$, $\rho_{\rm m}$ and $\rho_{\rm r}$ are the Hubble's parameter, energy density of matter and radiation respectively. We consider the Eqs. (5)–(7) in 4D by taking the limit $D \rightarrow 4$ and obtain the EoM as following,

$$3H^2 = \kappa^2\left(\rho_{\rm m} + \rho_{\rm r} + \frac{1}{2}\dot{\phi}^2 + V + 12\alpha\frac{\dot{\xi}}{\xi}H^3 + 3\alpha H^4\right), \tag{8}$$

$$\begin{aligned}2\dot{H} + 3H^2 = &-\kappa^2\left[\frac{\rho_{\rm r}}{3} + \frac{\dot{\phi}^2}{2} - V - 8\alpha\frac{\dot{\xi}}{\xi}H(\dot{H}+H^2)\right. \\ &\left. - 3\alpha H^4 - 4\alpha H^2\left(\dot{H} - \frac{\dot{\xi}^2}{\xi^2} + \frac{\ddot{\xi}}{\xi}\right)\right],\end{aligned} \tag{9}$$

$$\ddot{\phi} + V_\phi + 3\dot{\phi}H + 12\alpha\frac{\xi_\phi}{\xi}H^2\left(\dot{H} + H^2\right) = 0. \tag{10}$$

Obviously, the quartic term of Hubble parameter in Friedmann equations coming from a scaling coupling constant to GB term. However, the motivation from our investigation is the extension of regularized 4D EGB, which is the additional cubic term in Hubble parameter as a consequence of the non-minimal coupling function. Thus, both higher order terms in Hubble parameter might give interesting information about the evolution of universe. The continuity equation of radiation density can be written as,

$$\dot{\rho}_{\rm r} + 4H\rho_{\rm r} = 0, \tag{11}$$

we can rewrite Eqs. (8) and (9) as the following

$$3H^2 = \kappa^2 \left(\rho_{\rm m} + \rho_{\rm r} + \rho_\phi\right), \tag{12}$$

$$2\dot{H} + 3H^2 = -\kappa^2 \left(\frac{1}{3}\rho_{\rm r} + P_\phi\right), \tag{13}$$

where the definition of the energy density and the pressure of the scalar field are defined as,

$$\rho_\phi = \frac{1}{2}\dot{\phi}^2 + V + 12\alpha\frac{\dot{\xi}}{\xi}H^3 + 3\alpha H^4, \tag{14}$$

$$\begin{aligned} P_\phi =& \frac{\dot{\phi}^2}{2} - V - 8\alpha\frac{\dot{\xi}}{\xi}H(\dot{H} + H^2) - 3\alpha H^4 \\ & - 4\alpha H^2\dot{H} - 4\alpha\left(-\frac{\dot{\xi}^2}{\xi^2} + \frac{\ddot{\xi}}{\xi}\right)H^2. \end{aligned} \tag{15}$$

The equations of ρ_ϕ and P_ϕ satisfy the conservation equation and the Eq. (10) can be rewritten as

$$\dot{\rho_\phi} + 3H(1 + w_\phi)\rho_\phi = 0, \tag{16}$$

where the equation of state (EoS) of the scalar field is defined by $w_\phi \equiv P_\phi/\rho_\phi$ with the effective EoS also defined by,

$$w_{\rm eff} \equiv -1 - \frac{2\dot{H}}{3H^2}. \tag{17}$$

3 The Setup of Autonomous Equations

In order to consider the cosmological dynamics of the model, we apply DSA. We define the dimensionless variables to obtain the dynamics of these variables, then we find the fixed points and analyze its stability. Therefore, Eq. (8) can be rewritten as

$$\begin{aligned} 1 =& \frac{\kappa^2\rho_{\rm m}}{3H^2} + \frac{\kappa^2\rho_{\rm r}}{3H^2} + \frac{\kappa^2\dot{\phi}^2}{6H^2} + \frac{\kappa^2 V}{3H^2} + \frac{4\alpha\kappa^2\dot{\xi}H}{\xi} \\ & + \alpha\kappa^2 H^2, \end{aligned} \tag{18}$$

and we define the dimensionless variables as

$$\begin{aligned} & x = \frac{\kappa\dot{\phi}}{\sqrt{6}H}, \; y = \frac{\kappa\sqrt{V}}{\sqrt{3}H}, \; z = \sqrt{\alpha}\kappa H, \; \mu = -\frac{\sqrt{6}\xi_\phi}{\kappa\xi}, \\ & \lambda = -\frac{V_\phi}{\sqrt{6}\kappa V}, \; \epsilon = -\frac{\dot{H}}{H^2}. \end{aligned} \tag{19}$$

In terms of these new variables in Eq. (19), the Friedmann equation (8) can be simplified as

$$1 = \Omega_{\rm m} + \Omega_{\rm r} + \Omega_{\phi}, \tag{20}$$

$$\Omega_{\rm m} = \frac{\kappa^2 \rho_{\rm m}}{3H^2}, \tag{21}$$

$$\Omega_{\rm r} = \frac{\kappa^2 \rho_{\rm r}}{3H^2}, \tag{22}$$

$$\Omega_{\phi} = x^2 + y^2 + z^2 - 4xz^2\mu. \tag{23}$$

In addition, by introducing the dimensionless variable $N = \ln a$ and $dN = Hdt$ and take the derivative of with respect to N, we derive the autonomous equations as the following

$$\frac{dx}{dN} = \frac{\kappa\ddot{\phi}}{\sqrt{6}H^2} - \frac{\kappa\dot{\phi}\dot{H}}{\sqrt{6}H^3} = -\beta x + \epsilon x, \tag{24}$$

$$\frac{dy}{dN} = \frac{\kappa V_{\phi}\dot{\phi}}{2\sqrt{3}\sqrt{V}H^2} - \frac{\kappa\sqrt{V}\dot{H}}{\sqrt{3}H^3} = -3xy\lambda + y\epsilon, \tag{25}$$

$$\frac{dz}{dN} = \frac{\sqrt{\alpha}\kappa\dot{H}}{H} = -z\epsilon, \tag{26}$$

$$\frac{d\mu}{dN} = \frac{\sqrt{6}\xi_{\phi}^2\dot{\phi}}{\kappa\xi^2\, H} - \frac{\sqrt{6}\xi_{\phi\phi}\dot{\phi}}{\kappa\xi H} = x\mu^2(1-c), \tag{27}$$

$$\frac{d\Omega_{\rm r}}{dN} = -\frac{2\kappa^2\rho_{\rm r}\dot{H}}{3H^4} + \frac{\kappa^2\dot{\rho}_{\rm r}}{3H^3} = -4\Omega_{\rm r} + 2\Omega_{\rm r}\epsilon, \tag{28}$$

$$\frac{d\lambda}{dN} = \frac{V_{\phi}^2\dot{\phi}}{\sqrt{6}\kappa V^2\, H} - \frac{V_{\phi\phi}\dot{\phi}}{\sqrt{6}\kappa V H} = 6x\lambda^2(1-b), \tag{29}$$

where

$$b = \frac{V_{\phi\phi}V}{V_{\phi}^2}, \quad c = \frac{\xi_{\phi\phi}\xi}{\xi_{\phi}^2}, \quad \text{and} \quad \beta = -\frac{\ddot{\phi}}{\dot{\phi}H}. \tag{30}$$

From Eq. (10), we can write β as a function of ϵ,

$$\beta = -\frac{\ddot{\phi}}{\dot{\phi}H} = 3 - \frac{3y^2\lambda}{x} - \frac{2z^2\mu}{x} + \frac{2z^2\mu\epsilon}{x}, \tag{31}$$

in which according Eq. (9), we derive

$$\begin{aligned} \epsilon \equiv -\frac{\dot{H}}{H^2} = \frac{1}{2+4z^2(-1+2x\mu+2z^2\mu^2)}\Big[3-3z^2 \\ -4xz^2\mu + 8z^4\mu^2 + 3y^2(-1+4\lambda z^2\mu) + 3x^2 \\ +4x^2z^2\mu^2(1-c) + \Omega_{\rm r}\Big]. \end{aligned} \tag{32}$$

therefore we can rewrite the final autonomous equations as

$$\frac{dx}{dN} = -3x + 3y^2\lambda + 2z^2\mu - 2z^2\mu\epsilon + x\epsilon, \tag{33}$$

$$\frac{dy}{dN} = -3xy\lambda + y\epsilon, \tag{34}$$

$$\frac{dz}{dN} = -z\epsilon, \tag{35}$$

$$\frac{d\Omega_{\rm r}}{dN} = 2\Omega_{\rm r}(-2 + \epsilon), \tag{36}$$

$$\frac{d\mu}{dN} = x\mu^2(1 - c), \tag{37}$$

$$\frac{d\lambda}{dN} = 6x\lambda^2(1 - b). \tag{38}$$

From Eqs. (14), (15) and (17), we derive the EoS for the scalar field and the effective EoS as

$$\begin{aligned} w_\phi = & \frac{1}{3(x^2+y^2+z^2-4xz^2\mu)\{1-2z^2[1-2\mu(x+z^2\mu)]\}} \\ & \Big\{\Big(-3y^2+12y^2z^2\lambda\mu\Big)+z^2\Big[3+2\Omega_{\rm r}-4z^2\mu^2(1+\Omega_{\rm r})\Big] \\ & + x^2\Big[3+4z^2\mu^2(1-c)\Big]-4xz^2\mu(4+\Omega_{\rm r})\Big\}, \end{aligned} \tag{39}$$

$$w_{\rm eff} \equiv -1 - \frac{2\dot{H}}{3H^2} = -1 + \frac{2\epsilon}{3}. \tag{40}$$

Table 1. The fixed points and cosmological parameters

Pts	x	y	z	Ω_r	Ω_m	Ω_ϕ	Existence	$\omega_{\rm eff} = \omega_\phi$
$A_1^\pm$	$\frac{1}{2\lambda}$	$\pm\frac{1}{2\lambda}$	0	0	$1-\frac{1}{2\lambda^2}$	$\frac{1}{2\lambda^2}$	$\lambda^2 > 1/2$	0
$A_2^\pm$	$\frac{2}{3\lambda}$	$\pm\frac{\sqrt{2}}{3\lambda}$	0	$1-\frac{2}{3\lambda^2}$	0	$\frac{2}{3\lambda^2}$	$\lambda^2 > 2/3$	$1/3$
$A_3^\pm$	± 1	0	0	0	0	1	$\forall\lambda$	1
A_4	0	0	0	0	1	0	$\forall\lambda$	0
A_5	0	0	0	1	0	0	$\forall\lambda$	$1/3$
$A_6^\pm$	λ	$\pm\sqrt{1-\lambda^2}$	0	0	0	1	$\lambda^2 < 1$	$-1+2\lambda^2$
$A_7^\pm$	0	$-\sqrt{\frac{2\mu}{2\mu-3\lambda}}$	$\pm\sqrt{\frac{3\lambda}{3\lambda-2\mu}}$	0	0	1	$(\mu \geq 0, \lambda < 0)$ or $\mu \leq 0, \lambda > 0$	-1
$A_8^\pm$	0	$\sqrt{\frac{2\mu}{2\mu-3\lambda}}$	$\pm\sqrt{\frac{3\lambda}{3\lambda-2\mu}}$	0	0	1	$(\mu \geq 0, \lambda < 0)$ or $\mu \leq 0, \lambda > 0$	-1
$A_9^\pm$	$\frac{3-\sqrt{9-80\mu^2}}{20\mu}$	0	$\pm\sqrt{\frac{6}{3+\sqrt{9-80\mu^2}}}$	0	0	1	$-\frac{3}{4\sqrt{5}} \leq \mu \leq \frac{3}{4\sqrt{5}}, \mu \neq 0$	-1
$A_{10}^\pm$	$\frac{3+\sqrt{9-80\mu^2}}{20\mu}$	0	$\pm\sqrt{\frac{9+3\sqrt{9-80\mu^2}}{40\mu^2}}$	0	0	1	$-\frac{3}{4\sqrt{5}} \leq \mu \leq \frac{3}{4\sqrt{5}}, \mu \neq 0$	-1

4 The Fixed Points and Stability

In present work, we study the potential $V(\phi)$ and the coupling function $\xi(\phi)$ as the exponential functions of ϕ,

$$V(\phi) = V_0 e^{-\sqrt{6}\kappa\lambda\phi}, \text{ and } \xi(\phi) = \xi_0 e^{-\kappa\mu\phi/\sqrt{6}}, \tag{41}$$

where λ and μ are constants. The fixed points of the autonomous equations can be found by setting Eqs. (33)–(36) to zero and analyze its stability by looking at the sign of the real eigenvalues of the Jacobian matrix. One can notice that, we choose the values of $c = 1$ and $b = 1$ so that Eqs. (37) and (38) are automatically satisfied. Notice that if the real part of all eigenvalues are negative, the fixed point is stable. If all eigenvalues are positive, this indicates an unstable point, and it can be a saddle point if there are both positive and negative signs [32,33]. By solving the autonomous equations, we obtain ten fixed points. The expressions in Table 1 and 2 describe the ten fixed points in terms of their Eigenvalues, existent conditions, the values of the cosmological parameters, and the stability under specific conditions.

Points $\mathbf{A_1^\pm}$ to $\mathbf{A_6^\pm}$ have the fixed points $z = 0$ which imply that these are the GR fixed points, which is represented by the *quintessence* model with the canonical of scalar field minimally coupled to gravity and for the review, one is referred to [37]. In our solutions, we focus on GB fixed points where $z \neq 0$, so these $\mathbf{A^\pm_{7,8,9,10}}$ are the fixed points of interest that we can also relate to the accelerated expansion of the present universe with the EoS $\omega_{\text{eff}} = \omega_\phi = -1$. These fixed points also represent the scalar field dominated universe where ($\Omega_\phi = 1$). Notice that the GB fixed points must satisfy the conditions for the existence listed in Table 1.

$\mathbf{A^\pm_{7,8}}$: These fixed points are the same since the autonomous equations are invariant under the transformation $y \longrightarrow -y$, and they exist with the same conditions as listed in Table 1. Also, they represent the de Sitter solution with $\omega_{\text{eff}} = -1$ that allow the accelerating expansion in the scalar field domination phase. From conditions **A** and **B**, these fixed points can be stable with all negative signs of its eigenvalues and saddle points that contain three negative signs and one positive sign of the eigenvalues, respectively (Table 2).

$\mathbf{A^\pm_{9,10}}$: These fixed points can exist within certain range of μ as listed in Table 1. These points also produce the de Sitter solutions with $\omega_{\text{eff}} = -1$ which provide the stable and accelerating expansion with the negative signs of all its eigenvalues for condition **C** and saddle point with three negative and one positive Eigenvalues for condition **D**. Solutions that correspond to these points also correspond to the scalar field or the dark energy-domination phase.

$$\text{cond A} = \begin{cases} 0 < \mu \leq \frac{3\lambda^2}{-8\lambda^3+2\sqrt{16\lambda^6+17\lambda^4+\lambda^2}}, \ \lambda < 0 \\ -\frac{3\lambda^2}{8\lambda^3+2\sqrt{16\lambda^6+17\lambda^4+\lambda^2}} \leq \mu < 0, \ \lambda > 0 \end{cases} \tag{42}$$

$$\text{cond B} = \begin{cases} \mu > -\frac{3\lambda}{2\sqrt{1+9\lambda^2}}, \ \lambda < 0 \\ \mu < -\frac{3\lambda}{2\sqrt{1+9\lambda^2}}, \ \lambda > 0 \end{cases} \tag{43}$$

$$\text{cond C} = \begin{cases} -\frac{3}{4\sqrt{5}} \leq \mu < 0, \ \lambda < 0 \\ 0 < \mu \leq \frac{3}{4\sqrt{5}}, \quad \lambda > 0 \end{cases} \tag{44}$$

Table 2. The eigenvalues and the stability

Pts	Eigenvalues	Stability
$A_1^{\pm}$	$\left\{-\frac{3}{2},-1,-\frac{3}{4}\left(1+\sqrt{\frac{4}{\lambda^2}-7}\right),-\frac{3}{4}\left(1-\sqrt{\frac{4}{\lambda^2}-7}\right)\right\}$	stable for $\frac{1}{2}<\lambda^2\leq\frac{4}{7}$, stable spiral for $\lambda^2>\frac{4}{7}$
$A_2^{\pm}$	$\left\{-2,+1,-\frac{1}{2}\left(1+\sqrt{\frac{32}{3\lambda^2}-15}\right),-\frac{1}{2}\left(1-\sqrt{\frac{32}{3\lambda^2}-15}\right)\right\}$	saddle
$A_3^{\pm}$	$\{-3,+3,+2,3\pm 3\lambda\}$	saddle
A_4	$\{-\frac{3}{2},-\frac{3}{2},+\frac{3}{2},-1\}$	saddle
A_5	$\{-2,+2,-1,+1\}$	saddle
$A_6^{\pm}$	$\{-3\lambda^2,3\left(\lambda^2-1\right),6\lambda^2-3,6\lambda^2-4\}$	stable for $0<\lambda^2<\frac{1}{2}$, saddle if $\frac{1}{2}<\lambda^2<1$
$A_7^{\pm}$	$\left\{-4,-3,-\frac{3}{2}\left(1-\sqrt{1+\frac{32\lambda^2\mu^2-48\lambda^3\mu}{4\mu^2-\lambda^2(9-36\mu^2)}}\right),-\frac{3}{2}\left(1+\sqrt{1+\frac{32\lambda^2\mu^2-48\lambda^3\mu}{4\mu^2-\lambda^2(9-36\mu^2)}}\right)\right\}$	stable for condition A, saddle for condition B
$A_8^{\pm}$	$\left\{-4,-3,-\frac{3}{2}\left(1-\sqrt{1+\frac{32\lambda^2\mu^2-48\lambda^3\mu}{4\mu^2-\lambda^2(9-36\mu^2)}}\right),-\frac{3}{2}\left(1+\sqrt{1+\frac{32\lambda^2\mu^2-48\lambda^3\mu}{4\mu^2-\lambda^2(9-36\mu^2)}}\right)\right\}$	stable for condition A, saddle for condition B
$A_9^{\pm}$	$\left\{-4,-3,-3,-\frac{3\lambda}{20\mu}\left(3-\sqrt{9-80\mu^2}\right)\right\}$	stable for condition C, saddle for condition D
$A_{10}^{\pm}$	$\left\{-4,-3,-3,-\frac{3\lambda}{20\mu}\left(3+\sqrt{9-80\mu^2}\right)\right\}$	stable for condition C, saddle for condition D

$$\text{cond D} = \begin{cases} 0<\mu\leq\frac{3}{4\sqrt{5}}, & \lambda<0 \\ -\frac{3}{4\sqrt{5}}\leq\mu<0, \lambda>0 \end{cases} \tag{45}$$

5 Conclusions

In this present work, we have investigated the extension of the 4D EGB theory by introducing the non-minimal coupling function to the GB gravity. When we introduce these coupling function together with the dimensional regularization scheme proposed in [18], the scaling factor $1/(D-4)$ appears in the terms of $\xi(\phi)$. As a result, we redefine or regularize coupling function $\xi(\phi)\rightarrow\xi^{(D-4)}(\phi)$ and consider a spatially-flat FLRW metric where limit $D\rightarrow 4$ has been exploited to study 4D EGB gravity. Then, we derived a well-defined EoM of the model and obtained the extended version of dynamical equations as a result of regularizing both the coupling constant α and the non-minimal coupling function $\xi(\phi)$. From this model, it might be interesting to study further on the time evolution of the universe and constraining it with the current observational data.

Furthermore, We applied DSA to the model and chose the exponential form for the potential and coupling function as shown in (41) to examine the stability using fixed points. We have found and demonstrated more fixed points beyond GR or GB fixed points ($z\neq 0$) that can exist under the specific conditions listed in Table 1. In addition, the GB fixed points can present the stable and accelerating universe with the scalar field dominated. Therefore, our results show that the system is not only stable but also explain the late-time acceleration. However, the stable and accelerating solutions exist under specific conditions.

Acknowledgments. BB is partially supported by the project "Extension of the Standard Model and Cosmology, heavy meson study" by the Mongolian Ministry of Education and Science and Mongolian Science and Technology Foundation. SK is supported by the Higher Education Improvement Project (HEIP) funded by the Cambodian Government (IDA Credit No. 6221-KH) and PR is supported by the Swedish International Development Cooperation Agency (SIDA) through Sweden and the Royal University of Phnom Penh (RUPP)'s Pilot Research Cooperation Programme (Sida Contribution No. 11599). GT was supported by Basic Science Research Program through the National Research Foundation of Korea (NRF) funded by the Ministry of Education(grant number) (NRF-2022R1I1A1A01053784) and (NRF-2021R1A2C1005748).

References

1. Riess, A.G., et al.: Supernova Search Team. Astron. J. **116**, 1009–1038 (1998)
2. Perlmutter, S., et al.: Supernova cosmology project. Astrophys. J. **517**, 565–586 (1999)
3. Ade, P.A.R., et al.: Planck. Astron. Astrophys. **594**, A13 (2016)
4. Aghanim, N., et al.: [Planck]. Astron. Astrophys. **641**, A6 (2020). [erratum: Astron. Astrophys. **652**, C4 (2021)]
5. Sahni, V., Starobinsky, A.A.: Int. J. Mod. Phys. D **9**, 373–444 (2000)
6. Clifton, T., Ferreira, P.G., Padilla, A., Skordis, C.: Phys. Rept. **513**, 1–189 (2012)
7. Chevallier, M., Polarski, D.: Int. J. Mod. Phys. D **10**, 213–224 (2001)
8. Linder, E.V.: Phys. Rev. Lett. **90**, 091301 (2003)
9. Li, M.: Phys. Lett. B **603**, 1 (2004)
10. Zhang, X., Wu, F.Q.: Phys. Rev. D **72**, 043524 (2005)
11. Zhang, X., Wu, F.Q.: Phys. Rev. D **76**, 023502 (2007)
12. Khimphun, S., Lee, B.H., Tumurtushaa, G.: JHEP **10**, 232 (2021)
13. Cai, R.G., Khimphun, S., Lee, B.H., Sun, S., Tumurtushaa, G., Zhang, Y.L.: Phys. Dark Univ. **26**, 100387 (2019)
14. Dvali, G.R., Gabadadze, G., Porrati, M.: Phys. Lett. B **485**, 208–214 (2000)
15. Kamenshchik, A.Y., Moschella, U., Pasquier, V.: Phys. Lett. B **511**, 265–268 (2001)
16. Bento, M.C., Bertolami, O., Sen, A.A.: Phys. Rev. D **66**, 043507 (2002)
17. Xu, Y.Y., Zhang, X.: Eur. Phys. J. C **76**(11), 588 (2016)
18. Glavan, D., Lin, C.: Phys. Rev. Lett. **124**(8), 081301 (2020)
19. Gürses, M., Şişman, T.Ç., Tekin, B.: Eur. Phys. J. C **80**(7), 647 (2020)
20. Gurses, M., Şişman, T.Ç., Tekin, B.: Phys. Rev. Lett. **125**(14), 149001 (2020)
21. Arrechea, J., Delhom, A., Jiménez-Cano, A.: Phys. Rev. Lett. **125**(14), 149002 (2020)
22. Arrechea, J., Delhom, A., Jiménez-Cano, A.: Chin. Phys. C **45**(1), 013107 (2021)
23. Shu, F.W.: Phys. Lett. B **811**, 135907 (2020)
24. Ai, W.Y.: Commun. Theor. Phys. **72**(9), 095402 (2020)
25. Mahapatra, S.: Eur. Phys. J. C **80**(10), 992 (2020)
26. Hennigar, R.A., Kubizňák, D., Mann, R.B., Pollack, C.: JHEP **07**, 027 (2020)
27. Fernandes, P.G.S., Carrilho, P., Clifton, T., Mulryne, D.J.: Phys. Rev. D **102**(2), 024025 (2020)
28. Kobayashi, T.: JCAP **07**, 013 (2020)
29. Lu, H., Pang, Y.: Phys. Lett. B **809**, 135717 (2020)
30. Easson, D.A., Manton, T., Svesko, A.: JCAP **10**, 026 (2020)
31. Aoki, K., Gorji, M.A., Mukohyama, S.: Phys. Lett. B **810**, 135843 (2020)
32. Wiggins, S.: Introduction to Applied Nonlinear Dynamical Systems and Chaos. Springer, New York (1990)
33. Boehmer, C.G., Chan, N.: Dynamical systems in cosmology, [arXiv:1409.5585 [gr-qc]]

34. Granda, L.N., Jimenez, D.F.: Eur. Phys. J. C **77**(10), 679 (2017)
35. Santos Da Costa, S., Roig, F.V., Alcaniz, J.S., Capozziello, S., De Laurentis, M., Benetti, M.: Class. Quant. Grav. **35**(7), 075013 (2018)
36. Bayarsaikhan, B., Koh, S., Tsedenbaljir, E., Tumurtushaa, G.: JCAP **11**, 057 (2020)
37. Bahamonde, S., Böhmer, C.G., Carloni, S., Copeland, E.J., Fang, W., Tamanini, N.: Phys. Rept. **775–777**, 1–122 (2018)
38. Bayarsaikhan, B., Khimphun, S., Rithy , P., Tumurtushaa, G.: Eur. Phys. J. C **83**(3), 238 (2023)

Calibration of Detector Time Constant with a Thermal Source for the POLARBEAR-2A CMB Polarization Experiment

S. Takatori[1(✉)], M. Hasegawa[2,3], M. Hazumi[2,3,4,5], D. Kaneko[2], N. Katayama[5], A. T. Lee[6,7], S. Takakura[8], T. Tomaru[9], T. Adkins[6], D. Barron[10], Y. Chinone[2,3,5], K. T. Crowley[6], T. de Haan[2,3], T. Elleflot[7], N. Farias[6], C. Feng[11], T. Fujino[12], J. Groh[13], H. Hirose[12], F. Matsuda[4], H. Nishino[14], Y. Segawa[15], P. Siritanasak[16], A. Suzuki[7], and K. Yamada[17]

[1] Research Institute for Interdisciplinary Science (RIIS), Okayama University, Tsushima-naka 3-1-1 Kita-ku, Okayama 700-8530, Japan
takatori@okayama-u.ac.jp

[2] International Center for Quantum-field Measurement Systems for Studies of the Universe and Particles (QUP), High Energy Accelerator Research Organization (KEK), Tsukuba, Ibaraki 305-0801, Japan

[3] Institute of Particle and Nuclear Studies (IPNS), High Energy Accelerator Research Organization (KEK), 1-1 Oho, Tsukuba, Ibaraki 305-0801, Japan

[4] Institute of Space and Astronautical Science, Japan Aerospace Exploration Agency (JAXA), Sagamihara 252-5210, Japan

[5] Kavli Institute for the Physics and Mathematics of the Universe (Kavli IPMU), University of Tokyo, 5-1-5 Kashiwa-no-Ha, Kashiwa, Chiba 277-8583, Japan

[6] Department of Physics, University of California, Berkeley, CA 94720, USA

[7] Physics Division, Lawrence Berkeley National Laboratory, Berkeley, CA 94720, USA

[8] CASA & Department of Astrophysical and Planetary Sciences, University of Colorado Boulder, Boulder, CO 80309, USA

[9] Gravitational Wave Project Office, National Astronomical Observatory of Japan (NAOJ), Mitaka, Tokyo 181-8588, Japan

[10] Department of Physics and Astronomy, University of New Mexico, Albuquerque, NM 87131, USA

[11] University of Science and Technology of China, Hefei 230052, People's Republic of China

[12] Yokohama National University, Yokohama, Kanagawa 240-8501, Japan

[13] Superconductive Electronics Group, National Institute of Standards and Technology, Boulder, CO 80305, USA

[14] Japan Synchrotron Radiation Research Institute (JASRI), Sayo-gun, Hyogo 679-5198, Japan

[15] School of High Energy Accelerator Science, The Graduate University for Advanced Studies, Sokendai, Kanagawa 240-0193, Japan

[16] National Astronomical Research Institute of Thailand, Chiangmai 50180, Thailand

[17] Department of Physics, The University of Tokyo, Tokyo 113-0033, Japan

Abstract. The Simons Array (SA) project is a ground-based Cosmic Microwave Background (CMB) polarization experiment. The SA observes the sky using three telescopes, and POLARBEAR-2A (PB-2A) is the receiver system on the first telescope. For the ground-based experiment, atmospheric fluctuation is the primary noise source that could cause polarization leakage. In the PB-2A receiver

H. J. Choi et al. (Eds.): APPC 2022, *Proceedings of the 15th Asia Pacific Physics Conference*, pp. 55–62, 2024.
https://doi.org/10.1007/978-981-96-0191-2_6

system, a continuously rotating half-wave plate (HWP) is used to mitigate the polarization leakage. However, due to the rapid modulation of the polarization signal, the uncertainty in the time constant of the detector results in an uncertainty in the polarization angle. For PB-2A, the time constant of each bolometer needs to be calibrated at the sub-millisecond level to avoid introducing bias to the polarization signal. We have developed a new calibrator system that can be used to calibrate the time constants of the detectors. In this study, we present the design of the calibration system and the preliminary results of the time constant calibration for PB-2A.

Keywords: CMB · TES · calibrator

1 Introduction

Precise measurements of Cosmic Microwave Background (CMB) polarization, especially detecting the odd-parity CMB polarization pattern – called the B-mode – is a unique and powerful way of exploring the early universe. The Simons Array (SA) project [1] is a ground-based CMB polarization experiment conducted in the Atacama desert, Chile, at an altitude of 5,200 m. SA aims to characterize B-mode signals from inflationary gravitational waves and gravitational lensing effects. The projected constraint on a tensor-to-scalar ratio r is at the level of $\sigma(r) = 0.006$ by the inflationary B-modes, and the sensitivity to the sum of neutrino masses is 40 meV (68% C.L.) by the lensing B-modes [2]. To improve the sensitivity and separation of foreground radiation, SA uses a total of over 20,000 Transition Edge Sensor (TES) bolometers [3] in four frequency bands (90, 150, 220, and 280 GHz). SA observes the sky using three telescopes. POLARBEAR-2A (PB-2A) [4] is the receiver system on the first of three telescopes, with 7588 TES bolometers in two observing bands of 90 and 150 GHz. PB-2A had its first light in 2019 [5] and is currently being used for observing the CMB.

In ground-based CMB experiments, the sensitivity is limited by low-frequency $1/f$ noise especially for observing large angular-scale B-modes caused by inflationary gravitational waves. In PB-2A, a continuously rotating half-wave plate (HWP) [6] is used as a polarization modulator in front of the receiver window to modulate the polarization signal and reduce the effect of the $1/f$ noise [7]. In modulating the polarization signal, the uncertainty of the detector's time constant (time delay in response) causes uncertainties in the observed polarization angle, further resulting in the appearance of a pseudo B-mode [8].

To calibrate the time constant and the relative gain of the detector, we use the blackbody emission from a thermal source that can cover two observing bands simultaneously as a reference source. In this study, we explore the new calibration system using this thermal source and demonstrate the measurement of the time constant during the initial site commissioning test of PB-2A.

2 System Design

TES bolometers were 'tuned' around the transition temperature before starting the daily observations. The time constant of each bolometer is dependent on the tuned status, which is affected by the focal plane temperature and loading (sky) conditions. Therefore, the time constant τ needs to be calibrated for each observation. In PB-2A, the HWP was continuously rotated at 2 Hz in front of the receiver to modulate the polarization signal of the incoming CMB light and shifted it to a higher frequency to avoid the effect of $1/f$ noise [6]. When calibrating the time constants, the uncertainty in the time constant led to uncertainty in the HWP angle for each observation, and ultimately, to variations in the absolute polarization angle.

The uncertainty of the measured time constant of the TES bolometer $\delta\tau$ became the uncertainty $\delta\theta$ in the polarization angle θ. If the rotational speed of the HWP was $f_{\rm HWP}$ Hz, then the uncertainty in the polarization angle would be $\delta\theta \approx 2 \cdot 2\pi f_{\rm HWP}\delta\tau$ [9]. PB-2A calibrated the absolute polarization angles by observing a polarized celestial source such as tau A. We assumed that the uncertainty in the absolute polarization angle may be attributed solely to the variation of polarization angle due to the uncertainty in the time constant. The polarization angle uncertainty $\delta\theta$ should be smaller than $0.1°$ [10], so that the following equation holds.

$$\delta\tau < \frac{\delta\theta}{2 \cdot 2\pi f_{\rm HWP}} \approx 0.069\,{\rm ms}\left(\frac{\delta\theta}{0.1°}\right)\left(\frac{f_{\rm HWP}}{2\,{\rm Hz}}\right)^{-1}. \quad (1)$$

The uncertainties in the time constant should be less than 0.07 ms in each observation.

A TES output I is modeled using a one-pole model as

$$I(f) = \frac{I_0}{\sqrt{1 + (2\pi f\tau)^2}}, \quad (2)$$

where f is a signal modulation frequency, τ is the time constant and I_0 is the signal amplitude. The expected time constant for PB-2A ranged from 1 to 5 ms [11]. To simplify the discussion, the uncertainty of the time constant may be defined as

$$\delta\tau \equiv \left(\frac{dI}{d\tau}\right)^{-1} \delta I, \quad (3)$$

where δI is uncertainty of the signal amplitude. From Eq. 2, the estimated uncertainty of the time constant is

$$\delta\tau = \frac{(1 + (2\pi\tau f_{\rm max})^2)^{\frac{3}{2}}}{(2\pi f_{\rm max})^2\tau} \frac{\delta I}{I_0}. \quad (4)$$

For example, when $\delta I/I_0 = o(10^{-3})$, $\tau = 1$ ms and $f_{\rm max} = 44$ Hz, then $\delta\tau \sim 0.01$ ms, which is small for our requirement.

The system, including the thermal source (1000 K), was mounted behind the telescope's secondary mirror, as shown in Fig. 1. The radiation was emitted through a 9-mm hole at the center of the secondary mirror through a light pipe into the receiver. Using the expected noise-equivalent temperature value, NET $= 360\ \mu{\rm K}\sqrt{\rm s}$ [2], the effective temperature of the thermal source $T_{\rm s} \sim 0.04$ K and the measurement time $t_{\rm mes} = 120$ s, $\delta I/I_0$ can be estimated as $\delta I/I_0 = {\rm NET}/T_{\rm s}\sqrt{t_{\rm mes}} \sim 8 \times 10^{-4}$.

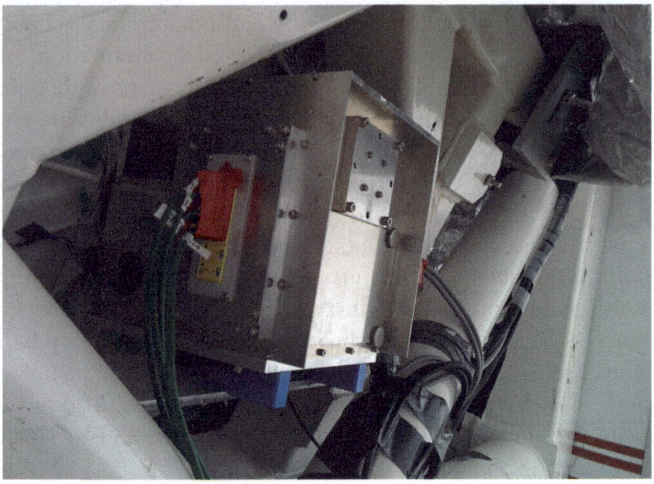

Fig. 1. Experimental system mounted behind secondary mirror. It emits a reference signal through a light-pipe in the 9-mm hole of the secondary mirror.

3 Overview of Calibration System

A side view of the calibration system is shown in Fig. 2. The entire system is a small box, approximately in the shape of a 20-cm cube, to be mounted in the limited space behind the telescope's secondary mirror, as shown in Fig. 1. The system employs a 25 mm × 25 mm ceramic heater (MS-1000 [12]) with a monitored temperature of approximately 1000 K, as the blackbody radiator. The nominal operating voltage of the heater is set as 40 V AC and the wattage of the heater is approximately 30 W. A thermocouple is attached with an alumina bond to monitor the heater temperature at 10-second intervals. The RMS value of the heater temperature stability over the entire season is 2 K, while it is 1 K after the seasonal variations are eliminated.

The system employs Winston cone horns, a type of compound parabolic concentrator (CPC) [13], to cover wide frequency bands. The horns are used to collect radiation from the heater surface and to improve the coupling efficiency before and after the optical chopper wheel. The geometry of the horn is designed to efficiently focus light into a light pipe with an inner diameter of 8 mm, which penetrates through the secondary mirror. The heater source section has a double-layered cover for improving the temperature stability. In the case of a heater breakdown under the harsh environment of the observation site, the first layer of the light source section alone can be replaced from behind in the form of a cartridge without dismantling the entire system. The system has remained operational throughout the past three years without any breakdowns.

The radiation from the heater is modulated by a rotating six-blade chopper to measure the time delay in the response of the detector. An encoder (photosensor, EE-SX1140) is mounted on the opposite side of the light pipe to monitor the timing of the opening and closing of the chopper blade. The signal from the encoder is synchronized with the time-ordered bolometer data, whose sampling rate is 152 Hz [14] for analysis.

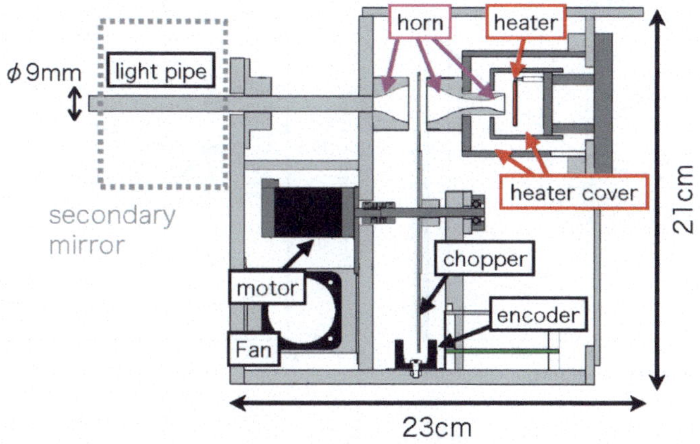

Fig. 2. Side view of the system. System has a small chopped thermal source, where a ceramic heater is used as a blackbody radiator to cover up observation frequency bands.

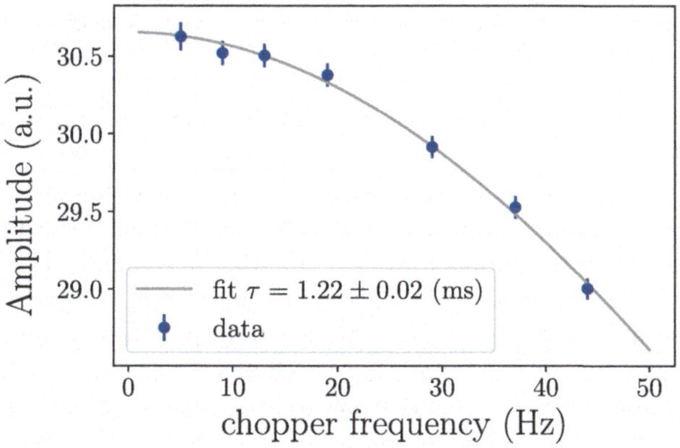

Fig. 3. Example of signal amplitudes from one calibration sequence in a 150 GHz bolometer. The time constant of the bolometer is measured by fitting the point of the signal amplitude at each frequency with the one-pole model.

The rotational stability of the chopper is typically 0.2%. It exhibits a good chopping stability, and the chopping frequency increases to the Nyquist frequency of the sample rate. The stepping motor for the rotating chopper, which needs to be cooled by fans, is placed in a separate room to improve the internal temperature stability of the box.

4 Site Commissioning Test

The PB-2A receiver and the developed thermal source system were installed on the telescope in late 2018 and early 2019. PB-2A had its first light in 2019 for a demonstration that took place during the early commissioning test in 2019.

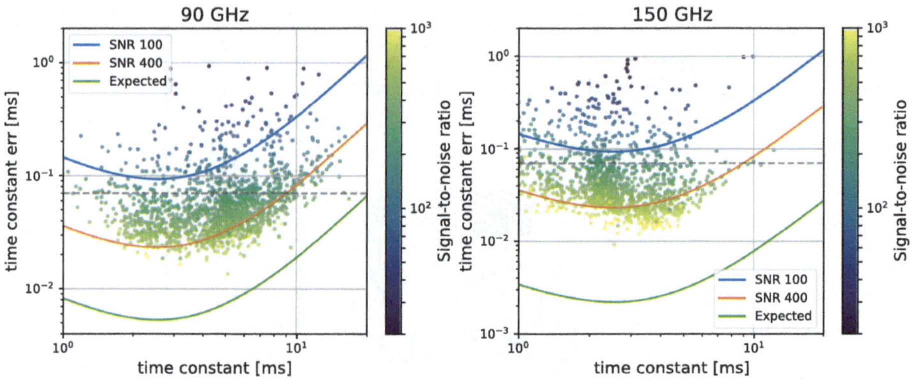

Fig. 4. Scatter plot of the time constants and their uncertainties. The signal-to-noise ratio of the signal from the thermal source is plotted using a color scale. The blue, orange, and green lines show $\delta\tau$ from Eq. 4 with signal-to-noise ratio, $I_0/\delta I$, of 100, 400, and that for the expected sensitivity, respectively. The bolometer with the higher signal-to-noise ratio meets the calibration accuracy requirement (< 0.07 ms) in the expected noise level. This data was taken in an initial commissioning phase, and sufficient calibration accuracy can be achieved by improving the expected noise level ($\mathrm{NET} = 360\mu\mathrm{K}\sqrt{\mathrm{s}}$).

The number of biased bolometers for which the signal from this system appears to have a signal-to-noise ratio of 20 or better is approximately 99 % across the entire focal plane of the PB-2A receiver. The time constant of the bolometers is measured by checking the detector time response to the modulated signal from this system. The rotational frequencies of the chopper are 5, 9, 13, 19, 29, 37, and 44 Hz, and the signal is observed for 120 s each, for a total of approximately 14 min. Before data analysis, the time-ordered bolometer data was passed through high-pass filters and the mean of the baseline was subtracted. In data analysis, we chunked and stacked the time-ordered bolometer data according to the encoder phase. Then, the amplitude of the modulated signal from the system was estimated by fitting the stacked data. We used the fitting function of the sum of the sinusoidal wave, $A(\theta) = \sum_{n=1}^{7} A_n \sin(n\theta + \phi_n)$ $[0 \leqq \theta < 2\pi]$, A_1 was assumed as an amplitude of the modulated signal. The one-pole time constant model (Eq. 2) was used to fit the point of the signal amplitude at each frequency. An example of the fitting is shown in Fig. 3. At the same time, the relative gain I_0 of the detector can also be obtained from the fit.

Figure 4 shows a scatter plot of the time constants and their uncertainties obtained by fitting the response of each bolometer. The signal-to-noise ratio based on the relative gain I_0 obtained by fitting is plotted using a color scale. As shown in Eq. 4, the magnitude of the time constant uncertainties is proportional to the noise level, and the trend of the experimental data is in accordance with the equation. The bolometer with the highest signal-to-noise ratio meets the time constant calibration accuracy requirement described in the above section. If the noise level is close to the design value, Eq. 4 suggests that, by improving the noise level, the uncertainty in the time constant can be reduced and the required calibration accuracy can be achieved. As this test was performed at the beginning of the commissioning test, the time constants were slower and

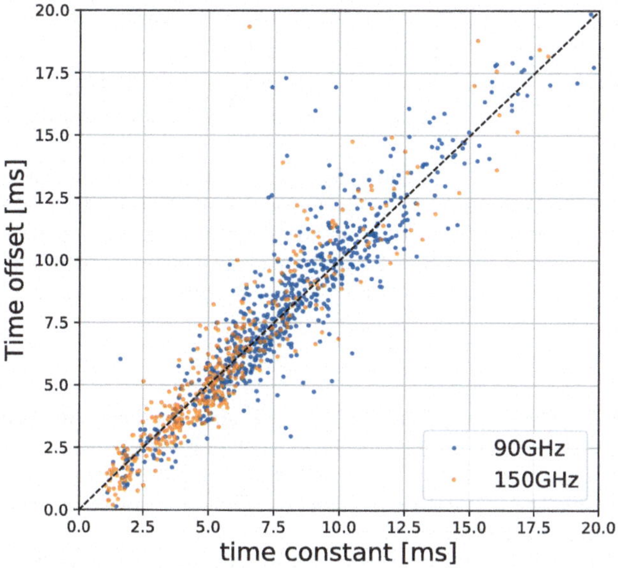

Fig. 5. Comparison of time constants between calibration system and astronomical point source. This scatter plot shows a strong correlation.

noise level were higher than expected because the bias tuning of the bolometers was not optimized.

The time constant was also obtained from the data of the astronomical point source (Jupiter) observation during commissioning and compared with the time constant obtained using this calibration system. In SA, the telescope observes the astronomical point source by changing the elevation slightly and scanning left and right in the azimuth direction. The effect of the time constant appears as a shift in the apparent position along the scanning direction. The position of the celestial body was obtained by fitting with a 2D Gaussian. The time constant value obtained from the difference in the position of point sources is shown in Fig. 5. We checked the consistency of the time constants obtained from our calibration system and from other calibration point sources. Note that as this data was acquired during the ongoing alignment of the receiver position, the variation in time offset was not directly indicative of the precision of the time constant measurement using the astronomical point source.

5 Conclusions

In the PB-2A receiver system, a continuously rotating HWP was used to mitigate the low-frequency noise. With rapid polarization modulation, the uncertainty in the time constant of the detector resulted in an uncertainty in the polarization angle. To calibrate the time constant of the detector, we developed a new calibration system with a thermal source. The system employed a thermal ceramic heater (1000 K) as a blackbody radiator to cover the entire frequency band of the PB-2A observation. In early commissioning tests at the observation site in Chile in 2019, we demonstrated that the time constant could be measured for several thousand detectors simultaneously and confirmed that our system could calibrate each bolometer at a sub-ms level. This system will contribute to the calibration of the detector time constant for PB-2A observation.

Acknowledgments. This work was supported by the Grant-in-Aid for JSPS Fellows (Grant Number 19J10652), Ozaki exchange program 2019 and SOKENDAI Advanced Research Course program. M.H. acknowledges the support from World Premier International Research Center Initiative (WPI,) MEXT, Japan and JSPS KAKENHI Grant Number 22H04945. I would like to thank T. Hamada and S. Kumazaki for their helpful comments and discussions.

References

1. Stebor, N., et al.: The Simons array CMB polarization experiment. In: Millimeter, Submillimeter, and Far-Infrared Detectors and Instrumentation for Astronomy VIII, vol. 9914, p. 99141H. International Society for Optics and Photonics. SPIE (2016)
2. Suzuki, A., et al.: The POLARBEAR-2 and the Simons array experiments. J. Low Temp. Phys. **184**(3–4), 805–810 (2016)
3. Irwin, K.D., Hilton, G.C.: Transition-Edge Sensors, pp. 63–150. Springer, Heidelberg (2005)
4. Hasegawa, M., et al.: POLARBEAR-2: a new CMB polarization receiver system for the Simons Array. In: Millimeter, Submillimeter, and Far-Infrared Detectors and Instrumentation for Astronomy IX, vol. 10708. International Society for Optics and Photonics, SPIE (2018)
5. Kaneko, D., et al.: Deployment of POLARBEAR-2A. J. Low Temp. Phys. **199**(3–4), 1137–1147 (2020)
6. Hill, C.A., et al.: Design and development of an ambient-temperature continuously-rotating achromatic half-wave plate for CMB polarization modulation on the POLARBEAR-2 experiment. In: Millimeter. Submillimeter, and Far-Infrared Detectors and Instrumentation for Astronomy VIII, vol. 9914, pp. 699–716. International Society for Optics and Photonics, SPIE (2016)
7. Takakura, S., et al.: Performance of a continuously rotating half-wave plate on the POLARBEAR telescope. J. Cosmol. Astropart. Phys. **2017**(05), 008–008 (2017)
8. Shimon, M., Keating, B., Ponthieu, N., Hivon, E.: CMB polarization systematics due to beam asymmetry: Impact on inflationary science. Phys. Rev. D **77**(8), 103002 (2008)
9. Abitbol, M.H., Alonso, D., Simon, S.M., Lashner, J., Crowley, K.T., et al.: The Simons Observatory: gain, bandpass and polarization-angle calibration requirements for B-mode searches. J. Cosmol. Astropart. Phys. **2021**(05), 032 (2021)
10. Navaroli, M.F., et al.: Design and characterization of a ground-based absolute polarization calibrator for use with polarization sensitive CMB experiments. In: Millimeter. Submillimeter, and Far-Infrared Detectors and Instrumentation for Astronomy IX, vol. 10708, pp. 274–288. International Society for Optics and Photonics, SPIE (2018)
11. Barron, D., Mitchell, K., Groh, J., et al.: Integrated electrical properties of the frequency multiplexed cryogenic readout system for POLARBEAR/Simons Array. IEEE Trans. Appl. Supercond. **31**(5), 1–5 (2021)
12. http://sakaguchi-dennetsu.co.jp/lineup/heater/ceramic_heater_index/
13. Winston, R.: Light collection within the framework of geometrical optics. J. Opt. Soc. Am. **60**(2), 245–247 (1970)
14. Nishino, H., et al.: Data acquisition and management system for the CMB polarization experiment: Simons Array. In: Millimeter. Submillimeter, and Far-Infrared Detectors and Instrumentation for Astronomy X, vol. 11453, pp. 328–340. International Society for Optics and Photonics, SPIE (2020)

Atomic and Molecular Physics

Photoionization Spectra of Atomic Sulfur ^1S in Transitions to Autoionizing Rydberg States in the Range of 73400–85300 cm^{-1}

Tzu-Ping Huang, Wan-Chun Pan, and Yin-Yu Lee(✉)

National Synchrotron Radiation Research Center, Hsinchu, Taiwan
tphuang@nsrrc.org.tw

Abstract. The photoionization spectra of sulfur atoms in transitions from state ^1S were recorded with synchrotron radiation (NSRRC, TLS BL21B2 beamline U9CGM) at resolution of up to 3 cm^{-1} in the region of 73400–85300 cm^{-1}. The photolysis of carbonyl sulfide (OCS) after 157 nm was used to produced the singlet (^1S) sulfur atoms that were then excited the neutral superexcited states with synchrotron radiation. Rydberg series $3s^2 3p^3(^2P_0)ns\ ^1P$ and $3s^2 3p^3(^2P_0)nd\ ^1P$ with *n* from 5–14 for the former and 3–12 for the latter are assigned.

Keywords: photoionization · Rydberg state · atomic sulfur

1 Introduction

Sulfur is one of the more abundant elements in the universe [1]. The relative high abundance of sulfur in the solar atmosphere has made the investigation of the ultraviolet (UV) and vaccum UV (VUV) spectroscopic lines of S important in the modeling of stellar atmosphere [2–5]. In planetary nebulae, the analysis of nebular emission from elements that have experienced nucleosynthesis in the parents star provides valuable information for stellar models [6,7]. Sulfur ion (S^+) has been discovered in the torus plasma of the Jupiter satellite Io [8]. The VUV photoionizaiton studies of the S atoms is also important for the fundamental understanding of photoionization dynamics [9, 10]. The photoionization studies of Rydberg lines include the photoabsorption spectra and photoionization spectra.

The ground state electronic configuration of atom sulfur is $1s^2 2s^2 2p^6 3s^2 3p^4$, which forms the three electronics terms, $^3P_{0,1,2}$, $^1D_2$ and $^1S_0$. Similarly, the ground state electronics configuration of sulfur cation is $1s^2 2s^2 2p^6 3s^2 3p^3$, forming the three electronics terms, $^4S^0_{3/2}$, $^2D_0$ and $^2P_0$. The photoionization spectrum of sulfur has been studied experimentally. Tondello reported cross sections of single-photon Rydberg lines in absorption from 10.38–13.8 eV using the flash-pyrolysis method, i.e., irradiating the sample with a strong flux of light from discharge flash lamp [11]. Sarma and Joshi modified measured the relative intensities for absorption in Rydberg series $3s^2 3p^3(^2D)nd$ ^3S, ^3D, ^3P, and ^3F from S(^3P) and series $3s^2 3p^3(^2P)ns\ ^1P$ and $nd\ ^1D$ from S($^1D_2$) [12]. Gibson et al. observed that the autoionization features of the $3s^2 3p^3(^2D_0)nd$, ^3D levels

H. J. Choi et al. (Eds.): APPC 2022, *Proceedings of the 15th Asia Pacific Physics Conference*, pp. 65–70, 2024.
https://doi.org/10.1007/978-981-96-0191-2_7

are broad, while that of the $3s^23p^3(^2D_0)nd$ ^3S, ^3P levels are narrow [13]. Innocenti et al. reported the partial photoionization cross sections and asymmetry parameters in the photon energy range of 10.0–30.0 eV; they measured the transitions from S(^3P) up to the fourth ionization limit [14].

In 2008, Yang et al. measured the single-photon excitation spectra from the lowest $3s^23p^4$ $^1D_2$ and $3s^23p^4$ $^1S_0$ metastable levels of sulfur atoms recorded with a tunable VUV laser served as the ionization source. The photoionization efficiency (PIE) spectra of metastable sulfurs (S) atoms have been recorded in the range 73350–84950 cm^{-1} [15]. Our previous studies have found that the photoionization spectra of sulfur atoms in transitions from the $3s^23p^4$ $^1D_2$ state in the energy range of 75800–89500 cm^{-1} [16]. We recorded and assigned the Rydberg series $3s^23p^3(^2D^0_{3/2})nd$[3/2] and $3s^23p^3(^2D^0_{5/2})nd$[5/2] with n extending to 16 and 32, respectively, to the $^2D_0$ series limit. In addition, Rydberg series $3s^23p^3(^2D^0_{3/2})nd$[1/2], $3s^23p^3(^2D^0_{5/2})nd$[5/2] and $3s^23p^3(^2D^0_{3/2})nd$[5/2] with n from 5–9 for the former two series and 7–13 for the latter, were able to be assigned. Inspired by these results, we have found that high quantum yields of $3s^23p^4$ $^1S_0$ metastable level of sulfur atoms produced by the photodissociation of OCS at 157 nm [17]. Here, we focus on autoionizing Rydberg states in transitions from $^1S_0$ in the range of 73400–85300 cm^{-1}.

2 Experiments and Discussion

A detailed experimental setup has been described previously [16]. The OCS used in the experiment was purchased from Sigma-Aldrich with a stated purity of 97.5%. In the molecular-beam source chamber, there is an Even-Lavie valve operating at 75 Hz that is used to produce a supersonic molecular beam of OCS. A F_2 excimer laser (GAM laser, EX750) running on a 2.4k torr F_2-He mixture was using at a rate of repetition rate 75 Hz in the photolysis of this study and the laser power was 0.7 mJ/pulse. The S^+ signals were selected by setting a quadruple mass at m/z = 32 and scanning vuv wavelength with synchrontron radiation. The main chamber pressure was typically operated at less the 2×10^{-6} torr. The synchrotron radiation was provided by an undulator at NSRRC(TLS BL21B2 U9CGM) with resolving power of 20000 at 10 eV with slit width of 50 μm.

The VUV-photoionization spectrum for neutral sulfur atoms produced from OCS on photolysis at 157 nm in the region of 73400–85300 cm^{-1} is shown in Fig 1. The peaks with asterisk have been observed and assigned by others previously [16]. At the high-energy end, two assignments from S(^3P) at wavenumbers 85427, and 85685 cm^{-1} are broad, expected to be short autoionization lifetime. The narrow lines from S(^1D) have a width limited by instrumental resolution indicating that these resonances have few picoseconds lifetime [15].

The quantum defect of Rydberg series is calculated with Rydberg formula Eq. (1),

$$E_r = T - \frac{Z^2 R_s}{(n-\mu)^2} \tag{1}$$

(1) in which R_s is the Rydberg constant, 109735.44 cm^{-1} for atomic sulfur, and E_r is the resonance energy between the initial and final states. The denominator n-μ

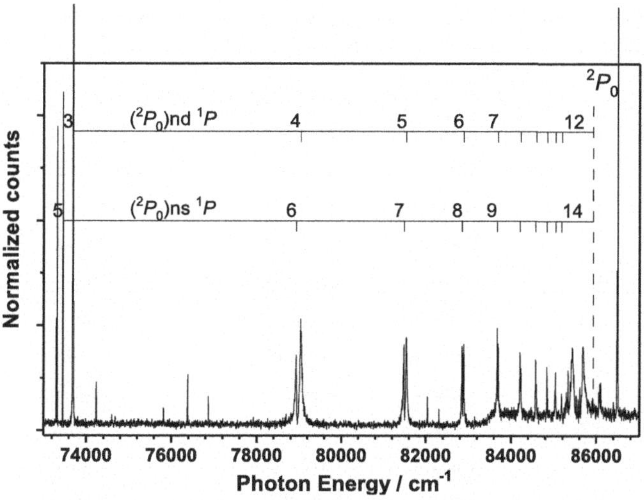

Fig. 1. The VUV-photoionization spectra of atomic sulfur $^1S_0$ in transitions to autoionizing Rydberg states $3s^23p^3(^2P_0)ns\ ^1P$ and $3s^23p^3(^2P_0)nd\ ^1P$.

represents the observed principal quantum number n^*; parameter μ denotes a quantum defect depending on the angular momentum of the electronic orbital. T represents the convergence limit of a Rydberg series. The values of T for formation of $S^+\ ^2P_0$ is 85927.05 cm^{-1} relative to S($^3P_2$) [18, 19]. Z is the effective charge of the core(in this case Z = 1). The observed line positions and quantum defects are listed in Table 1 and Table 2.

The energy levels at 73457 cm^{-1} and 73692 cm^{-1} are assigned as $3s^23p^3(^2P_0)5s\ ^1P$ and $3s^23p^3(^2P_0)3d\ ^1P$ [15, 16, 20]. These are in good agreement with the values from the NIST tabulations of 73459 and 73693.6 cm^{-1} [21]. Theoretical calculations on Rydberg states of S with electron configuration and electronic terms are $(^2P_0)ns\ ^1P$ and $(^2P_0)nd\ ^1P$; the quantum defects of these states are predicted to be approximately 1.997 and −0.0675, respectively [22]. In present work the quantum defects obtained for the *ns* resonance in Table 1 and the nd resonance in Table 2 are seen to be good accordance with the values available from theoretical calculations. Extrapolating with the Rydberg formula using [see Eq. (1)] could straightforwardly locate further members of the Rydberg resonance series.

In Table 1, we list the experimental data for energy resonance (E_r) of $3s^23p^3(^2P_0)ns\ ^1P$ Rydberg lines relatively to the $^1S_0$ metastable states of S^+ ion. All our quantum defects are obtained using Eq. (1) and converted into (cm^{-1}) for direct comparison. We note the present quantum defects is almost constant and positive along the Rydberg series.

In Table 2, we quote the results for energy resonance (E_r) and quantum defects of $3s^23p^3(^2P_0)nd\ ^1P$ Rydberg lines relatively to the $^1S_0$ metastable states of S^+ ion. These resonance states extend quantum number up to 12. Quantum defects are positive from

Table 1. Energies of autoionization Rydberg states with quantum numbers and defects. The assignments of these transitions as the Rydberg series $3s^2 3p^3(^2P_0)ns\ ^1P \leftarrow$ S($^1S_0$) are shown in Fig. 1.

n^c	$E_r(cm^{-1})$	$E_r(cm^{-1})^a$	$E_{level}(cm^{-1})^b$	μ	μ^c
5	73459[e]	73457	95637	2.033[e]	2.034
6		78930	101110		2.04
7		81477	103657		2.034
8		82846	105026		2.032
9		83673	105853		2.023
10		84210	106390		2.006
11		84573	106753		1.998
12		84835	107015		1.976
13		85032	107212		1.927
14		85173	107353		1.937
⋮		⋮	⋮		⋮
∞		85927[d]	107727		

[a] E_r is the resonance energy between the initial S $^1S_0$ and final states with uncertainty of < 4 cm^{-1} to the absolute value.
[b] E_{level} is the energy relative to S $^3P_2$.
[c] n is the principal quantum number. μ is the quantum defect.
[d] Transitions found in NIST database.
[e] Transitions first observed or reassigned by Yang and co-workers.

$n = 3 - 5$ because the effective charges of these self-ionizing state are less than the charge of the S^+ ion and if not, they are negative from $n = 6 - 12$.

We note for the case of the $3s^2 3p^4\ ^1S_0 {-} 3s^2 3p^3(^2P_0)5s\ ^1P$ Rydberg state that lies below the $S^+(^2D_0)$ continuum. The spectral line shape of the $3s^2 3p^3(^2P_0)5s\ ^1P$ level showing that might be an interloping resonance [23], as illustrated in Fig. 1, which disrupts the regular Rydberg pattern. Above the convergence limit of $S^+(^2D_0)$ and below the convergence limit of $S^+(^2P_0)$ the $3s^2 3p^3(^2P_0)ns\ ^1P$ series follows normal Rydberg pattern are obtained. In the $3s^2 3p^3(^2P_0)nd\ ^1P$ Rydberg series only the lowest observed member $3s^2 3p^3(^2P_0)3d\ ^1P$ was found.

Table 2. Energies of autoionization Rydberg states with quantum numbers and defects. The assignments of these transitions as the Rydberg series $3s^23p^3(^2P_0)nd\ ^1P \leftarrow$ S($^1S_0$) are shown in Fig. 1.

n^c	$E_r(cm^{-1})$	$E_r(cm^{-1})^a$	$E_{level}(cm^{-1})^b$	μ	μ^c
3	73693.6[e]	73692	98572	0.004[e]	0.005
4		79039	101219		0.009
5		81538	103718		0
6		82891	105071		−0.012
7		83702	105882		−0.023
8		84230	106410		−0.041
9		84593	106773		−0.07
10		84847	107027		−0.08
11		85040	107220		−0.122
12		85190	107370		−0.202
⋮		⋮	⋮		⋮
∞		85927[d]	107727		

[a] E_r is the resonance energy between the initial S $^1S_0$ and final states with uncertainty of < 4 cm^{-1} to the absolute value.

[b] E_{level} is the energy relative to S $^3P_2$.

[c] n is the principal quantum number. μ is the quantum defect.

[d] Transitions found in NIST database.

[e] Transitions first observed or reassigned by Yang and co-workers

3 Conclusions

The single photon excitation spectra of sulfur atoms from metastable state $^1S_0$ produced by OCS after a F_2 laser was recorded with synchrotron radiation. We have assigned autoionizing Rydberg series $3s^23p^3(^2P_0)ns\ ^1P$ and $3s^23p^3(^2P_0)nd\ ^1P$ with *n* extending to 14 and 12, respectively. Energies of autoionization Rydberg states with quantum numbers and defects are tabulated. Further theoretical calculations are expected to explain these experimental results.

Acknowledgments. The authors thank the National Synchrotron Radiation Center and the financial support of National Science and Technology Council of the Republic of China.

References

1. Asplund, M., Grevesse, N., Sauval, A.J., Scott, P.: Ann. Rev. Astron. Astrophys. **47**, 481 (2009)
2. Broadfoot, A.L., et al.: J. Geophys. Res. **86**, 8259 (1981)

3. Shemansky, D.E., Smith, G.R.: J. Geophys. Res. **86**, 9179 (1981)
4. Federman, S.R., Cardelli, J.A.: Astrophys. J. **452**, 269 (1995)
5. Deb, N.C., Hibbert, A.: J. Phys. B **25**, 2279 (1992)
6. Sharpee, B., et al.: Astrophys. J. **659**, 1265 (2007)
7. Sterling, N.C., Dinerstein, H.L., Kallman, T.R.: Astrophys. J. Suppl. Ser. **169**, 37 (2007)
8. Summers, M.E., Strobel, D.F., Yung, Y.L., Trauger, J.T., Mills, F.: Astrophys. J. **343**, 468 (1989)
9. Altun, Z.: J. Phys. B. **25**, 2279 (1992)
10. Dill, D., Starace, A.F., Manson, S.T.: Phys. Rev. A **11**, 1596 (1975)
11. Tondello, G.: Astrophys. J. **172**, 771 (1972)
12. Sarma, V.N., Joshi, Y.N.: Physica C **123**, 349 (1984)
13. Gibson, S.T., Greens, J.P., Ruscic, B., Berkowitz, J.: J. Phys. B At. Mol. Phys. **19**, 2825 (1986)
14. Innocenti, F., et al.: J. Chem. Phys. **126**, 154310 (2007)
15. Yang, X., Zhou, J., Jones, B., Ng, C.Y., Jackson, W.M.: J. Chem. Phys. **128**, 084303 (2008)
16. Pan, W.C., Chen, I.C., Huang, T.P., Yuh, J.Y., Lee, Y.Y.: J. Chem. Phys. **129**, 134305 (2008)
17. Loo, R.O., Strauss, C.E., Haerri, H.P., Hall, G.E., Houston, P.L.: J. Chem. Soc. Faraday Trans. 2 **85**(8) 1185 (1989)
18. Sugar, J., Musgrove, A.: J. Phys. Chem. Ref. Data **20**, 859 (1991)
19. Saloman, E.B.: J. Phys. Chem. Ref. Data **36**, 215 (2007)
20. Gallagher, T.F.: Rydberg Atoms. Cambridge University Press, New York (1994)
21. Ralchenko, Y., Kramida, A.E., Reader, J., NIST ASD Team: NIST Atomic Spectra Database, version 5 (National Institute of Standards and Technology, Gaithersburg, MD)
22. Mendoza, C.: Atomic and molecular Data and their Applications, p. 313. AIP, New York (2000). http://cdsweb.u-strasbg.fr/topbase/topbase.html. Edited by K. A. Berrington and K. L. Bell
23. Balance, C.P., Berrington, K.A., McLaughlin, B.M.: Phys. Rev. A **60**, R4217 (1999)

Condensed Matter Physics

Highly Tunable Negative Poisson's Ratio in Two-Dimensional Penta-SiCN

Shambhu Bhandari Sharma(✉) and Issam Qattan(✉)

Department of Physics, Khalifa University of Science and Technology, P.O Box 127788, Abu Dhabi, UAE
{shambhu.sharma,issam.qattan}@ku.ac.ae, shambhubhandari789@gmail.com

Abstract. In this work, we utilize first-principles Density Functional Theory (DFT) calculations to predict the structural, and mechanical stability, and properties of new ternary penta-SiCN. Penta-SiCN is found to be a highly auxetic material exhibiting negative Poisson's ratio and Young's modulus value of –0.136, and 131.29 N/m, respectively. In addition, mechanical parameters are highly tunable, super-elastic, and isotropic against the biaxial strain, making penta-SiCN a premium material in nanoauxetic and nanomechanics applications.

1 Introduction

The identification of negative Poisson's ratio (NPR) in the pentagonal lattice-based two-dimensinal (2D), pentagraphene [1], has shed light on pentagonal materials for auxetic mechanical properties. Since then, several pentagonal binaries to ternary (penta-B_2N_4 and penta-B N_2 [2], pentagraphene [1], and penta-$PdSe_2$ [3], penta-CNP [4], penta-PBN [5]) iso-structures are reported to possess rare NPR. These materials have peculiar physical properties such as enhanced noise insulation efficiency with increased indentation, impact, and fracture transmission resilience [6,7], which are unique from other ordinary materials. These excellent properties can be utilized in a diverse range of applications in automobiles, aviation, naval, and some other commercial fields [8]. More specifically, they are highly desired in fasteners, bullet-proof outfits, sporting goods, bio-engineering, and smart technological devices [9,10].

The NPR properties of 2D materials were mainly investigated theoretically and computationally. However, the first experimental study to verify the NPR property was done on black phosphorene [11] employing Raman spectroscopy. Nevertheless, these materials are expected to be characterized thoroughly with the advancement of new experimental techniques and instrumentation.

The unique non-planar geometry, and tri-layered geometry of pentagon-based 2D materials are reported to possess higher intrinsic NPR compared to their planar forms [9,12]. In addition, the tunability of NPR with strain has enormous potential in a wide range of applications [12]. For example, the ternary lightweight penta-sheets are gaining popularity because of their distinct physiochemical features, as well as their potential utility [4,5,13–17]. Of particular interest, the lightweight materials comprised of group II of the periodic table because of their abundance, environment friendly, and energetically stable nature.

H. J. Choi et al. (Eds.): APPC 2022, *Proceedings of the 15th Asia Pacific Physics Conference*, pp. 73–79, 2024.
https://doi.org/10.1007/978-981-96-0191-2_8

Understanding the strong chemistry between Si, C, and N with equivalent atomic radii, which provides them tight inter-atomic binding to gain substantial mechanical stability and strength, in this work, we have used first-principles DFT calculations to identify and predict penta-SiCN, a new lightweight ternary pentagonal-based compound. To check the stability of penta-SiCN, we have investigated the cohesive energy (E_{coh}), as well as performed frozen phonon calculations. Furthermore, the mechanical strength and stability are tested by the Bohr criteria. Important elastic parameters such as the in-plane stiffness (2D Young's modulus), and the Poisson's ratio, are calculated in order to understand the mechanical response of penta-SiCN. Further, the modulation of these parameters are tested against the biaxial strain. Finally, the ultimate strength and deformation mechanism is verified by applying a higher mode of strain.

2 Computational Details

The Spanish Initiative for Electronic Simulations with Thousands of Atoms (SIESTA) code [18, 19] is used to perform all the calculations. The semi-local form of the norm-conserving pseudopotentials [20] and the norm-conserving Troullier-Martins pseudopotentials [20] are used to describe the interactions between electrons and core in the Kleinman-Bylander factorized form. The generalized Gradient Approximation (GGA) of the Perdew-Burke-Ernzerhof (PBE) parametrization [21] is utilized to incorporate the electronic exchange and correlation potential. The standard double-ζ polarized (DZP) basis set assembled from numerical atomic orbitals (NAO) of the finite range is used to describe the electronic distribution of each atom. The convergence criteria are set to be 20×20×1 k-points under the Monkhrost pack scheme [22], and 350 Ry of the cutoff energy. To optimize the geometry, atomic forces less than 0.01 eV/Å, and 10^{-6} eV self-consistent field (SCF), as integrated in the conjugate-gradient (CG) scheme, are employed. A thick empty space (25 Å) is created along the out-of-plane direction to remove the spurious interactions between the adjacent unit cells. The inter-atomic binding and stability is ensured by calculating the cohesive (E_{coh}) energy [23]:

$$E_{coh} = \left(E_{p-SiCN} - 2E_{Si} - 2E_C - 2E_N\right)/6, \tag{1}$$

where E_{p-SiCN}, E_{Si}, E_C, and E_N denote the energy of penta-SiCN, free and isolated Si, C, and N atoms, respectively. The kinetic stability of the penta-SiCN is tested by calculating the phonon bands employing the finite displacement method of PHONOPY package [24] as integrated in SIESTA. The in-plane stiffness (Y) and Poisson's ratio (ν) at various orientation angle θ are represented by the relations [25]:

$$Y(\theta) = \frac{[C_{11}C_{22} - C_{12}^2]}{C_{22}\cos^4(\theta) + A\cos^2(\theta)\sin^2(\theta) + C_{11}\sin^4(\theta)}$$

$$\nu(\theta) = \frac{-C_{12}[\cos^4(\theta) + \sin^4(\theta)] + B\cos^2(\theta)\sin^2(\theta)}{C_{22}\cos^4(\theta) + A\cos^2(\theta)\sin^2(\theta) + C_{11}\sin^4(\theta)}$$

where $A = \left(C_{11}C_{22} - C_{12}^2\right)/C_{66} - 2C_{12}$ and $B = C_{11} + C_{22} - \left(C_{11}C_{22} + C_{12}^2\right)/C_{66}$.

3 Results and Discussion

The unit cell of penta-SiCN consists of 2 atoms of each Si, C, and N atom in equal proportion forming 3 virtual layers composited in a monolayer. The fully relaxed penta-SiCN belongs to P-421 m symmetry (space group No. 113) of pentagonal 2D, and lattice parameter a = b = 4.31 Å (Fig. 1a, b).

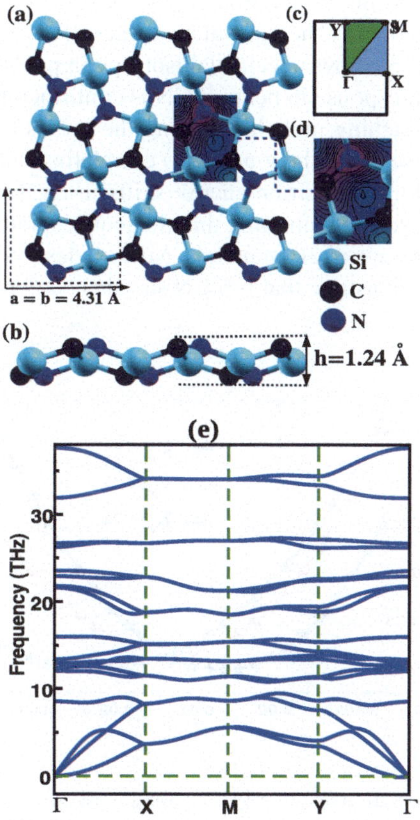

Fig. 1. (Color online) The fully optimized penta-SiCN (4 × 4 × 1 supercell): (a) top view, (b) side view, (c) first Brillouin zone, (d) charge density distribution, and (e) phonon dispersion band.

The bond lengths of Si–N, C–N and Si–C, are found to be 1.83, 1.38, and 1.85 Å, respectively. Similarly, the thickness of the monolayer (h) is 1.24 Å, which is equivalent to the pentagraphene [1] but contrasting with those of penta-BCN [13], penta-CNP [4], and penta-PBN [5]. The Mulliken charge density analysis shows that the higher charge density (pink color) is between the C–N, indicating covalent bonding, whereas other bonds appear to be ionic (Fig. 1d). In addition, the high negative cohesive energy of –4.36 eV/atom demonstrates strong inter-atomic binding in penta-SiCN. On the other hand, phonon calculations are calculated along the Brillouin zone path

$\Gamma - X - M - Y - \Gamma$ (Fig. 1c, e), which shows 3 acoustic and 15 optical frequencies with the absence of any imaginary band, and justifies its dynamical stability. Following the method of strain-energy [13,17], the calculated value of C_{11}, C_{22}, C_{12}, and C_{66} are found to be 132.16, 133.59, –17.44, and 74.80 N/m, respectively. The Born's criteria [26], $C_{11}C_{22} - C_{11}^2 > 0$, and $C_{66} > 0$ verifies the mechanical stability. The 2D Young's modulus [27,28] along the x-axis $Y_a = \left(C_{11}^2 - C_{12}^2\right)/C_{11}$, and along the y-axis $Y_b = \left(C_{22}^2 - C_{12}^2\right)/C_{11}$, are found to be 129.88 and 131.29 N/m, respectively. And the Poisson's ratios in these directions ($\nu_a = C_{12}/C_{11}$ and $\nu_b = C_{12}/C_{22}$) are -0.131 and -0.132, respectively. The equivalent values of $Y_{a/b}$, and $\nu_{a/b}$ with perfect circles in polar plots (Fig. 3) shows mechanial isotropy in penta-SiCN.

The remarkable NPR response in penta-SiCN is confirmed by using the methods of references [1,17,29]–stretching monolayer along the x-axis (ε_x = 4%, 6%, and 8%), and observing the response along the y-axis (ε_y) correlating with strain energy. In subsequent stretching, the minimum strain energy shifts in the forward direction (Fig. 2) justifying the NPR. It is worth mentioning that the tensile strain is sufficient enough to demonstrate the NPR behavior of 2D materials. Nevertheless, compressive strain is also applied in our work to demonstrate that p-SiCN sustains compression within the elastic range.

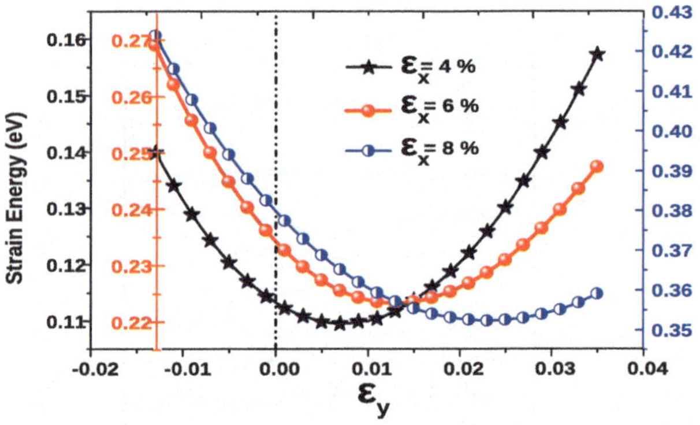

Fig. 2. (Color online) Confirmation test of NPR – variation of strain energy with different mode of strain.

Because the tunability of Y and ν against ε_{xy} impacts the piezoelectric property of materials [30], we load the equ-biaxial (compressive/tensile) strain ($-4\% \leq \varepsilon_{xy} \leq +4\%$) at an interval of 2% allowing atoms to fully relax (relaxed–ion). Applying compressive biaxial strain from equilibrium ($\varepsilon_{xy} = 0$) to $\varepsilon_{xy} = -4\%$, the NPR value decreases gradually and reaches –0.05. Conversely, for tensile strain ($\varepsilon_{xy} \leq +4\%$), the NPR increases significantly reaching –0.64, indicating high tunability of NPR in penta-SiCN. Further, the perfect circles displayed in the polar plot (Fig. 3a) of the Poisson's ratio, except at $\varepsilon_{xy} = -2\%$, demonstrates the mechanical isotropy of the monolayer even in strained case. On the other hand, applying compressive strain in the range $-4\% \leq \varepsilon_{xy} < 0$, the value of the Young's modulus increases monotonically reaching 145 N/m, and then decreases abruptly up to 42 N/m for $\varepsilon_{xy} = +4\%$. This indicates that penta-SiCN becomes soft and fragile with tensile strain, and the rigidity increases with compression. Further, the perfect circular distribution of the Young's modulus (Fig. 3b) for each strain step supports the mechanical isotropy. Consequently, the highly adjustable NPR and Young's modulus of the penta-SiCN generate a huge possibility for electromechanical devices in a variety of industrial domains such as health, military, and lightweight and versatile electronic gadgets [9, 12].

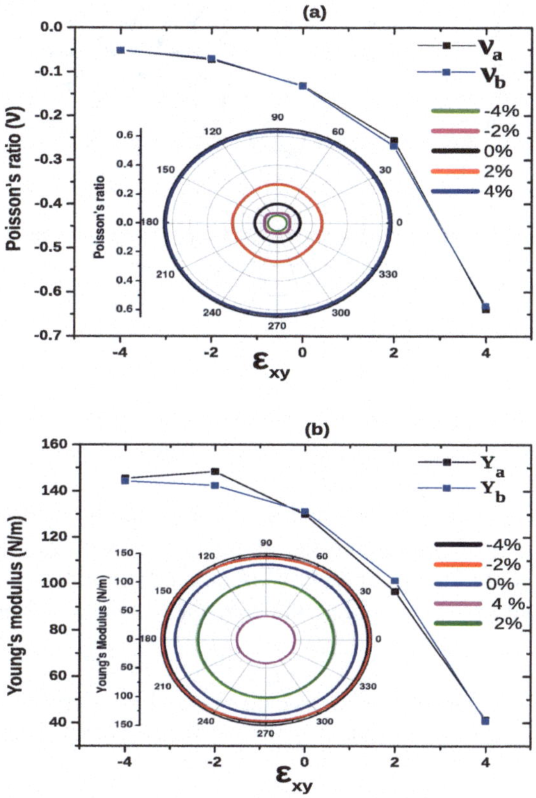

Fig. 3. (Color online) The tunablity of (a) Poisson's ratio and (b) Young's modulus as a function of biaxial strain. The inset demonstrates the orientation dependency.

Further, with an applied biaxial tensile strain above 16.0%, the monolayer completely breaks (Fig. 4). In addition, the stress and strain energy increase with increasing ε_{xy}. At $\varepsilon_{xy} = 16\%$, stress suddenly drops indicating the critical elastic point of penta-SiCN. The noted stress required to set the monolayer into that state, critical stress, is 12.14 N/m, which is relatively smaller than that of hexa-graphene (34 N/m) and puckkered-phosphorene (16 N/m) [31]. However, the strain energy continues to rise even beyond the critical point suggesting the possibilities of geometrical reconstruction.

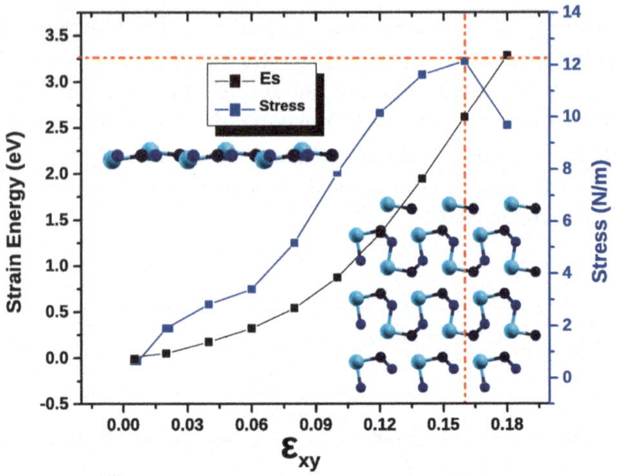

Fig. 4. (Color online) Stress and strain energies variation against the applied biaxial strain. The deformed structures beyond ε_c (dashed red line) are also shown.

4 Conclusions

We have performed first-principles DFT calculations to predict a new 2D ternary penta-SiCN, which is structurally and kinetically stable. Penta-SiCN is found to be mechanically isotropic and relatively softer material possessing a scare negative Poisson's ratio. Remarkably, both the Poisson's ratio and the Young's modulus of penta-SiCN can be modulated by stretching/compressing the monolayer. However, stretching the monolayer beyond 16% could lead to bond breaking, and deformation. Our results suggest that penta-SiCN could be a potential candidate for nanoauxetic and nanomechanical applications.

Acknowledgments. The authors acknowledge the financial and technical support provided by Khalifa University of Science and Technology (KU), and the KU-HPC facility.

References

1. Zhang, S., Zhou, J., Wang, Q., Chen, X., Kawazoe, Y., Jena, P.: Proc. Natl. Acad. Sci. **112**, 2372 (2015)
2. Yagmurcukardes, M., Sahin, H., Kang, J., Torun, E., Peeters, F.M., Senger, R.T.: J. Appl. Phys. **118**, 104303 (2015)
3. Liu, G., Zeng, Q., Zhu, P., Quhe, R., Lu, P.: Comput. Mater. Sci. **160**, 309 (2019)
4. Sun, W., Shen, Y., Guo, Y., Chen, Y., Wang, Q.: Appl. Surf. Sci. **554**, 149499 (2021)
5. Sharma, S.B., Qattan, I.A., KC, S., Alsaad, A.M.: ACS Omega **7**, 36235 (2022)
6. Alderson, A., Alderson, K.: Proc. Inst. Mech. Engineers Part G: J. Aeros. Eng. **221**, 565 (2007)
7. Yang, W., Li, Z.-M., Shi, W., Xie, B.-H., Yang, M.-B.: J. Mater. Sci. **39**, 3269 (2004)
8. Huang, C., Chen, L.: Adv. Mater. **28**, 8079 (2016)

9. Peng, R., Ma, Y., Wu, Q., Huang, B., Dai, Y.: Nanoscale **11**, 11413 (2019)
10. Shen, Y., Wang, Q.: Phys. Rep. **964**, 1 (2022)
11. Jiang, J.-W., Park, H.S.: Nat. Commun. **5**, 1 (2014)
12. Kong, X., et al.: Phys. Rev. B **98**, 184104 (2018)
13. Zhao, K., Guo, Y., Shen, Y., Wang, Q., Kawazoe, Y., Jena, P.: J. Phys. Chem. Lett. **11**, 3501 (2020)
14. Sharma, S.B., Santosh, K., Paudyal, D.: Appl. Surf. Sci. **593**, 153239 (2022)
15. Sharma, S.B., Bhatta, R., Adhikari, R., Paudyal, D.: Carbon Trends **7**, 100162 (2022)
16. Sharma, S.B., Qattan, I.: Appl. Surf. Sci. **599**, 153997 (2022)
17. Sharma, S.B., Qattan, I.A., Jaishi, M., Paudyal, D.: ACS Appl. Electron. Mater. **4**, 2561 (2022)
18. Soler, J.M., et al.: J. Phys.: Condens. Matter **14**, 2745 (2002)
19. Artacho, E., Sánchez-Portal, D., Ordejón, P., Garcia, A., Soler, J.M.: Physica Status Solidi (b) **215**, 809 (1999)
20. Troullier, N., Martins, J.L.: Phys. Rev. B **43**, 1993 (1991)
21. Perdew, J.P., Burke, K., Ernzerhof, M.: Phys. Rev. Lett. **77**, 3865 (1996)
22. Monkhorst, H.J., Pack, J.D.: Phys. Rev. B **13**, 5188 (1976)
23. Hou, C., Shen, Y., Sun, W., Chen, Y., Ni, D., Wang, Q.: J. Mater. Chem. C **10**, 10302 (2022)
24. Togo, A., Oba, F., Tanaka, I.: Phys. Rev. B **78**, 134106 (2008)
25. Sharma, S.B., Qattan, I., Kc, S., Abedrabbo, S.: Nanomaterials **12**, 4068 (2022)
26. Born, M., Misra, R.D.. Mathematical Proceedings of the Cambridge Philosophical Society, vol. 36, pp. 466–478. Cambridge University Press, Cambridge (1940)
27. Sharma, S.B., Paudel, R., Adhikari, R., Kaphle, G.C., Paudyal, D.: Physica E **146**, 115517 (2023)
28. Sharma, S.B., Bhatta, R., Sigdel, K.R., Adhikari, R.P., Kaphle, G.C.: Eur. Physical J. B **94**, 128 (2021)
29. Wang, B., Wu, Q., Zhang, Y., Ma, L., Wang, J.: ACS Appl. Mater. Interfaces. **11**, 33231 (2019)
30. Chen, S.-B., Zeng, Z.-Y., Chen, X.-R., Yao, X.-X.: J. Appl. Phys. **128**, 125111 (2020)
31. Hess, P.: Phys. Chem. Chem. Phys. **20**, 7604 (2018)

Anomalous Transport Properties in the Iron-Based Kagome Ferromagnet Fe_3Sn

Susumu Minami[1(✉)], Akito Sakai[1], Taishi Chen[1,2,3], Yangming Wang[1], Zili Feng[1,2], Rieko Ishii[2], Takuya Nomoto[4], Ryotaro Arita[4,5], and Satoru Nakatsuji[1,2,6,7]

[1] Department of Physics, The University of Tokyo, Hongo, Bunkyo-ku, Tokyo 113-0033, Japan
susumu.minami@phys.s.u-tokyo.ac.jp

[2] Institute for Solid State Physics, University of Tokyo, Kashiwa, Chiba 277-8581, Japan

[3] School of Physics, Southeast University, Nanjing 211189, China

[4] Research Center for Advanced Science and Technology, University of Tokyo, Komaba, Meguro-ku, Tokyo 153-8904, Japan

[5] Center for Emergence Matter Science (CEMS), RIKEN, Hirosawa, Wako, Saitama 351-0198, Japan

[6] CREST, Japan Science and Technology Agency (JST), 4-1-8 Honcho Kawaguchi, Saitama 332-0012, Japan

[7] Trans-scale Quantum Science Institute, University of Tokyo, Bunkyo-ku, Tokyo 113-8654, Japan

Abstract. Motivated by the discovery of the large anomalous transport responses from the Berry curvature in the noncollinear kagome antiferromagnet Mn_3X (X = Sn, Ge), we investigate the anomalous Hall conductivity σ_{ij} and anomalous thermoelectric conductivity α_{ij} of the iron-base kagome ferromagnet Fe_3Sn with the Curie temperature of 760 K. The maximum value of the transverse thermoelectric conductivity $|\alpha_{ji}|_{\max} \sim 2.5\ \mathrm{AK^{-1}m^{-1}}$ at 200 K is observed by using a polycrystalline sample of Fe_3Sn, which is comparable to the recently reported α_{ij} in other topological magnets. The magnetocrystalline anisotropy energy (MAE) and the magnetization direction dependence of the anomalous transport properties are also evaluated by the present theoretical calculations. We find the strong anisotropy in the chemical potential dependence of σ_{ij} and α_{ij}, and we conclude that the in-plane components produces the main component of α_{ij} in the polycrystalline sample.

Keywords: Anomalous Hall effect · Thermoelectrics · Magnetic anisotropy energy · Kagome lattice · First-principles calculation

1 Introduction

The large anomalous Hall effect (AHE) and anomalous Nernst effect (ANE) were observed in the noncollinear kagome antiferromagnet Mn_3X (X=Sn, Ge) [1–4]. The intrinsic contribution of AHE and ANE is caused by the finite Berry curvature due to time-reversal symmetry breaking [5,6]. The topological electronic structure, such as Weyl point and nodal lines, is a key feature to enhance AHE and ANE. In fact, the

H. J. Choi et al. (Eds.): APPC 2022, *Proceedings of the 15th Asia Pacific Physics Conference*, pp. 80–87, 2024.
https://doi.org/10.1007/978-981-96-0191-2_9

role of various topological electronic structures and their large or exotic responses have been theoretically and experimentally studied in various magnets [7–15].

Recently, the ANE, which has a transverse voltage induced by a longitudinal temperature gradient, is attracting renewed interest [16]. ANE can be exploited in developing high-efficiency energy-harvesting devices with a simple lateral structure, high flexibility, and low production cost. To realize a widespread use of thermoelectric devices based on the ANE, a high magnetic transition temperature is also a crucial condition except for a large ANE signal.

In this study, we investigate the electronic structure, magnetocrystalline anisotropy energy, and anomalous transport properties for the iron-based kagome ferromagnet Fe_3Sn with the Curie temperature of 760 K. Anomalous Hall conductivity (σ_{xy}) and transverse thermoelectric conductivity (α_{xy}) for Fe_3Sn with magnetization direction along $[2\bar{1}\bar{1}0]$, $[01\bar{1}0]$, and $[0001]$ are calculated. We find that the in-plane components of AHC show the sharp peak near the Fermi energy. We also compare the theoretical and experimental anomalous transport properties, and these results show good agreement with each other. Maximum value of transverse thermoelectric conductivity $|\alpha_{ji}|_{\max} \sim 2.5$ AK^{-1}m^{-1} at 200 K is observed. We can conclude that $|\alpha_{ji}|_{\max}$ originates from the intrinsic components of the anomalous Hall conductivities along with the in-plane magnetization direction based on the theoretical calculation results.

2 Methods

2.1 Computatinal Details

The electronic structure calculations were performed OpenMX code [17] based on the non-collinear density functional theory (DFT) [18]. DFT calculations are performed through the exchange-correlation functional within the generalized-gradient approximation and norm-conserving pseudopotentials [19]. The spin-orbit coupling (SOC) is included by using total angular momentum dependent pseudopotentials [20]. The wave functions are expanded by a linear combination of multiple pseudo-atomic orbitals [21]. A set of pseudoatomic orbital basis was specified as Fe6.0-$s3p3d3f1$, and Sn7.0-$s3p2d2f1$ where the number after each element stands for the radial cutoff in the unit of bohr and the integer after s, p, d, f indicates the radial multiplicity of each angular momentum component. The cutoff energy for charge density of 300 Ry and a k-point mesh of $32 \times 32 \times 32$ were used. The lattice constant of Fe_3Sn was set to $a = 5.46$ and $c = 4.33$ Å[22].

From the Bloch states obtained in the DFT calculation, a Wannier basis set was constructed by using the Wannier90 code [23]. The basis was composed of (s, d)-character orbitals at the Fe site and (s, p)-character orbitals at the Sn site. Therefore, we consider 76 orbitals per unit cell including the spin multiplicity. This basis set was extracted from 178 bands. The intrinsic contribution of the anomalous Hall conductivity σ_{xy} and the anomalous transverse TE conductivity α_{xy} can be calculated as $\sigma_{xy}(T, \mu) = e^2/(2\pi^3\hbar) \int d\mathbf{k} \sum_m \Omega_{m,z}(\mathbf{k}) f(\varepsilon_{m\mathbf{k}})$, $\alpha_{xy}(T, \mu) = -1/e \int d\varepsilon \sigma_{xy}(0, \varepsilon)(\varepsilon - \mu)/T \left(-\partial f/\partial \varepsilon\right)$, where $e, \hbar, \varepsilon, f, \mu$ are the elementary charge, the reduced Planck constant, the band energy, the Fermi-Dirac distribution function with the band index m and the wave vector $\mathbf{k}$, and the chemical potential, respectively. The

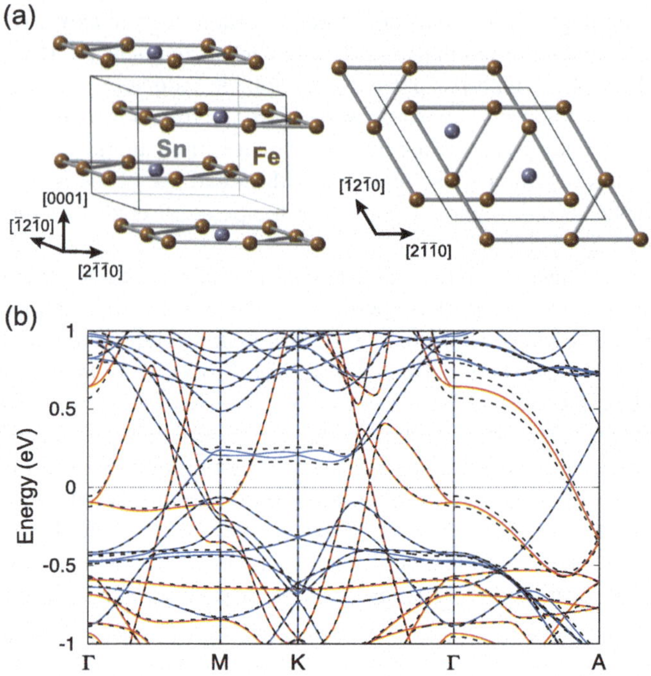

Fig. 1. Crystal structure of Fe_3Sn (a). Brown and gray atoms represent Fe and Sn, respectively. Electronic band structure of Fe_3Sn (b). Red (blue) lines correspond to majority (minority) bands calculated without SOC. Black dashed lines represent the bands with SOC. We assumed the magnetization direction as [0001].

Berry curvature can be determined as $\mathbf{\Omega}_m \equiv i\langle\nabla_{\mathbf{k}}u_m\mathbf{k}| \times |\nabla_{\mathbf{k}}u_m\mathbf{k}\rangle$. For the numerical integration, a k-point mesh of $300 \times\ 300 \times 300$ was used.

2.2 Sample Preparation and Thermoelectric Measurements of the Polycrystalline Fe_3Sn Sample

Polycrystalline Fe_3Sn ingots were made by arc melting the mixtures of iron chops (4 N) and tin grains (5 N) under an argon atmosphere for oxygen extinction. After being sealed into an ampoule and annealed for 7 days at 805 °C, the polycrystalline ingot was quenched into ice water [24]. The atomic ratio of iron to tin was confirmed to be 2.97:1.03, very close to stoichiometric 3:1, by the inductively coupled plasma (ICP) method.

The thermal transport properties were measured by a thermal transport option on a commercial system (PPMS, Quantum Design) with a rectangular bar-shaped sample ($\sim$ 6 mm in length by $\sim$ 1.3 mm in width by $\sim$ 0.5 mm in thickness). The Seebeck coefficient S_{ii} and Nernst signal S_{ij} were then estimated as $S_{ii} = E_i/\Delta T_i$ and $S_{ij} = E_j/\Delta T_i$, where ΔT_i, E_i, and E_j are the thermal gradient, the longitudinal, and transverse electric fields, respectively. The transverse thermoelectric conductivity was

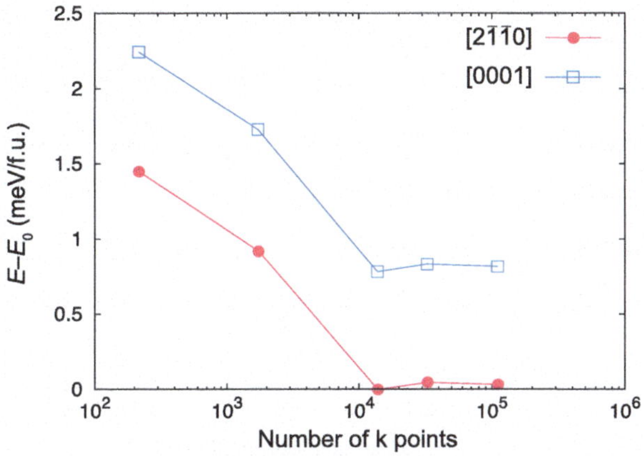

Fig. 2. k-points dependence of the total energy for Fe_3Sn. Pink circle-solid points and cyan open-square points represent the magnetization direction along $[2\bar{1}\bar{1}0]$ and $[0001]$ direction, respectively. E_0 is set to the converged total energy along the in-plane, that is $E_0 = -444.68932564$ hartree per formula unit (f.u.).

evaluated by using following equation as $\alpha_{ji} = 1/\rho S_{ji} - \rho_{ji}/\rho^2 S_{ii}$, where ρ, ρ_{ji} is the longitudinal and Hall resistivity, respectively. All parameters in the above equation can be obtained experimentally.

3 Results and Discussion

Figure 1 shows the schematic crystal structure and electronic band structure of Fe_3Sn. Fe_3Sn has the space group with $P6_3/mmc$. The Wyckoff position of the Fe (Sn) atom corresponds to $6h$ $(2d)$ site, respectively; therefore, Fe atoms construct the kagome lattice. The total magnetic moment of the unit cell is $\sim$ 14.48 μ_B, that is, the atomic magnetic moment is $\sim$ 2.41 μ_B/Fe atom which is consistent with reported values [25].

First, let us confirm the total energy and its convergence to clarify the magnetic direction and the magneto-crystalline anisotropy energy (MAE). Figure 2 shows the k-point dependence of the total energy. Here, we assumed the magnetization direction along $[2\bar{1}\bar{1}0]$ (in-plane) and $[0001]$ (out-plane) direction. We can see enough convergences were obtained above 10^4 k-points mesh in both magnetization directions, and it converged within less than 0.1%. We also find that in-plane anisotropy is realized in Fe_3Sn.

The MAE E_{MAE} was evaluated as follows: $E_{\mathrm{MAE}} = E_{[0001]} - E_{[2\bar{1}\bar{1}0]}$, where $E_{[0001]}$ and $E_{[2\bar{1}\bar{1}0]}$ are the total energy for Fe_3Sn with the magnetization direction along $[0001]$ $[2\bar{1}\bar{1}0]$, respectively. The evaluated AME is $E_{\mathrm{MAE}} = 0.782$ meV/f.u. (0.196 meV/atom), which leads to about 2.24×10^6 J/m^3 of the magnetic anisotropy constant. The absolute value of MAE in Fe_3Sn is comparable to the $L1_0$-ordered iron-based ferromagnets such as FePt ($E_{\mathrm{MAE}} \sim 2.15$ meV/atom), FePd ($E_{\mathrm{MAE}} \sim 0.15$ meV/atom), and FeNi ($E_{\mathrm{MAE}} \sim 0.10$ meV/atom), which show uniaxial anisotropy [26].

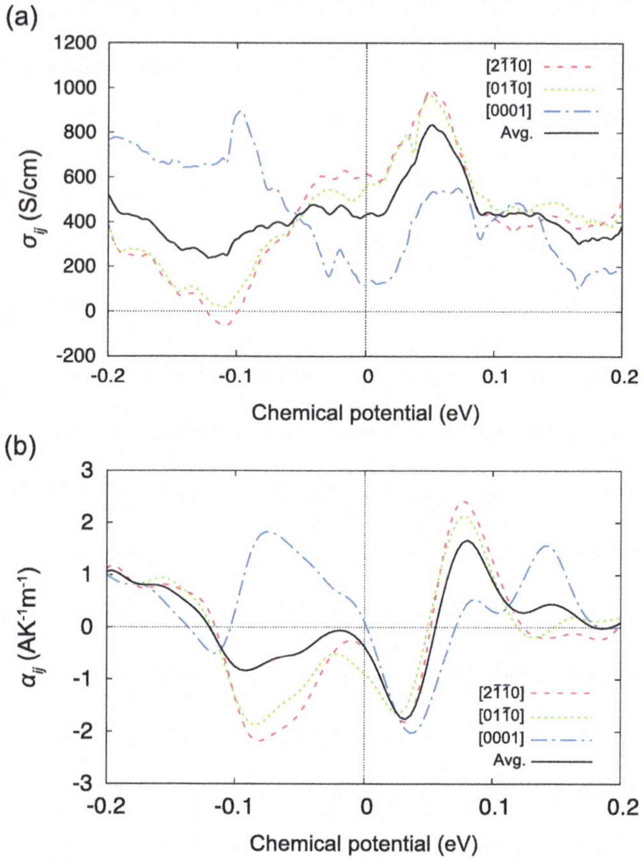

Fig. 3. Chemical potential dependence of anomalous Hall conductivity (σ_{ij}) at 0 K (a), and anomalous Nernst conductivity (α_{ij}) at 100 K (b). Pink dashed lines, green dotted lines, and cyan dash-dotted lines represent the magnetization direction along with $[2\bar{1}\bar{1}0]$, $[01\bar{1}0]$, and $[0001]$, respectively. Black solid lines correspond to the average of σ_{ij} and α_{ij}. Details see the main text.

Next, let us consider the anomalous transport properties based on the first-principles calculation. We consider the three magnetization directions; $[2\bar{1}\bar{1}0]$, $[01\bar{1}0]$, and $[0001]$ which correspond to x, y, and z direction in the cartesian coordinate, respectively. Figure 3 shows the chemical potential dependence of σ_{ij} and α_{ij} for Fe_3Sn. The in-plane components (pink dashed lines and green dotted lines) of σ_{ij} and α_{ij} as shown in Figs. 3(a) and 3(b) are almost consistent; on the other hand, the trend of the out-plane components (cyan dash-dotted lines) is different. This could also be understood from the strong magneto-crystalline anisotropy energy in Fe_3Sn.

The in-plane components of σ_{ij} show the sharp peak around the chemical potential is 50 meV [Fig. 3(a)]. The maximum value of σ_{ij} reaches 1000 (S/cm). This peak leads to the enhancement of α_{ij} around 50 meV as shown in Fig. 3(b). We can see the odd–

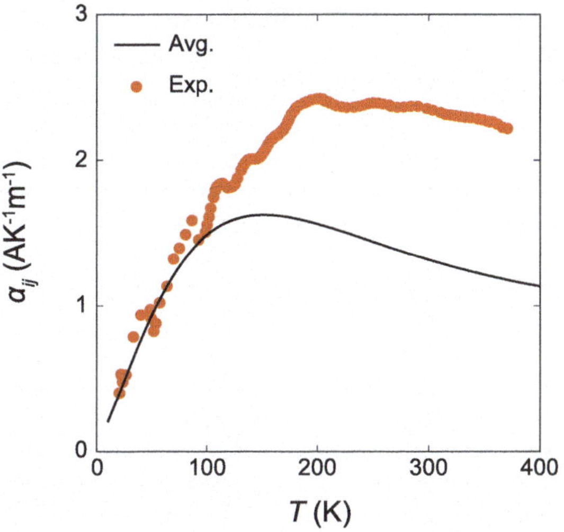

Fig. 4. Temperature dependence of anomalous Nernst conductivity α_{ij}. Black solid line represents the theoretical $\alpha_{ij}^{\mathrm{Adv.}}$ at $E = E_{\mathrm{F}} + 88$ meV. Orange solid square points correspond to the experimentally observed $\alpha_{ij}^{\mathrm{Exp.}}$.

function–like chemical potential dependence of α_{ij} before and after 50 meV because of the even–function–like chemical potential trend of σ_{ij} same as the conventional Seebeck effect. A sign change of α_{ij} appears before and after a peak of σ_{ij}, which gives crucial information to compare the experimental results.

To compare experimentally observed anomalous transport properties, we estimate σ_{ij} and α_{ij} for the polycrystalline Fe_3Sn. As a simplest assumption, we take an average of σ_{ij} as follows $\sigma_{ij}^{\mathrm{Avg.}} = 1/3(\sigma_{xy} + \sigma_{yz} + \sigma_{zx})$. The former theoretical work by E. Romam et al. shows that the average of the three high-symmetry axis $\sigma_{ij}^{\mathrm{Avg.}} = 1/3(\sigma_{xy} + \sigma_{yz} + \sigma_{zx})$ agrees well with the correct orientational average $\sigma_{\mathrm{Avg.}} = \int_0^{2\pi} \sigma_{ij}(\theta) \sin\theta d\theta$ and experimental results obtained for polycrystalline sample in hcp-Co [27]. The average of anomalous Nernst conductivity $\alpha_{ij}^{\mathrm{Avg.}}$ are evaluated as follows $\alpha_{ij}^{\mathrm{Avg.}} = 1/3(\alpha_{xy} + \alpha_{yz} + \alpha_{zx})$. $\sigma_{ij}^{\mathrm{Avg.}}$ and $\alpha_{ij}^{\mathrm{Avg.}}$ are shown in Figs. 3(a) and 3(b) as black solid lines. We can see the trend of $\sigma_{ij}^{\mathrm{Avg.}}$ and $\alpha_{ij}^{\mathrm{Avg.}}$ strongly depends on the in-plane components.

Let us finally move on to comparing the experimental and theoretical results. Figure 4 shows the temperature dependence of transverse thermoelectric conductivity α_{ij}. Black solid line and orange solid square points represent the theoretical $\alpha_{ij}^{\mathrm{Adv.}}$ at $E = E_{\mathrm{F}} + 88$ meV, and the experimentally observed $\alpha_{ij}^{\mathrm{Exp.}}$, respectively. The chemical potential shifting was estimated by counting the number of valence electrons difference from the Fermi energy, which was evaluated by using the slight off-stoichiometry of the composition (Fe : Sn $= 2.97 : 1.03$). We can see the trend of the temperature dependence of α_{ij} shows good agreement for each other, especially in less than 100 K regions.

The maximum value of the experimentally observed $\alpha_{ij}^{\mathrm{Exp.}}$ reaches 2.5 $\mathrm{AK^{-1}m^{-1}}$ at 200 K, comparable to the other recently reported topological magnets such as Co_2MnGa [$|\alpha_{ij}|_{\max} \sim$ 4 $\mathrm{AK^{-1}m^{-1}}$ (150 K)] [8], Fe_3Ga [$|\alpha_{ij}|_{\max} \sim$ 5.2 $\mathrm{AK^{-1}m^{-1}}$ (200 K)] [12], and $Co_3Sn_2S_2$ [$|\alpha_{ij}|_{\max} \sim$ 2.5 $\mathrm{AK^{-1}m^{-1}}$ (150 K)] [11]. We note that the single-crystalline sample of Fe_3Sn could be enhanced σ_{ij} and α_{ij} than the polycrystalline sample. The anisotropy of σ_{ij} and α_{ij} is much large than the cubic system which may come from the strong crystal anisotropy in the hexagonal system.

4 Conclusion

In summary, to investigate the electronic structure, MAE, and anomalous transport properties of the iron-base kagome ferromagnet Fe_3Sn, we performed the first-principles calculations. We also synthesized polycrystalline sample of Fe_3Sn and measured anomalous transport properties. The theoretical calculation results show in-plane anisotropy and its value reaches 2.2×10^6 J/m^3, comparable to the well-known iron-based ferromagnets. The experimentally observed maximum value of transverse thermoelectric conductivity is $\alpha_{ij}^{\mathrm{Exp.}} = 2.5$ $\mathrm{AK^{-1}m^{-1}}$ at 200 K. The trend of the temperature dependence of α_{ij} is consistent between the theoretical and experimental results. We can conclude that the main component of α_{ij} comes from the in-plane component of σ_{ij} from the Berry curvature.

Acknowledgments. This work was partly supported by JSPS KAKENHI Grants Numbers JP19H05825, JP20K22479, JP21H04437, JP22K14587, JST-PRESTO (JPMJPR20L7), JST-MIRAI Program (JPMJMI20A1), JST-ASPIRE(JPMJAP2317). The computations in this research were partially performed using the Fujitsu PRIMERGY CX400M1/CX2550M5 (Oakbridge-CX) in the Information Technology Center, The University of Tokyo and the facilities of the Supercomputer Center, the Institute for Solid State Physics, the University of Tokyo.

References

1. Nakatsuji, S., Kiyohara, N., Higo, T.: Nature **527**, 212 (2015)
2. Kiyohara, N., Tomita, T., Nakatsuji, S.: Phys. Rev. Appl. **5**, 064009 (2016)
3. Ikhlas, M., et al.: Nat. Phys. **13**, 1085 (2017)
4. Chen, T., et al.: Nat. Commun. **12**, 572 (2021)
5. Nagaosa, N., Sinova, J., Onoda, S., MacDonald, A.H., Ong, N.P.: Rev. Mod. Phys. **82**, 1539 (2010)
6. Xiao, D., Chang, M.-C., Niu, Q.: Rev. Mod. Phys. **82**, 1959 (2010)
7. Chang, G., et al.: Phys. Rev. Lett. **119**, 156401 (2017)
8. Sakai, A., et al.: Nat. Phys. **14**, 1119 (2018)
9. Minami, S., Ishii, F., Mizuta, Y.P., Saito, M.: Appl. Phys. Lett. **113**, 032403 (2018)
10. Liu, E., et al.: Nat. Phys. **14**, 1125 (2018)
11. Ding, L., et al.: Phys. Rev. X **9**, 041061 (2019)
12. Sakai, A., et al.: Nature **581**, 53 (2020)
13. Minami, S., Ishii, F., Hirayama, M., Nomoto, T., Koretsune, T., Arita, R.: Phys. Rev. B **102**, 205128 (2020)
14. Okamura, Y., et al.: Nat. Commun. **11**, 4619 (2020)

15. Nakamura, H., Minami, S., Tomita, T., Nugroho, A.A., Nakatsuji, S.: Phys. Rev. B **104**, L161114 (2021)
16. Nakatsuji, S., Arita, R.: Annu. Rev. Condens. Matter Phys. **13**, 119 (2022)
17. Openmx: Open source package for material explorer. http://www.openmx-square.org/
18. Kubler, J., Hock, K.H., Sticht, J., Williams, A.R.: J. Phys. F **18**, 469 (1988)
19. Morrison, I., Bylander, D.M., Kleinman, L.: Phys. Rev. B **47**, 6728 (1993)
20. Theurich, G., Hill, N.A.: Phys. Rev. B **64**, 073106 (2001)
21. Ozaki, T.: Phys. Rev. B **67**, 155108 (2003)
22. Chen, T., et al.: Sci. Adv. **8**, eabk1480 (2022)
23. Pizzi, G., et al.: J. Phys.: Condens. Matter **32**, 165902 (2020)
24. Kanematsu, K.: Trans. Jpn. Inst. Metals **27**, 225 (1986)
25. Sales, B.C., Saparov, B., McGuire, M.A., Singh, D.J., Parker, D.S.: Sci. Rep. **4**, 7024 (2014)
26. Kota, Y., Sakuma, A.: J. Phys. Soc. Jpn. **81**, 084705 (2012)
27. Roman, E., Mokrousov, Y., Souza, I.: Phys. Rev. Lett. **103**, 097203 (2009)

Magnetoresistance When Spin Effects on Conduction Are Weak

Vincent Sacksteder IV(✉)

Department of Physics and Astronomy, Rutgers University, New Brunswick, NJ, USA
vincent@sacksteder.com

Abstract. This paper considers certain materials, including topological insulators, where spin rotation symmetry is broken much more strongly than time reversal symmetry. When these materials are in the diffusive regime, i.e. when they have disorder that is strong enough to cause an electron to scatter many times while crossing a sample, electrons and holes move in pairs that have zero spin and are insensitive to spin physics. Working within this spinless scenario, we show that Fourier transforming the magnetoconductance with respect to external magnetic field obtains a curve describing the area distribution of loops traced by electrons and holes within the sample. We present loop area distributions of Landau levels, weak (anti)localization, conduction governed by Levy flights, and linear-in-field resistance. Of these four the last two are new results. Comparing these distributions, we argue that the linear-in-field resistance seen in some topological insulators is caused by the same diffusive scattering that causes weak antilocalization. The difference is that linear-in-field resistance materials retain a level of quantum coherence that is usually seen only on the surface of 2-D wires or in ring geometries. In an appendix we include some speculative material about linear-in-temperature resistance.

1 Introduction

In atomic units $m_e = e = \hbar = k_B = 1$ the magnetic field B has dimensions of inverse area, which is key to understanding the electrical conductance's dependence on B. Large fields probe areas at the scale of the unit cell, and small changes in field probe much larger areas. At large fields one finds Landau levels and Schubnikov-de Haas (SdH) oscillations, which are visible in the longitudinal conductance $G_{xx}(B)$ as peaks at characteristic field strengths determined by the Fermi energy E_F. The Fermi surface's cross section, an inverse area, can be read off from the peak positions. Landau levels and SdH oscillations are signals of ballistic physics, i.e. weak scattering, and are extinguished once scattering becomes too frequent.

At small fields one finds weak (anti) localization (WL/WAL), where scattering and quantum interference cooperate to cause G_{xx} to decrease or increase with field depending on whether spin rotation symmetry is broken faster or slower than time reversal symmetry. WL/WAL contrasts strongly against Landau levels and SdH oscillations. It is much more sensitive to magnetic field, is visible at much smaller field strengths, and varies smoothly with field rather than exhibiting oscillations. Moreover WL/WAL is

H. J. Choi et al. (Eds.): APPC 2022, *Proceedings of the 15th Asia Pacific Physics Conference*, pp. 88–116, 2024.
https://doi.org/10.1007/978-981-96-0191-2_10

caused by physics at length scales much longer than the scattering length. It is only weakly sensitive to the atomic unit cell and the Fermi wavelength.

Recent years have exposed mysteries which lie well beyond traditional SdH and WL/WAL physics. Many experiments report systems where the longitudinal resistance R_{xx} increases linearly with B and does not saturate [1–11]. This contrasts both with SdH oscillations which oscillate periodically in $1/B$ and with WL/WAL which is logarithmic. Explanations have been given for the cases of ballistic conduction when the Fermi surface has a cusp, of a 3-D Dirac cone in the presence of a strong magnetic field, of a density gradient across the sample, and of classical transport with strong sample inhomogeneities [12–16]. However this kind of case by case treatment is not entirely satisfactory given the wide range of experimental realizations.

The electrical conductance is determined by electrons and holes. Magnetic field affects electrons and holes in two ways. The first is the minimal coupling $\vec{p} - e\vec{R}$, where $\vec{R}$ is the gauge field determining the magnetic field by $\vec{B} = \vec{\nabla} \times \vec{R}$. This coupling is sensitive only to a fermion's kinetics, not its spin. In contrast, the $e\vec{\sigma} \cdot \vec{B}$ Zeeman term provides a very direct coupling between field and spin. The Zeeman term is accompanied by similar terms which are of higher order in momentum $\vec{p}$, electric field $\vec{E}$, and magnetic field [17].

In this paper we concern ourselves with materials where the Zeeman term's effect on conduction can be neglected, leaving in play only the minimal coupling. In other words, we confine ourselves to systems where electronic conduction is mediated by spin singlets; we explore where this simplification can lead us. Always working under the assumption that the Zeeman term can be neglected, we show that external magnetic field is conjugate to the area of the electron-hole loops inside a material. In non-interacting systems the conjugate area is that of a single loop, and in interacting systems it is the sum of the areas of all loops contributing to the conductance. Therefore if one analyzes the magnetoconductance by taking the Fourier transform with respect to external magnetic field, one can extract detailed information about the length and area scales in play in electronic conduction within the material. We apply this methodology to Landau levels, SdH oscillations, and weak antilocalization. We go on to see what this methodology says about materials with linear-in-field magnetoresistance, if their linear resistance is caused by the minimal coupling not the Zeeman term. Lastly we include some more speculative material concerning materials displaying linear-in-temperature resistance.

There exist many materials where electronic conduction is determined by spin singlets only; in these materials the spin triplet and the Zeeman effect have no influence. There are two ingredients for obtaining this scenario. First, the material should be in the diffusive regime, where the scattering length is much shorter than the sample size. Second, spin rotation symmetry must be broken more strongly than time reversal symmetry, as is the case with strong spin-orbit coupling.

The first ingredient for spin-singlet conduction is diffusive conduction, with many scatterings as electrons and holes traverse a sample. At lengths exceeding the scattering length l the electronic wave-function $|\psi\rangle$'s phase is randomized, rendering single wave-functions unable to mediate conduction. Therefore conduction over distances longer than l is mediated only by pairing between $|\psi\rangle$ and its complex conjugate $\langle\psi|$, whose random phases cancel. This pairing produces the single-particle density matrix $\rho(x)$, which is the key mediator of conduction at distances longer than the scattering length. This result is well known from the extensive work on sigma models and bosonization techniques in disordered systems [18, 19].

ρ's evolution is sensitive only to processes where $|\psi\rangle$ and $\langle\psi|$ undergo the same scattering events and follow the same path between scatterings, because only these processes satisfy the requirement that the disorder-induced random phases of $|\psi\rangle$ and $\langle\psi|$ must cancel. There are two such processes, the diffuson and the Cooperon. In the diffuson $|\psi\rangle$ and $\langle\psi|$ follow the same sequence in the same order, producing purely classical electronic motion. In contrast the Cooperon is a quantum interference process, distinct from classical motion. In the Cooperon $|\psi\rangle$ and $\langle\psi|$ follow the same sequence but in reversed order, and their phases seem to be randomized with respect to each other. It is not until their trajectories close a loop and come back to their origin that $|\psi\rangle$ and $\langle\psi|$'s phases suddenly cancel and $\rho(x)$ undergoes a revival. Comparing the diffuson and the Cooperon, the diffuson is purely classical and depends weakly on magnetic field, while the Cooperon is purely quantum mechanical and is strongly dependent on even very weak magnetic fields.

The main effect of spin on disordered transport is that $\rho(x)$ contains both a spin singlet (charge) and a spin triplet (spin polarization). Both the spin singlet and the spin triplet have characteristic relaxation times τ_s and τ_t that are determined by the material and by embedded impurities. When the spin singlet relaxes more slowly than the spin triplet, i.e. when $\tau_s \gg \tau_t$, then the spin singlet contribution dominates and the spin triplet's contribution to conduction is neglegible. This combination of lifetimes $\tau_s \gg \tau_t$ is found when spin rotation symmetry is broken more strongly than time reversal symmetry. This is the case in materials with strong spin-orbit coupling, including topological insulators. In particular, when topological insulators are in the diffusive regime and their charge carriers are from the Dirac cone not from other bands, then conduction is mediated only by spin singlets, not spin triplets.

If the Cooperon increases the conductance when the magnetic field is zero, then its effect is called weak antilocalization (WAL). If on the other hand the Cooperon decreases the conductance at zero field, then this is called weak localization (WL). WAL results in the resistance increasing with field, while WL results in the resistance decreasing with field. Whether weak localization is seen in a material, or instead weak antilocalization, is determined by whether the Cooperon triplet is present.

Magnetic field is a perfect diagnostic tool for distinguishing between systems where the Cooperon spin triplet is absent and ones where it is present. When the Cooperon spin singlet relaxes more slowly than the spin triplet (as in topological insulators), i.e. when $\tau_s \gg \tau_t$, the singlet produces weak antilocalization (WAL), which means that the resistance increases with field. On the other hand, when time reversal symmetry is not more favored than spin rotation, and the spin singlet and spin triplet relaxation time scales $\tau_s \sim \tau_t$ are similar, then the spin triplet dominates and the resistance decreases with field, which is called weak localization (WL) [20, 21]. In summary, WAL occurs when the spin triplet relaxes quickly so that conduction is mediated by charges which effectively have zero spin [22], while WL occurs when both the spin singlet and spin triplet contribute to conduction.

Experimentally, if the resistance increases with small fields in a diffusive material then only the Cooperon spin singlet is active in conduction, and not the Cooperon spin triplet. If, on the other hand, the resistance decreases with small fields then both the spin singlet and the spin triplet are in play.

Many materials displaying linear magnetoresistance do conduct only via the Cooperon spin singlet, not the spin triplet. Certainly topological insulators lie in this regime, as do Dirac semimetals with strong spin-orbit coupling [1–4]. Possibly other linear resistance materials, besides TIs and Dirac Weyl materials, may also have negligible spin triplet conduction. Experimental observations of linear magnetoresistance universally find that resistance increases rather than decreases with field, as seen in WAL. Observations of resistance increasing with field are consistent with a short spin triplet relaxation time τ_t and a significantly longer spin singlet lifetime τ_s, and with dominance of spin singlet conduction.

1.1 Loops and Magnetic Field

This paper develops a geometrical analysis of magnetotransport in terms of the loops traced by electrons and holes as they move through real space. In spin singlet materials where only the minimal coupling and not the Zeeman term couples magnetic field to electrons, electron-hole loops are controlled via the magnetic flux through them and by the Aharonov-Bohm effect. Because magnetic flux is equal to the product of field with loop area, field is conjugate to area. We will develop this fact into a methodology for analyzing experimental measurements of the conductance.

The starting point of our analysis is Feynman's formulation of quantum mechanics as a sum over paths [23]. In this formulation an electron follows not just one path, but instead many paths which fully explore the sample in which the electron is moving. All of these many paths are summed to determine the evolution of $|\psi\rangle$, the electronic state. In this paper our focus is not on the state $|\psi\rangle$, but on the many paths that contribute to $|\psi\rangle$. The two pictures are mathematically equivalent: starting from a knowledge of paths one can build the electronic propagator which controls the evolution of $|\psi\rangle$, and conversely $|\psi\rangle$'s evolution may be systematically decomposed into the paths which determine it.

To be clear, we are not discussing the semiclassical paths of an electron's average motion. For instance, in a magnetic field an electron's average position executes well-defined circular loops around the axis of the magnetic field, and one can measure this cyclotron motion. This is not the sort of path we are talking about. Instead we are talking about the infinitely many quantum mechanical paths, tracing many complex trajectories and fully exploring the sample, which sum up to produce the average cyclotron motion.

The paths traced by electrons never begin or end in isolation, since in this event electronic charge would not be conserved, i.e. charge would be generated or destroyed. Instead, the creation or annihilation of an electron is always accompanied by the creation or annihilation of an accompanying hole with opposite charge, and by this mechanism charge is conserved. The two paths of an electron and of its corresponding hole, followed from their origin together to their final disappearance together, form a loop.

It is important to be clear that the loops under discussion here are traced by bare electrons and holes, not by quasiparticles. The requirement to move in loops is an immediate consequence of charge conservation and gauge invariance, and does not require that interactions are weak, the existence of a Fermi liquid, or well-defined quasiparticles. Nor does it not assume any other sort of collective many-body behavior. The loop requirement applies independently of magnetic fields, scattering, and interactions. Analysis focused on electron-hole loops is therefore a very powerful way of understanding the mechanisms responsible for electronic transport.

It is also important to be clear that the loops discussed here move through real space, and occur even when disorder is very strong. Therefore they are not the same loops used in the analysis of Berry phases and topological invariants, which are usually calculated in momentum space and assume a well defined band structure. Moreover the loops discussed here concern the bare electron and hole carriers within a material; the U(1) symmetry which supports them comes from electromagnetism, not from effective interactions that occur in spin liquids and other interacting systems.

Experimental data on a sample's dependence on external magnetic field gives direct access to the areas of the electron and hole loops within the sample. Each loop couples to field via a phase which is proportional to the magnetic flux through the loop, and this phase is equal to the product of the loop area and the magnetic field strength. We neglect the Zeeman term, assuming that only the spin singlet contributes to conduction. Within the phase factor magnetic field has a conjugate relationship to area, with a proportionality constant that is determined by fundamental constants and can not be renormalized. Therefore measurements of a sample's dependence on external magnetic field give information about the area of the loops traced by electrons within the sample. More precisely, the area that one learns about is the sum of the areas of all loops that contribute to the conductance. In a non-interacting system this simplifies to the area of a single loop. The methodology of using magnetic fields to measure loops, which we explain further below, does not assume quasiparticle or Fermi liquid physics. It is a simple consequence of charge conservation. In interacting systems

In this article we apply this methodology focused on electron-hole loops and their areas to electrical conduction. We obtain results about the electron and hole loops responsible for Landau levels, SdH oscillations, WL/WAL, and linear magnetoresistance. Our rigor here is limited only by our omission of the Zeeman effect, which amounts to an assumption that only the spin singlet contributes to conduction. Where this assumption is valid, there is a firm link between experimental observables and loop areas. Details of the material's Hamiltonian, eigenstates, interactions, Fermi surfaces, or disorder do not affect this link, and are discussed very little in this paper. Instead we focus on reasoning about the electron and hole loops that are responsible for conduction, and on drawing conclusions about which physical mechanisms are responsible for producing those loops.

1.2 What We Can Learn From Loop Areas

Our methodology is to start with the conductance's dependence on external magnetic field, and then Fourier transform with respect to magnetic field. The resulting quantity depends on magnetic field's conjugate quantity, the total area of all loops that contribute to the conductance. We call this quantity the loop area distribution. The final step is to examine the weights of this distribution, i.e. the weight corresponding to a particular value of the total loop area, and allow these weights to inform us about what electrons and holes are doing inside the material.

One reference point for comparison is the Landau levels and SdH oscillations, which are phenomena that do not rely on scattering. The loop area distribution for these effects oscillates periodically with loop area, with a period of oscillation that is determined by the Fermi surface. It has both positive and negative signs. Both the presence of a particular area scale, and oscillations between negative and positive signs (i.e. ringing), are hallmarks of ballistic conduction.

A second reference point is weak (anti)localization (WL/WAL). Here the loop area distribution has a single sign, and does not show any concentration near any particular area scale (i.e. it is scale-free) [19,20,24]. The weights of the loops responsible for the WL/WAL signal decrease gradually and smoothly with area, and the decay follows a $1/A$ (where A is area) power law up to loop areas which are far larger than that determined by the Fermi surface. The very smooth distribution of loop areas and the very large area scales are caused by stochastic motion of the charge carriers, and are unmistakeable hallmarks of scattering-driven (diffusive) conduction.

We next compare the loop area distribution responsible for linear magnetoresistance to that of standard WL/WAL, Landau levels and SdH oscillations. We are making the assumption (explicit throughout this paper) that in the linear resistance materials of interest neither spin triplets nor the Zeeman term affect electronic conduction. This is certainly a reasonable assumption for those topological insulators which manifest linear magnetoresistance.

The theoretical predictions for Landau levels, SdH oscillations, and WL/WAL are all based on 2-D electron gas models with no interactions. Therefore the loop area distribution is that of single electron-hole loops. In principle a linear resistance material may be strongly interacting. In this case the loop area distribution would concern the total area of all loops contributing to conduction, and comparison to non-interacting loop area distributions like those of Landau levels or WL/WAL may be misleading. Fortunately this is not a concern in most topological insulators, which are only weakly interacting.

We find that the loop area distribution corresponding to linear magnetoresistance is very like the loop area distribution of WL/WAL. There are no changes in sign, there is no favored length scale, the area distribution decreases smoothly as area increases, and very large area scales are seen. This is evidence that scattering and stochastic motion are key to linear magnetoresistance. This is also evidence that linear magnetoresistance is caused by Cooperons, i.e. paired particles and holes, rather than by any species of single-particle motion. The phase of single electrons and holes oscillates vary rapidly at the Fermi wavelength, which is the ultimate reason why the loop area distribution of Landau levels exhibits an oscillating sign. In contrast the phase of a Cooperon varies at the much longer localization length, resulting in a very smooth single-sign loop area distribution.

Despite these similarities, there is one major difference between WL/WAL and linear magnetoresistance: the latter has a much longer tail than the former; the loop area distribution of linear magnetoresistance decays like $\ln A$ while WL/WAL decays like $1/A$. In other words, linear magnetoresistance has many more large loops, and much larger loops. This reweighting towards large loop areas is the opposite of what would be obtained by transitioning from 2-D conduction to 3-D conduction. It is impossible to explain if Cooperons move diffusively in 2-D, whether with ordinary random walks, or with Levy flights.

The extraordinarily long and fat tail seen in linear magnetoresistance shares something in common with Landau levels, whose loop area distribution is a simple cosine. In the absence of scattering and at zero temperature, this cosine extends to infinitely large areas, signifying that all sizes of electron-hole loops are in play. The reason is not, of course, that the electrons are exploring an infinite 2-D plane. Instead the repeated oscillations of the loop area distribution signify that if an electron can complete one cyclotron orbit, then it can complete any number of cyclotron orbits. The electron's ability to retrace its path around a loop is a manifestation of quantum coherence. If finite temperature disrupts quantum coherence, its effect is to cut off loops with large areas.

In two dimensional samples the Cooperons that produce WL/WAL do not repeat their loops; they are effectively random walkers moving according to purely classical dynamics, without quantum coherence. The reason is that scattering randomizes momentum, and therefore even when a random-walking Cooperon returns to its starting point, it starts off from there in a new direction. This is a peculiarity of 2-D planar materials. When Cooperons are confined to the 2-D surface of a wire, or move one dimensionally around a circle, then they do manifest quantum coherence and repeat their loops.

Linear magnetoresistance involves much larger loop areas than can be explained via Cooperons undergoing ordinary random walks. We therefore conclude that the Cooperons in linear magnetoresistance materials retain some degree of quantum coherence, allowing them to sometimes repeat their loops. In other words, linear magnetoresistance is caused by a combination of quantum coherence and scattering - a combination that is difficult to account for in planar 2-D materials. Further discussion of this topic is in Sect. 4.

Section 2 will begin with charge conservation and show that it implies that magnetic field is conjugate to area. This allows one to start with experimental measurements of the conductance's dependence on magnetic field and then obtain information about the areas of electron and hole loops. Next Sect. 3 applies this methodology to calculate and compare the loop area distributions of Landau levels, SdH oscillations, WL/WAL, linear magnetoresistance, and Levy flights. Section 4 discusses linear magnetoresistance and possible links to quantum coherence, and Sect. 5 wraps up with a few final thoughts. Appendix A discusses several ways of regularizing the Fourier transform from $1/A$ to the logarithm and vice versa, and Appendix B gives some more speculative comments on materials with linear-in-temperature resistance, such as bad metals.

2 Geometric Analysis of Electron and Hole Loops

Charge conservation is a fundamental feature of nature which has been subjected to intense scrutiny and has been found to be obeyed to extreme precision. Two strategies are available to engineer a theory to correspond to this experimental reality. One strategy is to select a set of quantum mechanical states available to the system, and a Hamiltonian which describes the system's ground state and dynamics. In the process of building the states and the Hamiltonian, one enforces a constraint: all states must have the same total charge, and the Hamiltonian may not change that charge. This approach is very powerful, and has shown extreme versatility in building models of the essential physics of many systems. These include simplified models where the electron's position is restricted to an integer number of discrete values and therefore electron paths are series of discrete jumps. However this approach has the weakness that charge conservation is essentially imposed by fiat, and that its fundamental connection to the geometry of electron and hole loops is lost.

To capture the geometry of electrons moving in the real world, it is necessary to start from the symmetries of the real world. These include continuous translation, i.e. the electrons' ability to move smoothly along continuous paths instead of jumping from position to position, and Poincare symmetry, the ability to smoothly change direction in the four dimensional manifold of space and time. The class of theories which unite these symmetries to quantum mechanics, relativistic quantum field theory, is our starting point for understanding the connection between charge conservation and geometry.

An early and essential result of relativistic quantum field theory was the Dirac equation, which predicted the existence of holes. Soon afterwards quantum electrodynamics was formulated, in which the principle of gauge invariance is key to guaranteeing charge conservation. Later the constraints of internal consistency (renormalizability, unitarity, etc.) and of experimental data resulted in the standard model, in which gauge invariance is elevated to a guiding principal that is key to all electronic interactions. The standard model incorporates quantum electrodynamics, and preserves its treatment of charge conservation of electrons and holes, which is based on gauge invariance.

Gauge invariant theories are built in two steps. The first step is adding a gauge field: a $U(1)$ phase $\phi(\vec{x})$ which depends on position. This phase modifies the momentum operator which describes translational motion, so that as an electron moves along its path, its phase is multiplied by the difference between the phases $\exp(\imath\phi)$ at the electron's final position and its initial position. The second and final step toward building a gauge invariant theory is the requirement that the theory be invariant under arbitrary changes of the gauge field. This is essentially a mathematical shorthand for imposing a restriction on the structure of every physical observable. The only way to build an observable $\mathcal{O}$ which is independent of $\phi(\vec{x})$ is by requiring each electron contributing to the observable to return to its starting position, forming closed loops. Because the final and initial positions are the same, the difference in phase $\exp(\imath\phi)$ between start and end is exactly zero, and therefore observables built from closed loops do not depend on $\exp(\imath\phi)$; they are gauge invariant. The actual shape and size of the loops, and also the number of loops, is completely unrestricted by gauge invariance; the emphasis is on their closed nature.

In relativistic quantum field theory the electron paths move through space-time, and the requirement to return to the origin does not mean only that an electron must reverse its motion in space. Each electron is also required to reverse its direction in time, so it can eventually return to its starting time. From one point of view this reversal of direction is simply a manner of pulling a loop's path back to its origin, but from another point of view the reversal of time-direction is the meeting (creation or annihilation) of a particle-hole pair. In other words, one can adopt a vocabulary that avoids discussing electrons moving backwards-in-time, but only at the cost of saying that for every electron [moving in forward in time] there is also a hole. The hole is thought of as moving forward in time, but it has exactly the same mathematical and physical consequences as an electron moving backward in time. The net effect is that instead of talking about electrons tracing loops in space-time, one requires that electrons and holes are always produced and destroyed in pairs. If one takes the hole and electron paths together one finds that they always add up to loops in space-time. This loop structure is an immediate and unavoidable consequence of combining gauge invariance with special relativity.

Gauge invariance ensures charge conservation in a way that is so powerful that it seems almost trivial, as follows. First it requires that observables be made out of space-time loops. Secondly we divide up each space-time loop into segments that move forward in time [electrons], and segments that move backward in time [holes]. Thirdly we assign charge $-e$ to electrons and charge $+e$ to holes. Fourthly we note that the loop structure implies that electrons are always created or destroyed with corresponding holes, in pairs. We thus arrive at charge conservation: this structure of pair-wise creation/destruction manifestly conserves total charge.

2.1 Coupling Magnetic and Electric Fields to Loops

In relativistic quantum field theory the effect of magnetic and electric fields on electrons is mediated by the minimal coupling: the momentum operator p^μ is replaced by $p^\mu - eR^\mu$, where the gauge potential $R^\mu = \partial^\mu \phi$ is the first derivative of the gauge phase $\phi(\vec{x})$. The electric and magnetic fields $\vec{E}, \vec{B}$ are encoded in $R^\mu = (\phi/c, \vec{R})$ via $\vec{E} = -\vec{\nabla}\phi - \partial\vec{R}/\partial t$ and $\vec{B} = \vec{\nabla} \times \vec{R}$. This minimal coupling is required to construct a gauge invariant continuum theory - it is necessary to ensure that when an electron moves from a starting position to a final position, it is multiplied by the difference in phases $\exp(\imath\phi)$ at the two positions.

The minimal coupling implies that as an electron or hole traces its path, the only effect of electric and magnetic fields is to introduce a multiplicative factor, the Wilson loop $W_\gamma = \exp(\imath \oint_\gamma ds_\mu R^\mu)$ [25,26]. The line integral $\oint_\gamma ds_\mu$ is taken along the electron/hole path, denoted by γ. It is a loop integral because electron and hole paths are required to form loops.

When several electron-hole loops occur the line integrals of each loop add up, and the multiplicative factors $\exp(\imath \oint_\gamma ds_\mu R^\mu)$ multiply each other. Therefore we shift the notation from γ which specifies a single loop to Γ, which specifies the paths traced by a set of one or more electron and hole loops. The total effect of all the loops together is to multiply by $W_\Gamma = \exp(\imath \oint_\Gamma ds_\mu R^\mu)$, where the line integral $\oint_\Gamma ds_\mu$ is the sum of each of the loop integrals $\oint_\gamma ds_\mu R^\mu$ in Γ, summing over the one or more loops specified by Γ. Gauge invariant observables are built from sums in which each term of the sum has a particular set of electron-hole loops specified by Γ and is weighted by the corresponding Wilson loop W_Γ.

By Stokes' theorem the phase $\oint_\Gamma ds_\mu R^\mu \equiv \Phi_\Gamma$ in the Wilson loop is equal to the electro-magnetic flux through the one or more loops specified by Γ. In other words, the line integral $\oint_\Gamma ds$ over loops Γ is equivalent to a flux integral $\int_\epsilon dS$ over the surfaces ϵ bounded by the loops. Mathematically $\oint_\Gamma ds_\mu R^\mu = \Phi_\Gamma = \int_\epsilon dS_{\mu\nu} F^{\mu\nu}$. Here $F^{\mu\nu} = \partial^\mu R^\nu - \partial^\nu R^\mu$ is the electromagnetic field tensor whose components are equal to the electric and magnetic fields, and $dS_{\mu\nu}$ is the infinitesimal surface tensor for a surface embedded in four-dimensional space-time. When the surfaces ϵ are all fixed to a particular time t this flux simplifies to the magnetic flux $\Phi_\Gamma = \int_\epsilon d\vec{S} \cdot \vec{B}$, where now $d\vec{S}$ is the surface infinitesimal in three-dimensional space and $\vec{B}$ is the magnetic field.

In atomic units where $e = \hbar = 1$ the phase Φ_Γ controlling the Wilson loop $W_\Gamma = \exp(\imath\Phi_\Gamma)$ is equal to the electromagnetic flux through all the loops specified by Γ. Because both the phase and the flux are strictly the argument of an exponential, they must be unit-free. In turn the units of electromagnetic field $F^{\mu\nu}$ must be the inverse of the units of the area infinitesimal $dS_{\mu\nu}$. This is the fundamental reason why the electromagnetic field tensor $F^{\mu\nu}$, and magnetic field $\vec{B}$ in particular, have units of inverse area.

We formalize these ideas using the loop representation of observables [27]. The previous discussion can be summarized by the statement that every physical observable $\mathcal{O}_i(\vec{B})$ must be written as

$$\mathcal{O}_i(\vec{B}) = \sum_\Gamma \langle \mathcal{O}_i | \Gamma \rangle \exp(\imath\Phi_\Gamma) \tag{1}$$

where Γ specifies the spatial trajectories of all the electron-hole loops, and $\sum_\Gamma$ is a sum over all the ways these electron-hole loops can be arranged - i.e. both their number and their individual trajectories. $\mathcal{O}_i$, is the observable, and $\langle \mathcal{O}_i | \Gamma \rangle$ is the weight with which a particular Wilson loop Γ contributes to $\mathcal{O}_i$. This weight is determined by both the observable and by the kinetics and interactions of the electrons and holes tracing the loops specified by Γ.

2.2 The Non-relativistic Limit

In relativistic quantum field theory the electron-hole pair follows a path which curves through space time and finally returns to its origin. The energy of the electron and the energy of the hole, taken separately, vary throughout their motion around the loop.

The non-relativistic limit is taken by assuming that the energies of the electron and hole are almost constant and remain very near their rest energy/mass. Obviously this requires a strict conceptual division between electron and hole. In this limit the Dirac equation, describing both electron and hole, simplifies to the Pauli equation which describes either the electron or the hole.

Because of the non-relativistic limit's special treatment of the time dimension, electric field drops out of the minimal coupling and takes a separate role in the Pauli equation, and is joined by the $e\vec{\sigma} \cdot \vec{B}$ Zeeman term. The remaining minimal coupling is $\vec{p} - e\vec{R}$, where $\vec{R}$ is the gauge field determining the magnetic field by $\vec{B} = \vec{\nabla} \times \vec{R}$. The minimal coupling produces Wilson loops $\exp(\imath\Phi_\Gamma)$ where the flux $\Phi_\Gamma = \int_\epsilon d\vec{S} \cdot \vec{B}$ depends only on magnetic field not electric field.

In the non-relativistic limit magnetic field affects electronic motion not only through Wilson loops but also via the Zeeman term. As we have argued earlier, in a class of systems which includes topological insulators in the diffusive regime, the spin triplet relaxation time τ_t is short compared to the spin singlet relaxation time τ_s. Therefore spin and the Zeeman term do not contribute to electronic transport over long distances [22], and observables measuring transport depend on magnetic field only via Wilson loops, not via the Zeeman term.

2.3 External Fields and the Loop Area Distribution

We specialize to the case of a uniform magnetic field $\vec{H}$ originating externally to the sample, which we will call $\vec{B}^{ext} \equiv \vec{H}$. We consider the effect of $\vec{B}^{ext}$ on the one or more electron-hole loops specified by Γ. Each electron-hole loop experiences not only the external field $\vec{B}^{ext}$, but also magnetic field generated by the electron-hole loops within the sample. The total field, including both external field and the field generated by the sample, is called $\vec{B}$. Total field $\vec{B}$ is related to external field $\vec{B}^{ext}$ via $\vec{B} = \mu_0 \vec{B}^{ext} + \mu_0 \vec{M}$, where $\vec{M}$ is the magnetization. μ_0 is the vacuum permeability, which we will drop for simplicity. In diamagnetic materials, including superconductors, $\vec{B}$ is smaller than $\vec{B}^{ext}$, while in paramagnetic materials $\vec{B}$ is larger than $\vec{B}^{ext}$. We separate the magnetic flux through a Wilson loop into two terms: a flux Φ_Γ^{int} caused by the field generated by the sample, and a flux $\Phi_\Gamma^{ext} = \int_\epsilon d\vec{S} \cdot \vec{B}^{ext}$ from external field. Therefore in an external magnetic field observables can be written as

$$\mathcal{O}_i(\vec{B}^{ext}) = \sum_\Gamma \langle \mathcal{O}_i | \Gamma \rangle \exp(\imath \Phi_\Gamma^{ext}) \tag{2}$$

The phase Φ_Γ^{int} from field generated within the sample is now packaged inside the weight $\langle \mathcal{O}_i | \Gamma \rangle$. The phase $\exp(\imath \Phi_\Gamma^{ext})$ depends only on external field, not on internally-generated field as well.

Because the external field $\vec{B}^{ext}$ has uniform direction and magnitude, it is useful to assign to each Wilson loop a vector-valued cross-section $\vec{A}_\Gamma = \int_\epsilon d\vec{S}$, which is a vector generalization of the Wilson loop's surface area. This area vector depends only on paths traced by the loop Γ, not on the shape of the surfaces ϵ which are bordered by Γ. The magnetic flux through Γ then simplifies to the dot product of the area $\vec{A}$ with the external field $\vec{B}^{ext}$; $\Phi_\Gamma = \vec{A}_\Gamma \cdot \vec{B}^{ext}$. This leads to our central result:

$$\begin{aligned} \mathcal{O}_i(\vec{B}^{ext}) &= \int d\vec{A}\, \mathcal{O}_i(\vec{A}) \exp(\imath \vec{A} \cdot \vec{B}^{ext}), \\ \mathcal{O}_i(\vec{A}) &\equiv \sum_\Gamma \langle \mathcal{O}_i | \Gamma \rangle \delta(\vec{A}_\Gamma - \vec{A}) \end{aligned} \tag{3}$$

In other words, every physical observable can be resolved into contributions corresponding to specific areas $\vec{A}$, each with weight $\mathcal{O}_i(\vec{A})$:

$$\mathcal{O}_i(\vec{A}) = \int \frac{d\vec{B}^{ext}}{(2\pi)^3} \mathcal{O}_i(\vec{B}^{ext}) \exp(-\imath \vec{A} \cdot \vec{B}^{ext}) \tag{4}$$

The area $\vec{A}$ is the sum of the areas of all the electron-hole loops that contribute to the observable.

In summary, Eqs. 3 and 4 state that measurements of an observable $\mathcal{O}_i(\vec{B}^{ext})$'s dependence on external magnetic field give direct and rigorous information about the sum of the areas traced out by electron loops within a sample, simply by performing a Fourier transform. The resulting loop area distribution $\mathcal{O}_i(\vec{A})$ completely describes how different loop areas contribute.

From a theorist's point of view, the problem of how to perform a regularized sum over loops $\sum_\Gamma$, i.e. how to control both long distance and short distance contributions to the sum, remains in general unsolved. This makes evaluation of the sum over loops $\sum_\Gamma$ in Eqs. 1 and 3 a difficult task. It is therefore both remarkable and encouraging that we can measure the sum experimentally and directly using Eq. 4; experimental measurements of the loop area distribution $\mathcal{O}_i(\vec{A})$ can lead the way in guiding theorists to correct procedures for performing the sum over loops.

3 Comparison of Loop Area Distributions

Throughout the remainder of this paper $\vec{B} = \vec{B}^{ext} = \mu_0 \vec{H}$ means the external magnetic field, and $B = |\vec{B}|$ is its magnitude.

3.1 Landau Levels and SdH Oscillations

To illustrate the transformation from the magnetoconductance to the loop area distribution, consider the case where the dominant energy scale is the Fermi level E_F, in which case a delta function - the first Landau level - will be found in $G_{xx}(B)$ at $B^{-1} = \frac{1}{2}A_F$. Here $2\pi/A_F \propto 2\pi E_F$ is the cross-section of the Fermi surface [28]. The loop area distribution is therefore a cosine $G_{xx}(A) \propto \cos(A/\frac{1}{2}A_F)$, which means that loops with every area contribute to the conductance and that there is ringing corresponding to the characteristic Fermi area A_F.[1]

The most interesting aspect of this result is that quantum coherence causes the first Landau level to be repeated exactly at characteristic areas $(N + 1/2)A_F$, with the N-th level's height proportional to N. The repetitions are manifested as a hierarchy of additional Landau level delta functions in $G_{xx}(B)$, as illustrated in Fig. 1a. Figure 1b shows the loop area distributions $G_{xx}(A)$ of the Landau levels. The N-th Landau level is a cosine $\cos(A/(N + \frac{1}{2})A_F)$ with period equal to $(N + 1/2)\ 2\pi$ times the characteristic area A_F. In mathematical terminology, the N-th Landau level is simply the N-th harmonic of the lowest Landau level. Speaking more plainly, quantum coherence ensures that if an electron can complete a loop once, then it can repeat that same loop any number of times.

The only limit to these repetitions (harmonics) is decoherence caused either by the scattering energy scale Γ or by the temperature T, which causes a decay in the loop area distribution [29]. Decoherence broadens the Landau levels into peaks with width Γ, T, which eventually merge into SdH oscillations. These oscillations have equal

[1] A slightly different result $\cos(A/NA^F)$ obtains for the N-th Landau level of Dirac fermions [61].

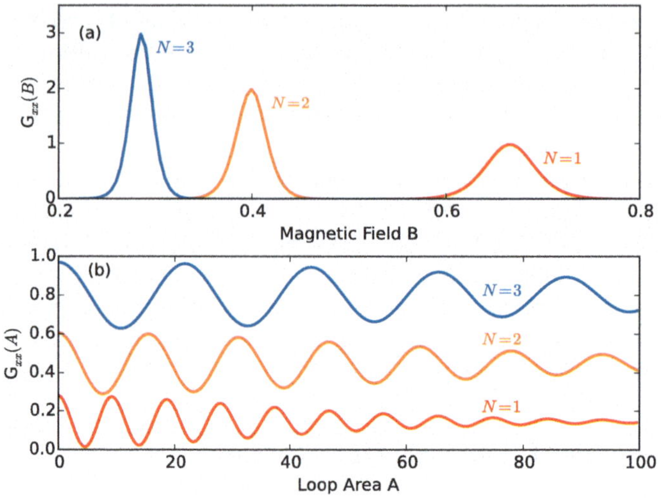

Fig. 1. (Color online.) Quantum coherence and the hierarchy of Landau levels. Panel (a) shows the magnetoconductance $G_{xx}(B)$ of the $N = 1, 2, 3$ Landau levels and panel (b) shows their loop area distributions $G_{xx}(A)$, which have been vertically shifted for clarity [29]. Quantum coherence is responsible for producing higher harmonics of the lowest Landau level, causing higher Landau levels at multiples $(N + \frac{1}{2})A_F$ of the characteristic area $A_F = E_F^{-1}$. Decoherence, i.e. loss of quantum coherence, is caused by either the scattering energy scale Γ or the temperature T. Decoherence widens peaks in $G_{xx}(B)$, suppresses $G_{xx}(A)$ at large A, and completely suppresses oscillations when the field B is as small as Γ, T. $E_F = 1,\ \Gamma = 0,\ T = 0.05$.

height instead of the height proportional to N seen in Landau levels. They decay exponentially when the value of $B = N^{-1}A_F^{-1}$ descends to Γ, T, and their loop area distribution also decays exponentially at areas larger than Γ^{-1}, T^{-1} [29][2].

3.2 Weak Antilocalization

We now investigate 2-D WL/WAL, whose magnetoconductance $G_{xx}^{WAL}(B)$ has been studied extensively. We will show that the WL/WAL loop area distribution varies inversely with area i.e. it obeys $G_{xx}^{WAL}(A) \propto A^{-1}$. At magnetic fields less than the inverse cutoff $q_C B < 1/L_{max}^2$ the conductance $G_{xx}^{WAL}(B)$ is controlled by the cutoff: it quickly transitions to zero, and this transition is quadratic in B because time reversal symmetry requires that G_{xx} be even under $B \rightarrow -B$. At larger fields $q_C B > 1/L_{max}^2$ the magnetoconductance of a 2-D spin singlet is a logarithm $G_{xx}^{WAL}(B) \propto -\ln B$. This logarithmic form, which will be our focus here, is robust and ubiquitous. If spin or orbital degrees of freedom are in play then the logarithmic conductance seen from the Cooperon spin singlet generalizes simply to a sum of logarithms corresponding to the Cooperon spin singlet, the Cooperon spin triplet, and

[2] To verify that loop area distribution for Landau levels and SdH oscillations decays exponentially, we numerically Fourier transformed the conductance formulas in Ref. [29], because analytical formulas for the Fourier transform of this signal are not available.

possibly other Cooperon sectors associated with valley degrees of freedom etc. The logarithmic conductance is seen for both Dirac and $p^2/2m$ dispersions, for both lattice and continuum models, for both short and long range scattering, and for a wide variety of quasi-2-D geometries [20,30–42][3]. The logarithm is found in all 2-D systems that are in the diffusive regime, i.e. all 2-D systems whose size is larger than the momentum relaxation length caused by scattering.

The reason that the same form of $G_{xx}^{WAL}(B)$ is universally found in diffusive 2-D systems is that at distances exceeding the momentum relaxation length the Cooperon loses all information about momentum and kinetics, and therefore its trajectory is simply a 2-D random walk. When spin or orbital degrees of freedom are in play the Cooperon carries several species of random walkers (e.g. a spin singlet and a spin triplet), but each of the species is still performing a simple random walk. The WL/WAL magnetoconductance signal is determined by Cooperon trajectories which return to their origin, forming loops. Quantum coherent repeats of loops do not contribute because the Cooperon's momentum is random and therefore even when a Cooperon returns to its origin it does not have the same momentum that it started with. Since the trajectories of these Cooperon loops are simply random walks, the magnetoconductance signal has the same form in all 2-D diffusive systems.

The fact that the universal form of the magnetoconductance $G_{xx}^{WAL}(B) \propto \ln B$ is logarithmic was first determined by Ref. [20]. Ref. [20] starts with a description of how scattering changes the motion of the electronic wave-function $|\psi\rangle$ and its complex conjugate $\langle\psi|$, and then uses this information to derive the behavior of the Cooperon, which is a pairing of $|\psi\rangle$ with $\langle\psi|$. It is found that the spin singlet component of the Cooperon is simply a random walker and that every random walk returning to the origin contributes with equal weight to the WAL conductance $G_{xx}^{WAL}(B)$. Mathematically this is expressed by writing the Cooperon Green's function as the inverse of a Laplacian operator which expresses how individual steps in the walk are taken. To derive from disordered scattering the Cooperon and its random walk, Ref. [20] uses the standard field theory of disordered scattering, bosonization, and sigma models [18, 19]. In this respect it is generally imitated by all other theoretical (non-computational) treatments of Cooperons and WL/WAL. Always the result is that WL/WAL is mediated by random walking Cooperons.

The WAL conductance is proportional to the probability that the Cooperon will return to its origin; mathematically it is the diagonal $\vec{x} = \vec{x}'$ matrix element of the Cooperon Green's function. At zero field $G_{xx}^{WAL}(B = 0)$ is simply the probability that random walkers will return to their origin, while non-zero field causes each walker trajectory to be reweighted by a Wilson loop phase factor. Calculation of the Cooperon spin triplet component is almost identical to that of the singlet - the main difference is that the triplet contribution to G_{xx} is multiplied by -1, producing weak localization instead of weak antilocalization.

We emphasize the fact established in Ref. [20] that all trajectories contributing to the Cooperon's spin singlet component contribute with the same weight, with no phase factor from kinetics, scattering, potentials, or time evolution. Likewise, all trajectories

---

[3] At very strong magnetic fields, when the magnetic length $\sqrt{\hbar c/eB}$ is compared with the mean free path, there is a further transition to $G_{xx} \propto 1/\sqrt{B}$ behavior [45–48].

contributing to the Cooperon's triplet component contribute with the same weight. In other words, although the Cooperon describes a special kind of quantum interference process that is exhibited by electrons, the actual behavior of a Cooperon is no different than that of a purely classical random walker. No phase information enters in because a Cooperon is a pairing of $|\psi\rangle$ with $\langle\psi|$, and their phases cancel each other. Therefore the Cooperon evolves as a classical random walker, and calculations of $G_{xx}^{WAL}(B)$ look purely classical. All Cooperon trajectories (of the singlet, or of the triplet) contribute with the same sign and weight. One very important consequence is that the loop area distribution $G_{xx}^{WAL}(A)$ of the Cooperon singlet (or triplet) trajectories has only one sign, like a purely classical probability distribution. Calculation of WL/WAL is isomorphic to calculating return probabilities of a purely classical walker.

This picture of a purely classical Cooperon is modified in samples with sizes approaching the localization length, where the conductance is reduced to the order of the universal conductance quantum G_0. At this long length scale processes coupling two or more Cooperons become important, as do processes coupling Cooperons with diffusons. However the systems which show the *weak* localization and antilocalization which is our focus here lie far from the localized limit, so that the WL/WAL contribution remains at the level of simple random walks without multi-Cooperon processes.

We demonstrate here that the Fourier transform of the logarithmic WL/WAL conductance $G_{xx}(B) \propto \ln B$ varies inversely with area, i.e. that the loop area distribution obeys $G_{xx}(A) \propto A^{-1}$. In other words the Fourier transform of $-\ln B$ is A^{-1} and, vice versa, the inverse Fourier transform of A^{-1} is $-\ln B$. The main difficulty is controlling the ultraviolet and infrared cutoffs, so we perform the demonstration three ways. First, we use purely dimensional analysis: $G_{xx}(A) = (2\pi)^{-1} \int dB \exp(-\imath AB) \ln B$. Therefore $G_{xx}(A)$ must carry units of inverse area; it must be equal to A^{-1} times some function that depends only on dimensionless quantities.

Second, we use specific functional forms for $G_{xx}(A)$ and carry out analytically the inverse Fourier transform from $G_{xx}(A)$ to $G_{xx}(B)$. For instance, we consider hard cutoffs at A_0 and A_{max}, with $G_{xx}^{WAL}(A) = A^{-1}$ for $A_0 < A < A_{max}$. With these cutoffs we arrive at $G_{xx}^{WAL}(B) \approx R_0 - \ln|q_C B| + (q_C B A_0)^2/4$ for $1/A_{max} < q_C B < 1/A_0$, where q_C is the Cooperon charge, $R_0 = -\gamma_E - \ln|A_0|$, and γ_E is the Euler-Mascheroni constant. Here R_0 should be adjusted to zero, as a regularization compensating for the hard cutoffs, to replicate the well-known fact that the WAL (Cooperon spin singlet) contribution to the conductance is always positive for all values of B, and that the triplet WL contribution is always negative [20]. The appendix imposes several other cutoffs on $G_{xx}^{WAL}(A) = A^{-1}$, in each case performs the Fourier transform exactly, and in each case obtains logarithmic forms for $G_{xx}^{WAL}(B) = \int dA \exp(\imath AB) A^{-1} \propto -\ln B$.

Lastly, Ref. [20] and many subsequent works have established that the magnetoconductance $G_{xx}^{WAL}(B)$ of a 2-D random walking Cooperon is logarithmic. Therefore we can show that the Fourier transform of $-\ln B$ is A^{-1} simply by showing that the loop area distribution of 2-D random walkers is $G_{xx}^{WAL}(A) \propto A^{-1}$. The reasoning is like a syllogism: random walkers have a logarithmic magnetoconductance, and random walkers have an A^{-1} loop area distribution. The loop area distribution is the Fourier transform of the magnetoconductance; therefore the Fourier transform of $-\ln B$ is A^{-1}.

Figure 2a summarizes the results of extensive numerical Monte Carlo simulations of random walks, all of which verify that the loop area distribution of random walks which return to their origin scales with A^{-1}. We use an infinite 2-D square lattice, and define the lattice spacing equal to one. The walkers take steps which are chosen from a Gaussian distribution with mean step length $l = 4\sqrt{2}, 8\sqrt{2}$, and an upper bound (IR cutoff) to the loop areas is supplied by limiting the walk length to 5×10^5. We count the walkers which return to their origin, compute the area of the loop traversed by each walker, and histogram the area distribution $G_{xx}(A)$. Panel (a) shows $G_{xx}(A)$, multiplied by A. The plateaus of constant $G_{xx}(A) \times A$ extending over several orders of magnitude verify that 2-D random walks follow an A^{-1} loop area distribution. We have verified that similar results are found with other cutoffs and with other probability distributions governing the step lengths, and a similar numerical Monte Carlo simulation of 2-D random walks has previously verified the A^{-1} loop area distribution [43]. These numerical demonstrations of the A^{-1} loop area law again verify that the Fourier transform of $-\ln B$ is $G_{xx}(A) \propto A^{-1}$, and vice versa.

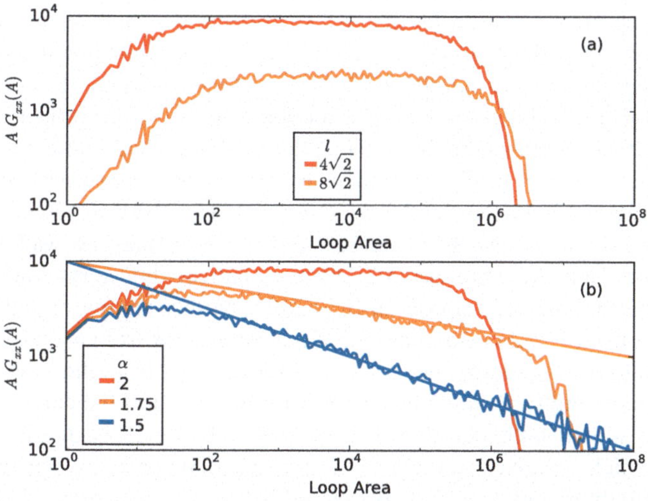

Fig. 2. (Color online.) The Cooperon's loop area distribution $G_{xx}(A)$ obtained by Monte Carlo simulations. Panel (a) shows $G_{xx}(A)$, multiplied by A, of standard 2-D random walks with Gaussian-distributed steps having average length $l = 4\sqrt{2}, 8\sqrt{2}$. The plateaus extending over several orders of magnitude verify that 2-D random walks follow an A^{-1} loop area distribution. The IR cutoff is caused by limiting the walk length to 5×10^5. Panel (b) shows that Levy flights have a loop area distribution which decays faster than A^{-1}. α is the stability parameter of the Levy alpha-stable distribution which controls the step lengths. $\alpha = 2$ produces Gaussian-distributed steps, and when α is decreased below 2 the distribution develops a heavy tail of very long steps. The straight lines are for $10^4 \times A^{-0.125}$ and $10^4 \times A^{-0.25}$, and their agreement with the Monte Carlo data shows that $G_{xx}(A) \propto A^{-2+\alpha/2}$. $l = 4\sqrt{2}$ and the IR cutoff is 5×10^5, 1.5×10^6, and 5×10^6 for the $\alpha = 2$, 1.75, 1.5 results.

3.3 Linear Magnetoresistance

This result can be reversed and applied to the case of linear magnetoresistance, where $G_{xx}(B) \propto B^{-1}$. We established above that the inverse Fourier transform of A^{-1} is $-\ln B$; interchanging A with B shows that the loop area distribution corresponding to linear magnetoresistance is logarithmic, $G_{xx}(A) \propto -\ln A$. This profile, whose broad features are general for any roughly linear resistance, is remarkably different from the A^{-1} distribution that governs loops generated by random walks. It has fewer small loops and many more large loops; a small head and a very long tail. We emphasize that the logarithmic loop area distribution is a rigorous result, as long as only spin singlets contribute to conduction and the Zeeman effect can be neglected. If this assumption holds, then linear magnetoresistance implies a logarithmic loop area distribution, no matter what physical mechanism is responsible for producing the loops.

3.4 Levy Flights and 3-D

Since linear magnetoresistance corresponds to a $\ln A$ distribution that decays very slowly with loop area, it would be of interest to determine how such a long tail can be produced. However the $1/A$ decay seen in WL/WAL seems to be the slowest decay and the biggest tail that can be produced by purely stochastic random walks moving on a 2-D lattice. Figure 2b shows the loop distribution that is obtained if the Cooperon does not follow a diffusive path where the step length distribution has a finite mean, variance, etc. Instead the Cooperon in Fig. 2b follows a Levy flight where step lengths follow a distribution including both short and very long steps. The result is a steeper decay than A^{-1}, because long steps tend to decrease the probability that the walker's path will complete a loop. Here we again use a 2-D square lattice, choose random steps from a probability distribution, and impose an IR cutoff l on the maximum path length. The only difference from Fig. 2a is that instead of choosing steps from a Gaussian probability distribution to produce a random walk, we choose steps from the Levy alpha-stable distribution, producing Levy flights where most steps are short but some steps are very long. The Levy alpha-stable distribution is controlled by the Levy distribution stability parameter α which we give values $\alpha = 2, 1.75, 1.5$. When $\alpha = 2$ it reduces to the Gaussian distribution, and when α is decreased below 2 the distribution develops a fat tail of very long steps. Panel 2b plots $G_{xx}(A) \times A$. At $\alpha = 2$ we see again the plateau signaling the $1/A$ area decay law of random walks. At $\alpha = 1.75, 1.5$ the loop area distribution decays faster than A^{-1}, tilting the plateau. We superimpose straight lines for $10^4 \times A^{-0.125}$ and $10^4 \times A^{-0.25}$, and their agreement with the $\alpha = 1.75, 1.5$ Monte Carlo data shows that $G_{xx}(A) \propto A^{-2+\alpha/2}$. In summary, Levy flights steepen the area law decay, taking it further from the very slow $\ln A$ decay necessary to produce linear magnetoresistance.

Increasing the dimensionality above two dimensions also produces a steeper decay, $A^{-3/2}$ in 3-D [44]. This is the opposite of what is wanted, so one might like to instead decrease the dimensionality below two dimensions. However in one dimension all loops have area identically equal to zero. Therefore neither changing the step distribution nor changing the dimensionality can reproduce the $\ln A$ tail seen in linear magnetoresistance systems.

We note that in very strong magnetic fields where the magnetic length $l_H = \sqrt{\hbar c/eB}$ is smaller than the mean free path a $\sqrt{A}$ area law is obtained at areas smaller than l_H^2 [45–48]. However this produces a $1/\sqrt{B}$ magnetoresistance, not a linear signal, and is strictly a large-field behavior, unlike experiments where linear magnetoresistance starts at small fields.

4 Linear Magnetoresistance and Quantum Coherence

Linear magnetoresistance's $-\ln A$ loop area distribution shows no characteristic scale, decreases uniformly and smoothly, and never changes sign. These facts are strong evidence that linear magnetoresistance is caused by Cooperons, i.e. by paired particles and holes scattering together, rather than by any species of single-particle motion.

The loop area distribution of linear magnetoresistance exhibits a long fat logarithmic tail that can not be explained by random walks or by Levy flights. Therefore we propose that the Cooperons responsible for linear magnetoresistance do follow random walks, but with the special feature that they maintain quantum coherence, allowing them to repeat their loops many times, in the same way that Landau levels are repeated in a hierarchy at $(N + \frac{1}{2})A_F$.

As a simple example, consider the case of 2-D random walks with hard cutoffs at A_0 and A_{max} and $G_{xx}^1(B, A_0, A_{max}) = \int_{A_0}^{A_{max}} dA \cos(q_C BA)/A$. The $1/A$ in this expression is the loop area distribution of standard random walks, and the cosine and integral perform the Fourier transformation from area to magnetic field. q_C is the Cooperon charge. At field strengths between the UV cutoff $1/A_0$ and the IR cutoff $1/A_{max}$ this magnetoconductance $G_{xx}^1(B, A_0, A_{max})$ follows the logarithmic form of standard WL/WAL.

Now take that expression, and allow loops to repeat up to N times, with a weight proportional to $1/N$. This procedure depletes the loop distribution's head, because if a loop area distribution has a UV cutoff at A_0, then its $N = 2$ first repetition will have a higher UV cutoff at $2A_0$, its second repetition will have its cutoff at $3A_0$, etc. Therefore between A_0 and $2A_0$ the total loop area distribution will have a contribution from only the base $N = 1$ loop distribution, while at larger areas higher and higher N will contribute. The net effect is to enlarge the tail of the loop area distribution and shrink its head.

Performing the sum over loop repetitions obtains $G_{xx}(B) = \gamma_E + \sum_{n=1}^{N} n^{-1} G_{xx}^1(B, nA_0, A_{max})$. We have added a regularization constant γ_E to compensate for cutoff effects. Figure 3a compares this loop area distribution to the diffusive $1/A$ profile produced by random walks without coherence. The new loop area distribution has the required small head and long tail; it is a logarithm. Figure 3b shows its Fourier transform, the magnetoresistance. It is linear, i.e. proportional to $R_{xx}(B) - R_{xx}(B = 0) \approx \pi q_C B A_0$, in the range $1/A_{max} < q_C B < 0.2/A_0$.

This simple example is a proof of principle about how to obtain linear magnetoresistance. We chose an abrupt IR cutoff at A_{max} and an abrupt UV cutoff at A_0, but the true forms of these cutoffs will be material dependent. These cutoffs can have visible effects. In Fig. 3b superimposed on the linear signal one sees a small amplitude fast ripple, which is ringing caused by the abrupt IR cutoff. A similar ripple has been seen

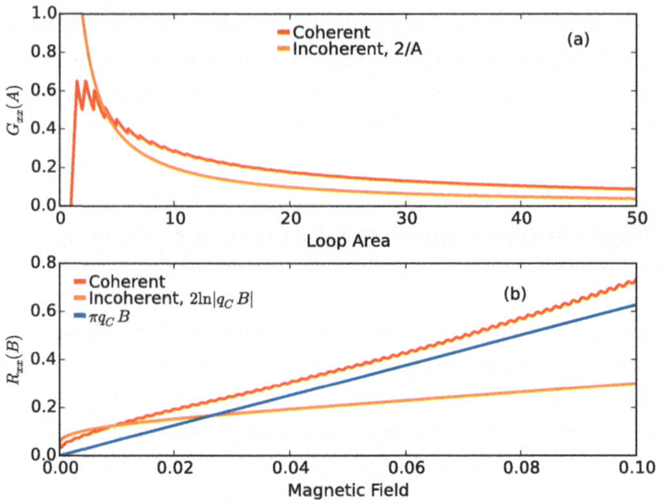

Fig. 3. (Color online.) Linear magnetoresistance from quantum coherence. The red lines show the results when quantum coherence allows loops to repeat up to $N = 100$ times. They can be compared to the orange lines which show standard behavior without quantum coherence. Panel (a) shows the loop area distribution $G_{xx}(A)$, where the quantum coherent distribution (red line) clearly has a small head and long tail as compared to the standard distribution (orange line). Panel (b) shows the resistance $R_{xx}(B)$. The quantum coherent result (red line) compares nicely to the blue straight line with slope πq_C. $A_0 = 1$, and the sharpness of the hard IR cutoff at $A_{max} = 2000$ causes the ripple in the quantum coherent result. The orange line gives a standard logarithmic curve, $R_{xx}(B) = (0.1 - 2\ln|q_C B|)^{-1}$.

in Ref. [1], signaling that a sharp IR cutoff was present in that experiment. However in most experimental data on linear resistance a ripple is not visible, indicating either a smooth IR cutoff or that magnetic field was not sampled at fine enough resolution. At large fields the form of the UV cutoff also manifests in the signal's behavior. Our sharp UV cutoff causes strong ringing at large fields $q_C B > 0.2/A_0$ [not shown in Fig. 3b] which overwhelms the linear magnetoresistance signal, but with a suitably chosen UV cutoff the linear growth will extend to larger B.

We note that linear magnetoresistance does not require the IR cutoff A_{max}, i.e. the area scale where phase coherence is suppressed, to be very large. For instance, a coherence area of $(100\,\text{Å})^2$ would allow linear magnetoresistance to extend down to ≈ 0.3T. Moreover, scatterers within a sample must not fluctuate or move more quickly than the time required for a Cooperon to complete its loop. However this time scale is extremely short. Using $T^{-1} = 4A$ from Appendix B and $\tau = \hbar/k_B T$, we find that τ is three picoseconds for loops of area $A = (100\,\text{Å})^2$.

We emphasize that a power law area distribution can always be given a depleted head and long tail by invoking quantum coherence and allowing loop repetitions. The form of the UV and IR cutoffs, the relative weight of each repetition, and even the base loop area distribution $G^1_{xx}(A)$ may be specific to the scattering source and to the mechanism responsible for Cooperon coherence. There are only two fundamental requirements: a scattering process producing an area distribution with broad support across a range of areas, and quantum coherence allowing Cooperon loops to repeat in a hierarchy like that of Landau levels.

The real difficulty here is that ordinarily Cooperons in 2-D planar materials do not display loop repetitions; they do not retrace their path two or more times. The fundamental reason is that in 2-D (or 3-D, 4-D, etc.) when a Cooperon returns to its starting point it does so with a momentum that is different from its original momentum. Therefore on the second time around the Cooperon leaves the origin moving in a direction that is different from its first time around. We now turn to the question of how 2-D Cooperons could retain a degree of quantum coherence.

There are two exceptional cases which prove that under the right circumstances Cooperons are able to maintain quantum coherence; they are able to show quantum loop repetitions similar to the higher harmonics seen in Landau levels and SdH oscillations. The most notable exception is 1-D systems with static disorder. In 1-D the Cooperon's particle and hole are constrained to follow the same path, and the static disorder means that their energies never change, and neither does the magnitude of their momenta. Therefore when a Cooperon returns to its starting point it can easily repeat the same exact loop that it has just ended, starting with a momentum that has the same magnitude and direction as before. As a result Cooperon-mediated quantum interference dominates the behavior of 1-D and quasi 1-D systems: once the system's length L increases enough so that the usual classical $1/L$ Ohm's law decay brings the conductance down to one conductance quantum G_0, quantum interference takes over. This can take two forms. In most systems the conductance decay transitions from the $1/L$ Ohm's law to the exponential decay characteristic of Anderson localization [49–51]. However in topologically protected systems the conductance plateaus at a constant value of G_0, manifesting a perfectly conducting channel. Both cases manifest complete dominance of quantum interference over classical conduction processes, and are fueled by Cooperons maintaining quantum coherence [52,53].

A second case where Cooperons retain quantum coherence is 2-D cylindrical geometries when a magnetic field is parallel to the cylinder. In this geometry the Cooperon produces Altshuler-Aronov-Spivak oscillations where the conductance oscillates as field strength is varied. Without Cooperon loop repetitions, i.e. without quantum coherence, the oscillations would have a simple cosine profile. With the loop repetitions the Altshuler-Aronov-Spivak oscillations deviate strongly from a simple cosine signal [54,55].

These two examples prove that the absence or presence of quantum coherence in Cooperons is determined by geometry. In 1-D systems and in 2-D cylindrical systems Cooperons can display loop repetitions. The remaining question is how Cooperon coherence and loop repetitions are preserved in the experiments that report linear magnetoresistance, which generally have non-cylindrical (simply connected) 3-D or 2-D geometries. The cause cannot be a non-Gaussian disorder distribution with large outliers, because non-Gaussian disorder penalizes higher-order processes with small weights.

We speculate, very briefly, that the resolution to this question is in geometry. It may be that in linear resistance materials transport may be locally one-dimensional, with carriers constrained along certain race-track like trajectories. This is seen in snake states in graphene, in edge states in the quantum Hall effect and in topological insulators, and might also be realized in C_4 symmetry-broken (nematic) states in underdoped cuprates [56–60]. When particle motion is locked to a locally one dimensional track, at any particular point the momentum is limited to only two values differing only in sign, and therefore the Cooperon's position and momentum are locked to each other, up to the same sign. This could allow Cooperons to retain some degree of quantum coherence and result in linear magnetoresistance.

The point of introducing quantum coherence into this discussion was the hope that it could explain the extraordinarily large areas that occur in the loop area distribution of materials whose resistance is linear in field. We close by discussing two other alternatives. First, if spin physics and the Zeeman term were to affect conduction, then the Fourier transform of the magnetoconductance, i.e. the quantity which we call the loop area distribution, will contain information not only about loop areas but also about spin physics. In this case the long tail of the Fourier transform may not say anything about loops but may instead be caused by some spin effect.

In topological insulators the only way that spin physics can affect conduction is if conduction is ballistic instead of diffusive; if scattering is weak enough or the sample small enough that an electron can transverse the sample with very few scatterings. It is a relatively simple matter for experimentalists to determine whether their sample is diffusive or ballistic. Moreover, the conductance in a ballistic sample would exhibit strong dependence on the geometry, leads, and Fermi energy, and could be expected to oscillate with field. There should be little doubt about whether a particular topological insulator sample is in the diffusive regime. In turn there should be little doubt about the validity of the loop area transform and its long tail.

Second, the area detected in the loop area distribution is the sum of the areas of all loops that contribute to the material's conductance. In non-interacting materials this sum includes only one loop. We found that in the diffusive regime random walkers completing a single loop cannot reproduce the area distribution's logarithmic tail, which motivated our discussion of quantum coherence. However in strongly interacting materials, where a sea of electron-hole loops determine the conductance, the story may run differently. In this case the logarithmic distribution's fat tail and depleted head could be caused by some multi-body effect and not by quantum coherence.

Most topological insulators, including the ones reported to exhibit linear magnetoresistance, are known to be weakly interacting [1–4]. Therefore it seems unlikely that the long tail of their loop area distribution could be caused by interactions, and we are left again with the same conundrum that led us to quantum coherence.

5 Final Comments

In our view one of the most important aspects of the present work is its methodology, which focuses on geometric analysis of electron and hole loops, and especially on the loop area distribution that can be obtained from Fourier transforms of magnetoconductance data. We have presented a non-perturbative framework for understanding electron and hole behavior that can give additional physical insight. This framework depends only on the assumption that neither the spin triplet nor the Zeeman term affect conduction, which is valid in topological insulators in the diffusive regime, and perhaps in other materials as well. We anticipate experiments focusing on the loop area distribution, with special attention to accessing large areas using carefully controlled small increments of the field, to removing leads effects, and to performing careful Fourier transforms. We also expect increased use of vector magnets and multi-dimensional Fourier transforms.

Acknowledgments. We gratefully acknowledge formative and stimulating discussions with S. Kettemann, A. Leggett, Y. Li, C. Lin, X. Dai, V. Dobrosavljevic, H.-J. Lee, Q. Wu, T. Ohtsuki, J. Zaanen, K. Schalm, T. Takimoto, K.-S. Kim, X. Wan, P. Niklowitz, A. Ho, J. Saunders, L. Levitin, J. Koelzer, T. Schapers, H. Luth, C. Weyrich, P. Hasnip, A. Kim, M. Ma, P. Coleman, S. Hayden, I. Bozovic, P. Abbamonte, and G. Parisi. We also thank especially A. Petrovic who discussed the manuscript during preparation.

Funding Information. We acknowledge support from EPSRC grant EP/M011038/1.

Appendix A: Alternate Regularizations

We demonstrate in this appendix that reasonable regularizations of the Fourier transform of x^{-1} produce a logarithm, and vice versa. In our first example we perform the Fourier transform of x^{-1}.

$$\begin{aligned}\mathcal{A}(\gamma, x_0, k) &= \int_0^\infty dx\, \cos(kx) \exp(-\gamma x) \frac{1}{x + x_0} \\ &= Re(\exp(x_0(\gamma + \imath k))\Gamma(0, x_0(\gamma + \imath k)))\end{aligned} \tag{A1}$$

$\mathcal{A}(\gamma = 0, x_0 = 1, k)$ is approximately logarithmic in the range $x = [0, 0.1] \times 2\pi$, as can be verified by plotting $\mathcal{A}(\gamma = 0, x_0 = 1, k)/\log(k)$.

Our second example again performs the Fourier transform of x^{-1}, but with a different UV cutoff.

$$\mathcal{B}(\gamma, x_0, k) = \int_0^\infty dx\, \cos(kx) \exp(-\gamma x) \frac{1}{\sqrt{1 + x^2/x_0^2}} \tag{A2}$$

$\mathcal{B}(\gamma, x_0, k)$ can be integrated exactly, and includes Bessel, logarithmic, Struve, and hypergeometric functions. $\mathcal{B}(\gamma = 0, x_0 = 1, k)$ is approximately logarithmic in the range $x = [0, 0.1] \times 2\pi$.

Our third example performs the Fourier transform of $\log |x|$.

$$\mathcal{C}(\gamma, x_0, k, \nu) = \int_0^\infty dx \cos(kx) \exp(-\gamma x) \log |1 + (x/x_0)^\nu| \tag{A3}$$

This integral can be performed analytically for many values of $\nu = 1,\ 3/2,\ 5/3,\ 7/4,\ 19/10,\ 2$, and is always proportional to k^{-1} plus corrections of order k at small k in the range $[0, 0.2]$. For $\nu = 3/2,\ 5/3,\ 7/4,\ 19/10$ it is written in terms of Meijer functions. The value of $k \times \mathcal{C}(\gamma = 0, x_0 = 1, k, \nu)$ at $k = 0$ is generally a rational fraction times π. For instance,

$$\begin{aligned} \mathcal{C}(\gamma, x_0, k, \nu = 1) &= Re(\omega^{-1} \exp(\omega x_0) \Gamma(0, \omega x_0)) \\ \omega &= -\imath k + \gamma. \end{aligned} \tag{A4}$$

Appendix B: Linear-in-Temperature Resistance

In this appendix we continue our discussion of systems whose resistance increases linearly with magnetic field. We make remarks of a more speculative nature arguing that these systems' resistance should increase linearly with temperature. We are continuing to assume that conduction is diffusive, with many scatterings, and that the spin triplet and the Zeeman term do not affect conduction. We are also continuing along the lines of Sect. 4, attributing the long $\ln A$ tail of the loop area distribution to quantum coherence. Within this framework we examine the resistance's dependence on temperature.

As we have seen, the loop area distribution has a smooth logarithmic form with only two characteristic area scales: the UV cutoff A_0, and the infrared cutoff A_{max} which regulates large loops. The ultraviolet cutoff A_0 may be controlled by many length scales: the scattering length, the lattice spacing, the scale of the Fermi surfaces, and the crossover from WAL to WL which is controlled by the scales at which spin rotation symmetry and time reversal symmetry are broken, etc. These mechanisms depend weakly on temperature.

The main source of temperature dependence comes instead from the infrared cutoff A_{max} which regulates large loops. The tail of the $\ln A$ loop area distribution reflects quantum coherence. Regardless of which short distance physics controls the loop area distribution without quantum coherence, this physics has no control over quantum coherence, and therefore cannot determine the infrared cutoff A_{max} regulating the loop area distribution. The *only* mechanism available to limit the loop repetitions and thus supply a cutoff A_{max} on the loop area distribution's tail is quantum decoherence, which is controlled by temperature.

How does the IR cutoff A_{max} depend on temperature? Since in atomic units inverse temperature has units of area, it would be natural for $A_{max} \propto T^{-1}$ to scale inversely with temperature. We have already noted that $A_{max} \propto T^{-1}$ does obtain in the specific case of SdH oscillations. [29] It also holds in 2-D diffusive systems when the dominant mechanism of decoherence is through electron-electron interactions. [64,65] In linear resistance systems $A_{max} \propto T^{-1}$ scaling should be very robust because the loop areas in the tail are so large that T^{-1} is the only area scale large enough match them.

If the decoherence area $A_{max} \propto T^{-1}$ scales inversely with temperature, it is then a simple matter to show that at $B = 0$ the resistance must be linear in temperature. Most simply, the linear magnetoresistance is $R_{xx}(B) = \pi q_C B A_0$. When $B = 0$ the inverse area B^{-1} diverges. The only scale available to take the place of B^{-1} is T^{-1}, and plugging this into the magnetoresistance formula immediately gives linear in temperature resistance $R_{xx}(T) = \pi q_C T A_0$.

Going into more detail, the logarithmic loop area distribution is $G_{xx}(A) \approx -(\pi A_0)^{-1} \ln |A|$. This should be regularized to give positive values because, as we have seen, in WL/WAL the Cooperon does not carry phase information and all Cooperon loop trajectories carry the same weight with the same sign. One possible regularization is to add $(\pi A_0)^{-1} \ln |A_{max}|$. Setting $B = 0$ and integrating with respect to area gives $R_{xx}(B = 0) \approx \pi A_0/A_{max} \propto \pi A_0 T$ for $A_0 \ll A_{max} \propto T^{-1}$, plus regularization-dependent terms which are sensitive to the cutoffs. When $A_{max} \propto T^{-1}$ is comparable to A_0 the linear dependence on T collapses. At smaller temperatures the linear in temperature resistance is robust because B^{-1} and T^{-1} both act at large length scales and therefore are roughly interchangeable.

We summarize these mathematical arguments in three experimental observables:

- α is the coefficient of linear temperature dependance, i.e. $R(B = 0, T) = \alpha k_B T$.
- β is the coefficient of linear field dependence, i.e. $R(T = 0, B) = \beta \mu_B B$.
- $\gamma = \alpha/\beta$ is the ratio of the coefficient of temperature dependence to the coefficient of field dependence.

These observables have the values

$$\begin{aligned} \alpha &= \pi A_0/(A_{max} T) \\ \beta &= \pi q_C \mu_B^{-1} A_0 \\ \gamma &= q_C^{-1} \mu_B/(A_{max} T) \end{aligned} \tag{B1}$$

where $\mu_B = 1/2$ and $q_C = 2$ are the Bohr magneton and the Cooperon charge.

We can go a step further by combining the Heisenberg uncertainty relation with a $p^2/2m$ dispersion: $A_{max} = \langle (\Delta x)^2 \rangle = \hbar^2/\langle p^2 \rangle$, where the momentum scale is determined by $\langle p^2/2(m_C/m_e) \rangle = T$, m_C is the Cooperon mass, and m_e is the electron mass. This produces $A_{max} T = m_e/2m_C$. The Heisenberg relation used here is a quantum mechanical upper bound on the decoherence scale which can be obtained at a given temperature.

Using this relation gives:

$$\begin{aligned} \alpha &= 4\pi \frac{m_C}{2m_e} A_0 = \gamma \times 4\pi A_0 \\ \beta &= \pi q_C \mu_B^{-1} A_0 = 4\pi A_0 \\ \gamma &= \frac{2m_C \mu_B}{q_C m_e} = \frac{m_C}{2m_e} \end{aligned} \tag{B2}$$

These arguments indicate that γ's physical meaning is simply the ratio of the Cooperon mass to that of two electrons, i.e. $\gamma = m_C/2m_e$. When $\gamma = 1$ the coefficients α and β of both the linear-in-temperature resistance and the linear magnetoresistance are direct measures of the UV cutoff A_0, with $R_{xx}(B = 0) = \alpha T = 4\pi A_0 T$ and $R_{xx}(T = 0) = \beta \mu_B B = 2\pi A_0 B$.

1. Relation to Bad Metals

Our derivation of linear-in-temperature resistance was made based on assumptions about diffusive scattering and the absence of spin physics. These assumptions hold in disordered topological insulators, but in many other materials they may not be valid. Nonetheless the linear-in-temperature resistance predicted here is also seen in high T_c superconductors (cuprates and pnictides) at temperatures above the superconducting phase. These strongly correlated materials display a resistance which increases linearly with temperature and does not saturate [66]. This linear dependence is inconsistent both with the WAL signal which increases much more slowly with T (logarithmically in 2-D and as a square root in 3-D) and with phonon based scattering which should saturate at a maximum determined by the atomic spacing. Materials whose resistance is linear in temperature are called bad metals and are understood to be strongly correlated, but the details of the conduction process responsible for linear resistance are not understood. It is widely believed that linear-in-temperature resistance is correlated with high-T_c superconductivity. Very recently the linear dependence on temperature has been linked to linear magnetoresistance found in the same bad metal regime [67–69].

Most work on linear resistance in bad metals neglects the quantum interference contributions to electronic conduction which are responsible for WL/WAL. Linear resistance in bad metals extends up to temperatures of 1000 K, where it seems unreasonable for quantum coherence to play a role. Most work lays the responsibility for linear resistance with a momentum relaxation time scale τ which varies inversely with temperature. At large temperatures this time scale becomes very short, opening up a range of questions about correlation effects at short time scales [66], as well as the puzzling consequence that τ can be smaller than $\hbar/E_F$, where E_F is the Fermi energy [70]. Explanations of linear in temperature resistance have been given using resonating-valence-bond theory [71], marginal Fermi liquids [72], hydrodynamics of quantum critical liquids [73], the SYK model with effective medium theory [74], and bounds on diffusion in incoherent metals [70]. In Refs. [71–73] the momentum relaxation time $\tau \propto T^{-1}$ scales inversely with temperature, and therefore simple formulas such as the Drude formula which have $\rho \propto \tau^{-1}$ produce linear in temperature resistance. Quantum interference is omitted from these simple estimates of the resistance. Ref. [70] follows a similar argument but instead of τ uses a time scale $\tau_{eff} \propto T^{-1}$ which controls diffusion. Ref. [74] computes the resistance from the leading order term in Boltzmann theory, with a self-energy $\Sigma \propto \tau^{-1}$ which is linear in T. The one thing in common between these works and the results presented in the present paper is that the source of linear resistance is not attributed to quantities at the atomic length scale, such as the band structure.

We list some arguments for not using the present paper's calculations as an explanation of linear resistance in bad metals:

- Many believe that scattering in bad metals is so strong, and the temperatures of up to 1000 K are so large, that linear resistance cannot rely on quantum interference or quantum coherence. Instead bad metals are in a regime of incoherent, short lived excitations, and strong dephasing.
- Many believe that bad metals are not Fermi liquids; that their carriers and excitations are not fundamental (bare) electrons and holes. Nor can they be (adiabatically) mapped to bare fermions. Therefore it would be surprising to many if bare electrons and holes and their pairs (Cooperons) were to play a determining role in conduction.

- Given this doubt, it is difficult to distinguish between spin-singlet and spin-triplet components of the Cooperon, or to use reasoning about the singlet and triplet to decide whether spin has a role in conduction.
- The spin physics of bad metals may be extremely rich, both in experimental signatures such as magnon spectra, and in the many theories of bad metals and high-T_c superconductivity. It is difficult to exclude the possibility that spin has a role in conduction in bad metals.
- The theory of diffusons and Cooperons, of weak localization and weak antilocalization, generally assumes that the scattering centers are static and that scattering is elastic; it conserves energy. Moreover the diffusive conduction discussed in the present paper requires a substantial density of scatterers. In contrast, static disorder does not seem to be a salient feature of bad metals. Instead inelastic scattering caused by interactions is the key.

In the face of these arguments it is necessary to review the assumptions that went into the present paper. Certainly scattering is strong in bad metals, putting them in the diffusive regime - though the scatterers are dynamic not static. The real problem with this paper's assumptions is the requirement that spin physics does not play a role in conduction. Because of this difficulty, we can not be sure of the transformation from field to loop area distribution; we can not be sure that the loop areas in bad metals follow a $\ln A$ distribution.

If we were able to exclude spin effects on conduction, then the arguments of this paper would proceed inexorably, with all the rigor of charge conservation, gauge invariance, and quantum field theory, and arrive at the $\ln A$ loop area distribution. At this point we would have to confront the difficulty that the A figuring in the $\ln A$ is the sum of areas of a vast number of strongly interacting electron-hole pairs; it is not the area of a single non-interacting random walker. Therefore perhaps the $\ln A$ loop distribution is caused by strongly interacting physics, not by simple random walks. This difficulty in interpreting the $\ln A$ distribution may complicate the relation between temperature and the IR cutoff, and thus impede our arrival at linear in temperature resistance. On the other hand $A_{max} \propto T^{-1}$ is dimensionally correct, and a large area scale like T^{-1} is necessary to explore the tail of the $\ln A$ distribution. Therefore we may arrive at linear-in-temperature resistance, even if the $\ln A$ loop area distribution is caused by strong interactions and not Cooperons.

2. Experimental Results in Bad Metals

Bearing in mind the difficulties discussed in the previous paragraphs, we comment on a few interesting experiments. First, Hayes et al. measured $\gamma = \alpha/\beta$, where α is the coefficient of linear-in-temperature dependence and β is the coefficient of linear-in-field dependence. [67] γ was measured in the high-T_c pnictide superconductor $BaFe_2(As_{1-x}P_x)_2$ at dopings ranging between 0.31 and 0.41. It was found to be identical to one in atomic units, within the experimental error bar of 7%. Further studies of the cuprate $La_{2-x}Sr_xCuO_4$ and of $Yb_{1-x}La_xRh_2Si_2$ at various dopings near optimal doping have found constants between 0.7 and 2.3. [67,75] We conclude that γ does depend on the host material, but only mildly, and that it is of order 1. If our formula

$\gamma = m_C/2m_e$ is correct, then experimental observation of $\gamma = 1$, i.e. a $p^2/2m_e$ dispersion, implies that in the compounds studied by Refs. [67,75] the carriers responsible for linear resistance have mass $m_C = 2m_e$ equal to twice the bare electron mass m_e, and are insensitive to mass renormalization via band structure and other effects.

We examine the UV cutoff A_0's scaling in several cuprates analyzed by Ref. [76], which determined the linear coefficient α in $La_{2-x}Sr_xCuO_4$ (LSCO), $YBa_2Cu_3O_{6+\delta}$ (YBCO), $Tl_2Ba_2CuO_{6+\delta}$ (Tl2201), and $HgBa_2CuO_{4+\delta}$ (Hg1201). Ref. [76] found that α - the coefficient of linear-in-temperature resistance - is the same in all four compounds if one uses the sheet resistance per CuO_2 plane, not per unit cell. They also found that α is inversely proportional to the doping p for $p \leq 0.20$, where p is the number of holes per unit cell, not per CuO_2 plane. After conversion to atomic units one finds that $\alpha = \pi \times 64\, a_0^2 \times p^{-1}$.[4]

We surmise on dimensional grounds that α should not depend on the doping, but instead on the 2-D carrier density ρ_{2D}. Using $p = \rho_{2D} \times \mathcal{A}$ where $\mathcal{A} \approx 53a_0^2$ is the cross-section of the unit cell in the cuprates' copper oxide plane, we arrive at $R_{xx}(B = 0) = 0.30 \times 4\pi \times T \times \rho_{2D}^{-1}$.

Comparison to our formula where $R_{xx}(B = 0, T) = 4\pi\gamma \times A_0 T$ suggests that in these compounds the UV cutoff A_0 of the loop area distribution is the inverse of the carrier density, i.e. $A_0 = \rho_{2D}^{-1}$.[5] This offers the possibility of determining the charge carrier density directly from either the linear magnetoresistance or the linear in temperature resistance, without any speculation about the compound's chemistry or band structure.

References

1. Zhang, S.X., et al.: Appl. Phys. Lett. **101**, 202403 (2012)
2. Zhang, W., et al.: Phys. Rev. Lett. **106**, 156808 (2011)
3. Wang, X., Du, Y., Dou, S., Zhang, C.: Phys. Rev. Lett. **108**, 266806 (2012)
4. Feng, J., et al.: Phys. Rev. B **92**, 081306 (2015)
5. Kisslinger, F., et al.: Nat. Phys. **11**, 650 (2015)
6. Zhang, C., et al.: Phys. Rev. B **92**, 041203 (2015)
7. Wang, K., Graf, D., Petrovic, C.: Phys. Rev. B **89**, 125202 (2014)
8. Xu, X., et al.: J. Phys.: Condens. Matter **27**, 335701 (2015)
9. Sun, Y., Pyon, S., Tamegai, T.: Phys. Rev. B **93**, 104502 (2016)
10. Tanabe, Y., et al.: Phys. Rev. B **84**, 100508 (2011)
11. Bhoi, D., Mandal, P., Choudhury, P., Pandya, S., Ganesan, V.: Appl. Phys. Lett. **98**, 172105 (2011)
12. Fenton, J., Schofield, A.J.: Phys. Rev. Lett. **95**, 247201 (2005)
13. Koshelev, A.E.: Phys. Rev. B **88**, 060412 (2013)
14. Abrikosov, A.A.: Phys. Rev. B **58**, 2788 (1998)

[4] Up to the multiplicative constant, this formula $R \propto T/p$ is the same as Ref. [71]'s RVB result. See also Ref. [62] for discussion of the resistance per CuO_2 plane.

[5] Ref. [63] has argued that the linear coefficient of the resistivity dR/dT is proportional to λ_L^2 across a range of cuprate, pnictide, and heavy fermion materials, where λ_L is the London penetration depth. Since both Ref. [63] and Ref. [76] analyze the same data on LSCO, this suggests a scaling relationship between λ_L^2 and the inverse carrier density ρ_{2D}^{-1}.

15. Khouri, T., et al.: Phys. Rev. Lett. **117**, 256601 (2016)
16. Parish, M.M., Littlewood, P.B.: Nature **426**, 162 (2003)
17. Winkler, R.: Spin-orbit coupling effects in two-dimensional electron and hole systems, vol. 191. Springer, Heidelberg (2003)
18. Schäfer, L., Wegner, F.: Zeitschrift für Physik B Cond. Matter **38**, 113 (1980)
19. Efetov, K.: Supersymmetry in Disorder and Chaos. Cambridge University Press, Cambridge (1999)
20. Hikami, S., Larkin, A.I., Nagaoka, Y.: Progress Theoret. Phys. **63**, 707 (1980)
21. Bergman, G.: Phys. Rev. Lett. **48**, 1046 (1982)
22. Chen, J., et al.: Phys. Rev. B **83**, 241304 (2011)
23. Feynman, R.P., Hibbs, A.R.: Quantum Mechanics and Path Integrals. McGraw-Hill, New York (1965)
24. Rammer, J.: Quantum Transport Theory. CRC Press, Boca Raton (2018)
25. Wilson, K.G.: Phys. Rev. D **10**, 2445 (1974)
26. Feynman, R.P.: Phys. Rev. **80**, 440 (1950)
27. Ashtekar, A., Rovelli, C.: Class. Quant. Gravity **9**, 1121 (1992)
28. Onsager, L.: Lond. Edinburgh Dublin Phil. Maga. J. Sci. **43**, 1006 (1952)
29. Laikhtman, B., Altshuler, E.L.: Ann. Phys. **232**, 332 (1994)
30. Altshuler, B.L., Aronov, A.G.: JETP Lett. **33**, 499 (1981)
31. Dugaev, V.K., Khmelnitskii, D.E.: Sov. Phys. JETP **59**, 1038 (1984)
32. Bergmann, G.: Phys. Rev. B **39**, 11280 (1989)
33. Raichev, O.E., Vasilopoulos, P.: J. Phys.: Condens. Matter **12**, 589 (2000)
34. Meyer, J.S., Altland, A., Altshuler, B.L.: Phys. Rev. Lett. **89**, 206601 (2002)
35. Garate, I., Glazman, L.: Phys. Rev. B **86**, 035422 (2012)
36. Beenakker, C.W.J., van Houten, H.: Phys. Rev. B **38**, 3232 (1988)
37. Lin, C.J., et al.: Phys. Rev. B **88**, 041307 (2013)
38. Sacksteder, V.E., Arnardottir, K.B., Kettemann, S., Shelykh, I.A.: Phys. Rev. B **90**, 235148 (2014)
39. Wu, Q., Sacksteder, V.E.: Phys. Rev. B **90**, 045408 (2014)
40. Nomura, K., Koshino, M., Ryu, S.: Phys. Rev. Lett. **99**, 146806 (2007)
41. Bardarson, J.H., Tworzydło, J., Brouwer, P.W., Beenakker, C.W.J.: Phys. Rev. Lett. **99**, 106801 (2007)
42. Lewenkopf, C.H., Mucciolo, E.R., Neto, A.H.C.: Phys. Rev. B **77**, 081410 (2008)
43. Minkov, G.M., Germanenko, A.V., Larionova, V.A., Negashev, S.A., Gornyi, I.V.: Phys. Rev. B **61**, 13164 (2000)
44. Baxter, D.V., Richter, R., Trudeau, M.L., Cochrane, R.W., Strom-Olsen, J.O.: J. de Physique **50**, 1673 (1989)
45. Dmitriev, A.P., Kachorovskii, V.Y., Gornyi, I.V.: Phys. Rev. B **56**, 9910 (1997)
46. Gasparyan, V., Zyuzin, A.Y.: Sov. Phys. Solid State **27**, 999 (1985)
47. Dyakonov, M.: Solid State Commun. **92**, 711 (1994)
48. Cassam-Chenai, A., Shapiro, B.: J. Phys. I **4**, 1527 (1994)
49. Ishii, K.: Prog. Theor. Phys. Suppl. **53**, 77 (1973)
50. Simon, B., Spencer, T.: Commun. Math. Phys. **125**, 113 (1989)
51. Beenakker, C.W.J.: Rev. Mod. Phys. **69**, 731 (1997)
52. Ando, T., Suzuura, H.: J. Phys. Soc. Jpn. **71**, 2753 (2002)
53. Takane, Y.: J. Phys. Soc. Jpn. **73**, 1430 (2004)
54. Altshuler, B.L., Aronov, A.G., Spivak, B.Z.: JETP Lett. **33**, 94 (1981)
55. Sacksteder, V.E., Wu, Q.: Phys. Rev. B **94**, 205424 (2016)
56. Oroszlány, L., Rakyta, P., Kormányos, A., Lambert, C.J., Cserti, J.: Phys. Rev. B **77**, 081403 (2008)

57. Ghosh, T.K., De Martino, A., Häusler, W., Dell'Anna, L., Egger, R.: Phys. Rev. B **77**, 081404 (2008)
58. Williams, J.R., Marcus, C.M.: Phys. Rev. Lett. **107**, 046602 (2011)
59. Fujita, K., et al.: Science **344**, 612 (2014)
60. Wu, Q., Du, L., Sacksteder, V.E.: Phys. Rev. B **88**, 045429 (2013)
61. Mikitik, G.P., Sharlai, Y.V.: Phys. Rev. Lett. **82**, 2147 (1999)
62. Leggett, A.J.: Quantum Liquids: Bose Condensation and Cooper Pairing in Condensed-Matter Systems. Oxford University Press, Oxford (2006)
63. Hu, T., Liu, Y., Xiao, H., Mu, G., Yang, Y.-F.: Sci. Rep. **7**, 9469 (2017)
64. Altshuler, B.L., Aronov, A.G., Khmelnitskii, D.E.: J. Phys. C: Solid State Phys. **15**, 7367 (1982)
65. Matsuo, S., et al.: Phys. Rev. B **88**, 155438 (2013)
66. Hussey, N.E., Takenaka, K., Takagi, H.: Phil. Mag. **84**, 2847 (2004)
67. Hayes, I.M., et al.: Nat. Phys. **12**, 916 (2016)
68. Giraldo-Gallo, P., et al.: Science **361**, 479 (2018)
69. Kumar, R., Singh, S., Nair, S.: arXiv preprint arXiv:1801.03768 (2018)
70. Hartnoll, S.A.: Nat. Phys. **11**, 54 (2015)
71. Nagaosa, N., Lee, P.A.: Phys. Rev. Lett. **64**, 2450 (1990)
72. Varma, C.M., Littlewood, P.B., Schmitt-Rink, S., Abrahams, E., Ruckenstein, A.E.: Phys. Rev. Lett. **63**, 1996 (1989)
73. Davison, R.A., Schalm, K., Zaanen, J.: Phys. Rev. B **89**, 245116 (2014)
74. Patel, A.A., McGreevy, J., Arovas, D.P., Sachdev, S.: Phys. Rev. X **8**, 021049 (2018)
75. Giraldo-Gallo, P., et al.: arXiv preprint arXiv:1705.05806 (2017)
76. Barišić, N., et al.: Proc. Natl. Acad. Sci. **110**, 12235 (2013)

Room Temperature In-Plane Magnetotransport Properties in the Chiral Antiferromagnet Mn_3Sn

Shunichiro Kurosawa[1], Mingxuan Fu[1,2], Akito Sakai[1], Muhammad Ikhlas[2], and Satoru Nakatsuji[1,2,3,4,5(✉)]

[1] Department of Physics, Faculty of Science and Graduate School of Science, The University of Tokyo, Hongo, bunkyo-ku, Tokyo 113-0033, Japan
[2] Institute for Solid State Physics, University of Tokyo, Kashiwa, Chiba 277-8581, Japan
[3] CREST, Japan Science and Technology Agency (JST), 4-1-8 Honcho Kawaguchi, Saitama 332-0012, Japan
[4] Institute for Quantum Matter and Department of Physics and Astronomy, Johns Hopkins University, Baltimore, MD 21218, USA
[5] Trans-scale Quantum Science Institute, University of Tokyo, Hongo, bunkyo-ku, Tokyo 113-0033, Japan
satoru@phys.s.u-tokyo.ac.jp

Abstract. Chiral antiferromagnet Mn_3Sn exhibits large Berry-curvature-induced anomalous Hall and Nernst effect despite its vanishing net magnetization, attracting wide, interdisciplinary attention. This material is suited to investigate the interplay between the magnetic Weyl semimetal state and the underlying spin structure. Here, we report in-plane angular dependence of magnetotransport properties at room temperature. The anisotropic magnetoresistance and planar Hall effect show nearly two-fold periodicity under low fields, whereas higher-order oscillations grow with increasing field. Our findings suggest that the interplay between chiral anomaly and field-induced spin texture deformation may lead to complex magnetotransport anisotropy beyond the typical two-fold oscillation, offering valuable insight that helps understand the chiral-anomaly-driven magnetotransport in Weyl magnets.

Keywords: Weyl semimetal · chiral antiferromagnet · anisotropic magnetoresistance · planar Hall effect · chiral anomaly

1 Introduction

The chiral antiferromagnet Mn_3Sn exhibits large anomalous Hall effect (AHE) [1], anomalous Nernst effect (ANE), [2] and magneto-optical Kerr effect [3] at room temperature, which are comparable to that found in typical ferromagnets. These large responses stem from the intrinsic contribution governed by the Berry curvature and are believed to be signatures of topological band properties. Angle resolved photoemission spectroscopy (ARPES) measurements have confirmed the existence of Weyl nodes, the major source of the enhanced Berry curvature that is responsible for the large AHE and

H. J. Choi et al. (Eds.): APPC 2022, *Proceedings of the 15th Asia Pacific Physics Conference*, pp. 117–122, 2024.
https://doi.org/10.1007/978-981-96-0191-2_11

ANE. The topological properties and the resulting large responses in Mn_3Sn are not only of profound fundamental interest but also hold promising technological applications [4]

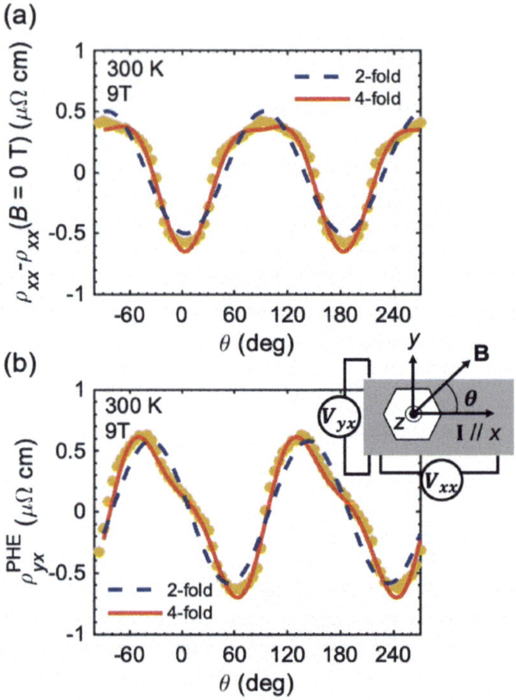

Fig. 1. Angular dependence of AMR (a) and PHE (b) measured under $B = 9$ T at room temperature. The broken blue lines represent two-fold fitting. The solid red lines include two- and four-fold components. Inset: the schematic of the measurement configuration

The Weyl nodes tie closely with the underlying magnetic structure. The crystal structure of Mn_3Sn consists of Mn kagome lattices stacking along the [0001] direction (z-axis). Below the Néel temperature ($T_N \sim 430$ K), the Mn moments form an inverse triangular spin structure, breaking the time reversal symmetry and establishing the Weyl nodes.

The chiral anomaly is a transport phenomena characteristic of the Weyl semimetal state; namely, the number imbalance of electrons between the right- and left-handed Weyl nodes happens when applying an electric field parallel to the magnetic field, resulting in an additional electric current and thereby a negative longitudinal magnetoresistance (nLMR) [5,6]. More recent theoretical works suggest that the chiral anomaly induces anisotropic MR (AMR) and planar Hall effect (PHE), which are given by the following equations according to the simplest model comprising a single pair of Weyl

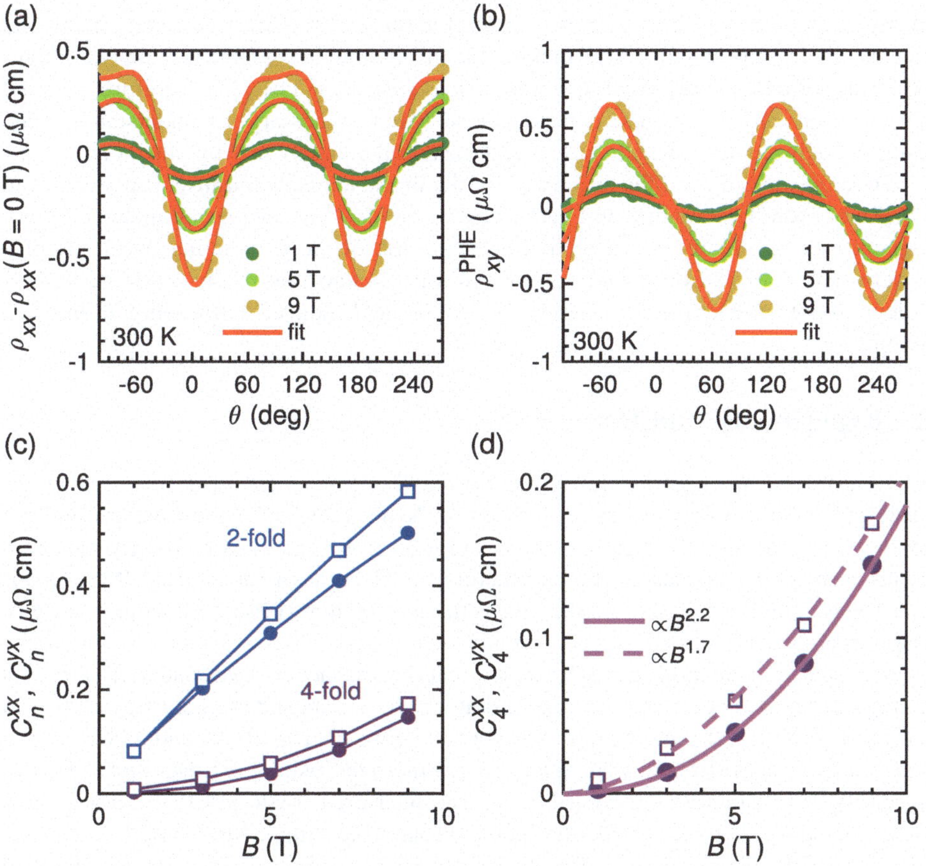

Fig. 2. (a),(b) Angular dependence of AMR (a) and PHE (b) obtained under several fields at room temperature. The red lines are four-fold fitting. (c) Field dependence of the amplitude of the two- and four-fold components. The close and open symbols represent fitting results of AMR and PHE, respectively. (d) Power-law fitting (aB^n) of the field dependence of the four-fold components. The solid and broken lines represent fits to AMR and PHE data, respectively.

nodes [7,8]:

$$\rho_{xx} = \frac{\rho_\perp + \rho_\parallel}{2} - \frac{\rho_\perp - \rho_\parallel}{2}\cos 2\theta \tag{1}$$

$$\rho_{yx}^{\mathrm{PHE}} = -\frac{\rho_\perp - \rho_\parallel}{2}\sin 2\theta \tag{2}$$

where $\rho_\parallel$ ($\rho_\perp$) and θ represent the longitudinal (transverse) MR and the angle between the current and the field, respectively (Fig. 1(b) inset). In the case of the chiral anomaly, the nLMR appears for parallel applied current and magnetic field while the transverse MR (TMR) is positive, leading to substantial amplitude of of AMR and PHE oscillations. The nLMR, the AMR and PHE have been observed in the inverse triangular state of Mn_3Sn, considered evidence of the chiral anomaly [9]. However, magnetic effects can also induce the negative MR, making the discussion on the chiral anomaly ambiguous, as reported for other Weyl magnets such as $Co_3Sn_2S_2$ [10, 11]. Therefore, detailed investigation into the angular dependence of MR is necessary for understanding the interplay between the magnetic effect and the chiral anomaly in Weyl magnets.

Here, we investigated the anisotropy of the in-plane magnetotransport properties in Mn_3Sn at room temperature and under various fields. With increasing magnetic field, higher-order oscillatory components appear on top of the dominating two-fold component typical of chiral anomaly. We discuss the connection of the observed higher-order component with the chiral anomaly and the field-induced deformation of the spin texture.

2 Experiments and Discussion

Single-crystal Mn_3Sn was synthesized by the Bridgman method as reported previously [2]. The sample was oriented based on X-ray Laue diffraction pattern and cut by the spark erosion. The magnetization and four-probe electrical resistivity measurements were performed using the commercial SQUID magnetometer (MPMS, Quantum Design) and physical properties measurement system (PPMS, Quantum Design), respectively.

We measured the room-temperature angular dependence of AMR and PHE at $B = 9$ T (Figs. 1(a)(b)). The current was applied along the x-axis and the field was rotated in the xy-plane (Fig. 1(b), inset). In this paper, we define $[2\bar{1}\bar{1}0]$, $[01\bar{1}0]$ and $[0001]$ as x, y and z axis. The AMR and PHE data are symmetrized using $+B$ data and $-B$ data to remove the ordinary Hall contribution. The measured AMR and PHE curves show clear deviations from two-fold oscillation (broken blue line). Motivated by analyses of angular dependence reported for other topological semimetals [12–14], we included higher order terms in the fitting functions, such that

$$\rho_{xx} = C_0^{xx} + \sum_{\substack{n=2 \\ n:\text{even}}} C_n^{xx} \cos(n\theta - \phi_n^{xx}) \tag{3}$$

$$\rho_{yx}^{\text{PHE}} = - \sum_{\substack{n=2 \\ n:\text{even}}} C_n^{yx} \sin(n\theta - \phi_n^{yx}) \tag{4}$$

The solid red lines in Figs. 1(a)(b) show the fitting result that consists of both two- and four-fold components, which describe the data better than the simple two-fold oscillation (broken blue line).

Based on this analysis, we carried out the fitting up to the four-fold oscillation for other fields (Fig. 2(a)(b)). Figure 2(c) shows the resulting field dependence of the amplitude of each oscillation. At all measured fields, the AMR and PHE oscillations are predominately two-fold, showing negative sign on approaching the parallel current and magnetic field configuration (i.e., near $\theta = 0°$ and $180°$). Such behavior is consistent with the prediction of the chiral anomaly, as described by Eqs. (1) and (2). On the other hand, the four-fold component gradually increases with increasing field. These higher-order oscillation is likely relate to the enhanced magnetic anisotropy. Recent magnetic torque measurements on Mn_3Sn reveal six- and twelve-fold oscillations together with the two-fold one at room temperature, which are attributed to the spin texture deformation resulting from competition among the Dzyaloshinskii-Moriya interaction, single-ion anisotropy, exchange couplings and Zeeman effect [15, 16]. Such spin texture deformation might also affect the electrical transport, leading to higher-order oscillations in addition to the two-fold one arising from the chiral anomaly.

On the other hand, we note that the Weyl nodes should survive up to the highest measured in-plane field of 9 T because the large AHE of topological origin persists up to least 10 T [17]. This fact gives rise to another possibility that the four-fold oscillation is intrinsic to the chiral anomaly. A recent theoretical work on $Mn_3X(X = \mathrm{Ge, Sn})$ systems predicts that the chiral anomaly arising from the multiple pairs of Weyl nodes can induce four- and six-fold oscillations, with the four-fold oscillation increasing quadratically with field [18]. Our observed field dependence of the four-fold amplitude agrees with this prediction (Fig. 2(d)). At this stage, it is difficult to distinguish between the effects of spin texture deformation and the chiral anomaly on the angular dependence of magnetotransport. Further investigation into the Mn doping dependence of the AMR and PHE will help determine the origin of the higher order oscillations because changing Mn composition can effectively tune the relative position of the chemical potential to the Weyl nodes, and thus affect the chiral-anomaly-driven transport phenomena [9].

3 Conclusions

To conclude, we report the angular dependence of the in-plane magnetotransport properties in Mn_3Sn at room temperature and carried out analysis including two-fold and higher-order oscillations. The AMR and PHE data are dominated by two-fold oscillations, consistent with the chiral anomaly. High-order component becomes noticeable beyond 3 T, whose magnitude increases with field. Such higher-order components are likely associated with the interplay of field-induced spin texture deformation and chiral anomaly. To identify their origin, comprehensive studies of Mn doping effects on transport phenomena will be essential.

Acknowledgments. This work was partially supported by JST-Mirai Program (JPMJMI20A1), JST-CREST (JPMJCR18T3) from Japan Science and Technology Agency, and Grants-in-Aid for Scientific Research (19H00650). The work at the Institute for Quantum Matter, an Energy Frontier Research Center was funded by DOE, Office of Science, Basic Energy Sciences under Award # DE-SC0019331. M.F. acknowledges support from the Japan Society for the Promotion of Science (Postdoctoral Fellowship for Research in Japan (Standard)). The use of the facilities of the Materials Design and Characterisation Laboratory at the Institute for Solid State Physics, the University of Tokyo, is gratefully acknowledged.

References

1. Nakatsuji, S., Kiyohara, N., Higo, T.: Nature **527**, 212 (2015)
2. Ikhlas, M., et al.: Nat. Phys. **13**, 1085 (2017)
3. T. Higo, et al., Nat. photonics **12**, 73 (2018)
4. Nakatsuji, S.: AAPPS Bull. **32**, 1 (2022)

5. Nielsen, H.B., Ninomiya, M.: Phys. Lett. B **130**, 389 (1983)
6. Son, D.T., Spivak, B.Z.: Phys. Rev. B **88**, 104412 (2013)
7. Nandy, S., Sharma, G., Taraphder, A., Tewari, S.: Phys. Rev. Lett. **119**, 176804 (2017)
8. Burkov, A.A.: Phys. Rev. B **96**, 041110 (2017)
9. Kuroda, K., et al.: Nat. Mater. **16**, 1090 (2017)
10. Shama, R., Gopal, Singh, Y.: J. Magn. Magn. Mater. **502**, 166547 (2020)
11. Yang, S.-Y., et al.: Nano Lett. **20**, 7860 (2020)
12. Yang, S.-Y., Chang, K., Parkin, S.S.P.: Phys. Rev. Research **2**, 022029 (2020)
13. Dai, Y., et al.: Phys. Rev. Lett. **128**, 247202 (2022)
14. Wu, M., et al.: Nano Lett. **22**, 73 (2022)
15. Duan, T.F., Ren, W.J., Liu, W.L., Li, S.J., Liu, W., Zhang, Z.D.: Appl. Phys. Lett. **107**, 082403 (2015)
16. Li, X., et al.: Phys. Rev. B **106**, L020402 (2022)
17. Sung, N.H., Ronning, F., Thompson, J.D., Bauer, E.D.: Appl. Phys. Lett. **112**, 132406 (2018)
18. Chaudhary, G., Burkov, A.A., Heinonen, O.G.: Phys. Rev. B **105**, 085108 (2022)

Magnetic-Field Effect on Transport Properties of the Multipolar Kondo Metal $PrTi_2Al_{20}$

Takachika Isomae[1,2], Akito Sakai[1], Mingxuan Fu[1,2], and Satoru Nakatsuji[1,2,3,4,5(✉)]

[1] Department of Physics, Faculty of Science and Graduate School of Science, University of Tokyo, Hongo, Bunkyo-ku, Tokyo 113-0033, Japan
[2] Institute for Solid State Physics, University of Tokyo, Kashiwa, Chiba 277-8581, Japan
[3] CREST, Japan Science and Technology Agency (JST), 4-1-8 Honcho Kawaguchi, Saitama 332-0012, Japan
[4] Institute for Quantum Matter and Department of Physics and Astronomy, Johns Hopkins University, Baltimore, MD 21218, USA
[5] Trans-scale Quantum Science Institute, University of Tokyo, Bunkyo-ku, Tokyo 113-0033, Japan

satoru@phys.s.u-tokyo.ac.jp

Abstract. $PrTi_2Al_{20}$ is a model multipolar Kondo lattice system featuring a non-magnetic crystal-electric-field (CEF) ground state with active multipolar moments. In this study, we perform transverse magnetoresistance (MR) measurements on $PrTi_2Al_{20}$ and its non-4f analog $LaTi_2Al_{20}$ with a magnetic field applied along the [001] direction. In the Fermi liquid state above the ferro-quadrupolar (FQ) transition, the transverse MR of $PrTi_2Al_{20}$ shows strong temperature dependence, forming a maximum at about 10 K and undergoing a sign change from positive to negative on approaching the FQ ordering. By contrast, the transverse MR of $LaTi_2Al_{20}$ increases toward an extremely large value of $\sim 10^3\%$ on cooling and remains positive down to the lowest measured temperature of about 2 K. The qualitatively different MR behavior of the two systems suggests that the multipolar moments and their fluctuations play a central role in the low-T electrical transport of $PrTi_2Al_{20}$.

Keywords: Heavy fermion system · Magnetotransport · Magnetoresiatance · Multipolar moments · Multipolar Kondo effect

1 Introduction

Strong electron correlations can break down the paradigmatic quasi-particle picture, leading to various exotic quantum phenomena such as non-Fermi liquid (NFL) and unconventional superconductivity (SC) [1–4]. An extensively explored playground of electronic correlations is the magnetic heavy fermion metal, where NFL and unconventional SC stem from quantum critical point (QCP) due to competition between the magnetic Kondo effect and the Ruderman-Kittel-Kasuya-Yosida (RKKY) [5–8]. On the other hand, strong hybridization between quadrupolar moments and conduction electrons can induce a distinct type of Kondo entanglement, so-called quadrupole Kondo

H. J. Choi et al. (Eds.): APPC 2022, *Proceedings of the 15th Asia Pacific Physics Conference*, pp. 123–129, 2024.
https://doi.org/10.1007/978-981-96-0191-2_12

effect [9–12], offering an alternative route to the NFL ground state without involving magnetic criticality. Pr-based cubic heavy fermion systems can host a Γ_3 CEF ground doublet with no magnetic dipolar moment but active quadrupolar and octupolar moments, well suited for investigating the novel physics arising from the quadrupolar Kondo effect.

Recent experimental and theoretical works reveal that the Pr-based cage compounds PrTr_2Al$_{20}$ (Tr = Ti, V) are suitable for studying the quadrupole Kondo effect[13, 14]. The cubic T_d symmetry of Pr^{3+} site in a $CeCr_2Al_{20}$-type crystal structure stabilizes the non-magnetic Γ_3 ground state, which is well isolated from the excited state $\Delta \sim 60$ K (Tr = Ti) and 40 K (Tr = V)[13, 15–22]. Moreover, strong c-f hybridization effect is confirmed by various experiments [13,21,23,24]. While PrV_2Al_{20} shows a two-stage multipolar transition at $\sim$ 0.75 K and 0.6 K, and NFL behavior above the transition, $PrTi_2Al_{20}$ shows a single ferroquadrupolar (FQ) transition at $T_Q \sim 2.0$ K and Fermi liquid (FL) behavior above T_Q [13, 17, 24, 25]. Moreover, superconductivity appears at $T_c \sim 0.2$ K, inside the FQ ordered state [17,26]. More interestingly, T_c of $PrTi_2Al_{20}$ enhances strongly under hydrostatic pressure, reaching $\sim$ 1.1 K under $P \sim 8.8$ GPa, accompanied by suppression of T_Q and the enhancement of the effective mass, implying the presence of a quadrupolar QCP [16]. The behavior of electrical transport under magnetic fields is crucial for understanding the physical properties of the quadrupolar Kondo lattice system, yet such a study is so far lacking.

In this paper, we report the study of the transverse magnetoresistance (MR) of $PrTi_2Al_{20}$ and its non-4f analog $LaTi_2Al_{20}$. Transverse MR in $PrTi_2Al_{20}$ shows the positive MR below T = 20 K in the para-quadrupolar phase and negative MR on approaching the FQ transition. On the other hand, the transverse MR in $LaTi_2Al_{20}$ shows the large positive MR (MR $\sim$ 1200%) at T = 2 K under B = 9 T. These differences indicate that the 4f multipolar moments strongly influence the transport properties under magnetic fields.

2 Experiments and Discussions

Single crystal samples of $PrTi_2Al_{20}$ and $LaTi_2Al_{20}$ were prepared by the Al self-flux method [13]. We used high-quality samples with the residual resistivity ratio RRR = ρ (300 K) / ρ (0.3 K) $\sim$ 100 for the transport measurements using a ^{3}He-^{4}He dilution refrigerator (Kelvinox 25, Oxford instruments). The isofield resistivity curves for $B \parallel$ [001] and $I \parallel$ [011] were measured using the standard dc four-probe method with a physical property measurement system (PPMS, Quantum Design).

Figure 1 (a) shows the temperature dependence of resistivity ρ (T, B) (left axis) and 4f-electron contribution $\rho_{4f}(T, B)$ (right axis) of $PrTi_2Al_{20}$ obtained at various magnetic fields applied along the [001] direction. The $\rho_{4f}(T, B)$ is obtained by subtracting the resistivity curve of $LaTi_2Al_{20}$ from that of $PrTi_2Al_{20}$. The behavior of zero-field $\rho_{4f}(T, 0)$ of $PrTi_2Al_{20}$ can be divided into three temperature regimes separated by characteristic energy scales such as the crystalline-electric-field (CEF) gap and the FQ ordering transition. The $\rho_{4f}(T, 0)$ shows -ln$T$ behavior and forms a peak at about $T_{peak} \sim 50$ K. This behavior is due to the magnetic Kondo effect originating from the magnetic dipole moments of the excited triplet, with T_{peak} corresponding to

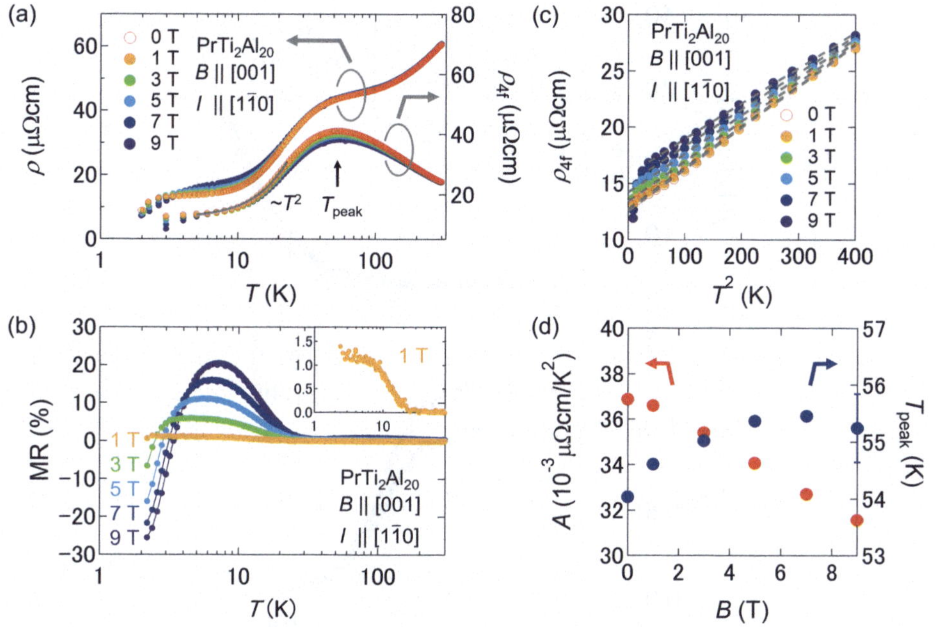

Fig. 1. (a) Temperature dependence of resistivity $\rho(T,B)$ (left axis) and $4f$-electron contribution $\rho_{4f}(T,B)$ (right axis) in $PrTi_2Al_{20}$ under various magnetic field along $B \parallel [001]$. Black arrow shows peak temperature in $\rho_{4f}(T,B)$. Black curve indicates the FL phase with $\rho \propto T^2$. (b) Temperature dependence of the transverse magnetoresistance MR in $PrTi_2Al_{20}$ under various magnetic fields (main panel) and $B = 1$ T (inset) along $B \parallel [001]$. (c) Squared-temperature T^2 dependence of the $4f$-electron contribution $\rho_{4f}(T,B)$ in $PrTi_2Al_{20}$. The gray dashed line show the FL behavior $\rho \propto T^2$. (d) Field dependence of A coefficient (left axis) and the peak temperature T_{peak} in the $4f$- electron contribution (right axis).

the CEF gap between the nonmagnetic Γ_3 ground-state doublet and the first-excited magnetic triplet. With further cooling, a FL phase with $\rho_{4f}(T,0) \sim T^2$ appears below $T_{\rm FL} \sim 20$ K (Figs. 1(a) and (c)). This regime is strongly influenced by the interaction of the $4f$ multipolar moments of the non-magnetic CEF ground state with conduction electrons. Around $T_Q = 2$ K, $\rho_{4f}(T,0)$ shows a downturn due to the FQ transition and the associated suppression of the multipole-induced scattering. The observed temperature evolution of $\rho_{4f}(T,0)$ is consistent with previous studies [13].

With a magnetic field B applied along the [001] direction, the high-temperature behavior of $\rho(T,B)$ is nearly unchanged, yet deviation from the zero-field $\rho(T,0)$ curve occurs below $T_{\rm FL} \sim 20$ K, which can be clearly seen in the transverse magnetoresistance (MR) plot shown in Fig. 1(b). The transverse MR is defined as $\mathrm{MR}(\%) \equiv (\rho_{xx}(B) - \rho_{xx}(0))/\rho_{xx}(0) \times 100$. As shown in Fig. 1(b), the transverse MR is negligible above $T_{\rm FL} \sim 20$ K for all measured field values. Below $T_{\rm FL}$, the MR curve measured at $B = 1$ T shows a mild upturn and saturates at a small magnitude of $\sim 1\%$ near $T_Q \sim 2$ K (Fig. 1 (b), inset). For $B = 3$ T and above, the magnitude of low-T MR enhances sharply and peaks around $T \sim 10$ K. The maximum magnitude is MR $\sim 20\%$ at B

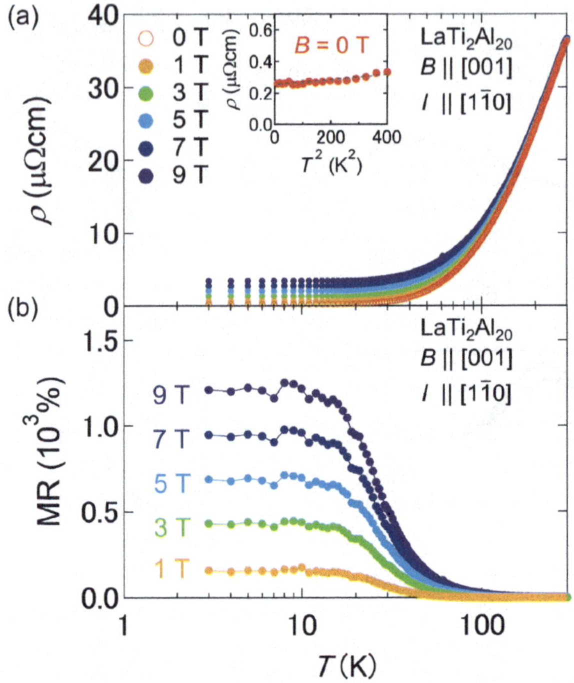

Fig. 2. (a) Temperature dependence of the resistivity $\rho(T, B)$ in $LaTi_2Al_{20}$ under various magnetic fields along $B \parallel [001]$. Squared temperature T^2 dependence of the resistivity under zero-field $\rho(T, 0)$ (inset). Gray dashed line indicates the linear fitting. (b) Temperature dependence of the transverse magnetoresistivity MR in $LaTi_2Al_{20}$.

= 9 T in the para-quadrupolar phase. Interestingly, the sign of the MR changes from positive to negative on approaching the FQ transition at $T_Q \sim 2$ K. The negative MR is common in the magnetic Kondo lattice systems, typically due to suppressed magnetic scattering of the localized f-electrons, namely, the development of a coherent Kondo state [27,28]. Moreover, the sister compound PrV_2Al_{20} exhibits negative MR around $T \sim 10$ K, accompanied by the emergence of non-Fermi liquid behavior characteristic of a quadrupolar Kondo lattice. The negative MR is therefore attributed to the coherence effect of the quadrupolar Kondo lattice [13,29]. Although $PrTi_2Al_{20}$ also shows the negative MR at low-Ts, its origin seems to depart from that typically observed in Kondo lattice systems. Unlike PrV_2Al_{20}, the negative MR takes place not within the FL phase but around the lower T_Q, suggesting that it ties in with the FQ ordering instead of the multipolar Kondo interaction. Indeed, the second-order FQ transition turns into a smooth crossover under $B \geq 2$ T, revealed by transport, thermodynamic, and NMR studies [30,31]. The resulting suppression of the resistivity anomaly near T_Q is likely the origin of the negative MR.

We next turn to the magnetic-field effect on the CEF scheme and the FL behavior of $PrTi_2Al_{20}$. Figure 1(c) shows the $\rho_{4f}(T, B)$ as a function of T^2 measured at various B values. The isofield resistivity curves all follow FL behavior with $\rho(T) \sim T^2$; the dashed lines in Fig. 1(c) represent fits to $\rho_{4f}(T, B) = \rho_0 + AT^2$, where ρ_0 is the residual resistivity. This indicates that the FL behavior is robust under B. However, the A coefficient, which measures the c-f hybridization strength, monotonically decreases with increasing B (Fig. 1(d), left axis). The reduction of c-f hybridization might stem from the competition between the magnetic Zeeman effect and the quadrupolar Kondo effect, the field-induced splitting of the ground-state doublet, or both. By contrast, the peak temperature, T_{peak}, in $\rho_{4f}(T, B)$ that represents the CEF gap remains constant within experimental uncertainties under $B \parallel [001]$ up to 9 T (Fig. 1(d), right axis). This result is consistent with previous specific heat measurements that report nearly unchanged CEF scheme under $B \parallel [001]$ [30].

To further investigate the transport properties in the para-phase of $PrTi_2Al_{20}$, we measure the temperature dependence of $\rho(T, B)$ in the non-4f analog $LaTi_2Al_{20}$ for $B \parallel [001]$ (Fig. 2(a)). The isofield resistivity curve in $LaTi_2Al_{20}$ shows a typical FL behavior below T = 20 K. A coefficient of $LaTi_2Al_{20}$ under zero-field is about 1.5×10^{-4} ($\mu\Omega$cmK^{-2}), which is two orders of magnitude smaller than that of $PrTi_2Al_{20}$ (Fig. 2(a), inset) and is nearly field independent. Figure 2 (b) shows the temperature dependence of transverse MR in $LaTi_2Al_{20}$. This system shows drastic enhanced MR below $T \sim 90$ K. Surprisingly, maximum MR reaches MR $\sim 1200\%$ at $T \sim 2$ K under $B = 9$ T.

We then discuss the possible causes of the different MR behavior in $PrTi_2Al_{20}$ and its non-4f analog $LaTi_2Al_{20}$. Although both materials show the FL behavior below 20 K, the FL state in $PrTi_2Al_{20}$ is likely stabilized with competing magnetic and multipolar Kondo effect [32], which is absent in $LaTi_2Al_{20}$. The presence of multipolar Kondo interaction and fluctuations of multipolar moments might result in the distinct behavior of MR in the FL state of $PrTi_2Al_{20}$. In $LaTi_2Al_{20}$, MR shows strong enhancement below about 90 K and saturates below 20 K, mainly due to the suppression of inelastic phonon scattering on cooling. We anticipate that the peak and sign change of MR observed in $PrTi_2Al_{20}$ stem from a combination of the following two components: (1) a positive MR component due to the suppressed multipolar-fluctuation-induced scattering as the multipolar moments are polarized by the Zeeman field; (2) a negative MR component due to the field-induced suppression of the resistivity anomaly at the FQ ordering transition [33]. This interpretation is supported by the fact that the magnitude of MR in the FL state of $PrTi_2Al_{20}$ enhances with increasing magnetic field (Fig. 1(b)).

3 Conclusions

In conclusion, our transverse MR study on the para-quadrupolar phase of $PrTi_2Al_{20}$ reveals the role of the magnetic-field effect on transport properties in the quadrupolar Kondo metals. The temperature dependence of transverse MR in $PrTi_2Al_{20}$ shows the drastic increasing at T = 20 K and the negative MR originated from the anomaly due to the quadrupolar ordering around T_Q. The behavior of the transverse MR in $PrTi_2Al_{20}$ follows the emergence of three distinct regimes: magnetic Kondo regime, para-quadrupolar FL regime, and the quadrupolar ordered phase. On the other hand, $LaTi_2Al_{20}$ shows large positive MR at low-Ts. The qualitatively different behavior of MR in the FL state of $PrTi_2Al_{20}$ and $LaTi_2Al_{20}$ suggests that multipolar fluctuations above the FQ ordering provide an essential electron scattering mechanism. Our study offers valuable insight into the response of quadrupolar Kondo systems under applied magnetic fields, which may help understand their exotic emergent behavior.

Acknowledgments. This work is partially supported by JST-ASPIRE (JPMJAP2317), CREST (JPMJCR18T3), Japan Science and Technology Agency, by Grants-in-Aids for Scientific Research on Innovative Areas (15H05882 and 15H05883) from the Ministry of Education, Culture, Sports, Science and Technology of Japan, and by Grants-in-Aid for Scientific Research (19H00650, 20K03829, 23K03298) from the Japanese Society for the Promotion of Science (JSPS). T.I. is supported by JST SPRING, Grant Number JPMJSP2108. M.F. acknowledges support from the Japan Society for the Promotion of Science Postdoctoral Fellowship for Research in Japan (Standard).

References

1. Keimer, B., Kivelson, S.A., Norman, M.R., Uchida, S., Zaanen, J.: Nature **518**, 179 (2015)
2. Stewart, G.R.: Rev. Mod. Phys. **73**, 797 (2001)
3. Gegenwart, P., Si, Q., Steglich, F.: Nat. Phys. **4**, 186 (2008)
4. Si, Q., Yu, R., Abrahams, E.: Nat. Rev. Mater. **1**, 16017 (2016)
5. Yuan, H., Grosche, F., Deppe, M., Geibel, C., Sparn, G., Steglich, F.: Science **302**, 2104 (2003)
6. Custers, J., et al.: Nature **424**, 524 (2003)
7. Nakatsuji, S., et al.: Nat. Phys. **4**, 603 (2008)
8. Matsumoto, Y., et al.: Science **331**, 316 (2011)
9. Cox, D.L.: Phys. Rev. Lett. **59**, 1240 (1987)
10. Cox, D.L., Zawadowski, A.: Adv. Phys. **47**, 599 (1998)
11. Tsuruta, A., Miyake, K.: J. Phys. Soc. Jpn. **84**, 114714 (2015)
12. Patri, A.S., Khait, I., Kim, Y.B.: Phys. Rev. Res. **2**, 013257 (2020)
13. Sakai, A., Nakatsuji, S.: J. Phys. Soc. Jpn. **80**, 063701 (2011)
14. Onimaru, T., Kusunose, H.: J. Phys. Soc. Jpn. **85**, 082002 (2016)
15. Kangas, M.J., Schmitt, D.C., Sakai, A., Nakatsuji, S., Chan, J.Y.: J. Solid State Chem. **196**, 274 (2012)
16. Matsubayashi, K., Tanaka, T., Sakai, A., Nakatsuji, S., Kubo, Y., Uwatoko, Y.: Phys. Rev. Lett. **109**, 187004 (2012)
17. Tsujimoto, M., Matsumoto, Y., Tomita, T., Sakai, A., Nakatsuji, S.: Phys. Rev. Lett. **113**, 267001 (2014)
18. Sato, T.J., et al.: Phys. Rev. B **86**, 184419 (2012)
19. Ito, T.U., et al.: J. Phys. Soc. Jpn. **80**, 113703 (2011)
20. Ito, T.U., Higemoto, W., Sakai, A., Tsujimoto, M., Nakatsuji, S.: Phys. Rev. B **92**, 125151 (2015)
21. Taniguchi, T., et al.: J. Phys. Soc. Jpn. **85**, 113703 (2016)
22. Nakanishi, Y., et al.: Physica B: Cond. Matt. **536**, 125 (2018)
23. Matsunami, M., et al.: Phys. Rev. B **84**, 193101 (2011)
24. Machida, Y., et al.: J. Phys: Conf. Ser. **592**, 012025 (2015)
25. Tokunaga, Y., Sakai, H., Kambe, S., Sakai, A., Nakatsuji, S., Harima, H.: Phys. Rev. B **88**, 085124 (2013)
26. Sakai, A., Kuga, K., Nakatsuji, S.: J. Phys. Soc. Jpn. **81**, 083702 (2012)
27. Fert, A., Levy, P.M.: Phys. Rev. B **36**, 1907 (1987)

28. Sumiyama, A., Oda, Y., Nagano, H., Ō nuki, Y., Shibutani, K., Komatsubara, T.: J. Phys. Soc. Jpn. 55, 1294 (1986)
29. Fu, M., Sakai, A., Sogabe, N., Tsujimoto, M., Matsumoto, Y., Nakatsuji, S.: J. Phys. Soc. Jpn. **89**, 013704 (2020)
30. Kittaka, S., et al: J. Phys. Soc. Jpn. **89**, 043701 (2020)
31. Taniguchi, T., et al.: J. Phys. Soc. Jpn. **88**, 084707 (2019)
32. Kusunose, H., Onimaru, T.: J. Phys: Conf. Ser. **592**, 012099 (2015)
33. Isomae, T., Sakai, A., Fu, M., Taniguchi, T., Takigawa, M., Nakatusji, S.: Extremely large magnetoresistance and anisotropic transport in the multipolar Kondo system PrTi2Al20. Phys. Rev. Res. **6**, 013009 (2024). https://journals.aps.org/prresearch/abstract/10.1103/PhysRevResearch.6.013009

Nuclear Physics

New Prediction of the Positron-Proton to Electron-Proton Elastic Scattering Ratio $R_{e^+e^-}$ from *ep* Elastic Scattering Data

I. A. Qattan(✉)

Department of Physics, Khalifa University of Science and Technology, P.O Box 127788, Abu Dhabi, United Arab Emirates
issam.qattan@ku.ac.ae

Abstract. I present a new extraction of the positron-proton to electron-proton elastic scattering cross sections ratio $R_{e^+e^-}(\varepsilon, Q^2)$. Contrary to several previous phenomenological studies and extractions, which predicted the ratio $R_{e^+e^-}$ using model-independent parametrization of $F(\varepsilon, Q^2)$, the hard two-photon exchange (TPE) correction to the electron-proton (*ep*) elastic scattering cross section σ_R, I do not assume prior knowledge of the functional form of $F(\varepsilon, Q^2)$, but rather I use combined unpolarized, and polarized *ep* elastic scattering experimental data to extract the TPE contributions to σ_R, and provide a simple parametrization of $F(\varepsilon, Q^2)$ along with the uncertainties associated with the model dependence of the extractions. I use the new parametrization of $F(\varepsilon, Q^2)$ to predict the ratio $R_{e^+e^-}$, and then compare my extractions of $R_{e^+e^-}$ to several previous phenomenological extractions, TPE-hadronic calculations, and world data emphasizing mainly the kinematics range of the recent precise direct measurements from the OLYMPUS, VEPP-3, and CLAS experiments.

Keywords: electron-proton · positron-proton · elastic scattering · two-photon-exchange · cross-sections

1 Introduction

The proton's form factors (FFs) ratio $\mu_p R_p = \mu_p G_E(Q^2)/G_M(Q^2)$ as extracted using the Rosenbluth separation technique [1], and the recoil polarization technique [2] are significantly different, and differ almost by a factor of three for $Q^2 > 1.0$ (GeV/c)^2, where Q^2 is the four-momentum transferred squared. The ratio $\mu_p R_p$ as extracted using the Rosenbluth technique shows FFs scaling, while the recoil polarization technique predicts a linearly decreasing ratio with increasing Q^2, followed by flattening out for $Q^2 > 5.0$ (GeV/c)^2. While it is still a speculation, the source of the discrepancy is believed to be attributed to a missing higher order radiative corrections to the reduced cross section $\sigma_R(\varepsilon, Q^2)$, mainly the hard two-photon-exchange (TPE) correction. To correct σ_R for TPE corrections, the real function $F(\varepsilon, Q^2)$, which accounts for the interference of the one-photon-exchange (OPE) and TPE amplitudes, is added to the Born reduced cross-section σ_{Born}:

$$\sigma_R(\varepsilon, Q^2) = \left[G_M(Q^2)\right]^2 + \frac{\varepsilon}{\tau}\left[G_E(Q^2)\right]^2 + F(\varepsilon, Q^2), \tag{1}$$

H. J. Choi et al. (Eds.): APPC 2022, *Proceedings of the 15th Asia Pacific Physics Conference*, pp. 133–140, 2024.
https://doi.org/10.1007/978-981-96-0191-2_13

where $G_E(Q^2)$, and $G_M(Q^2)$ are the true electric and magnetic FFs of the proton, respectively, ε is the virtual photon longitudinal polarization parameter, $\tau = Q^2/4M_p^2$ is a kinematics factor, and M_p is the mass of the proton.

The TPE effect and its impact on ep scattering observables was studied thoroughly theoretically [3,4], phenomenologically [5–18], and experimentally [19–21]. However, measuring the positron-proton to electron-proton cross sections ratio $R_{e^+e^-}(\varepsilon, Q^2)$ [22–24] is by far the most precise and direct technique used to measure the size of TPE effect, as the function $F(\varepsilon, Q^2)$ changes sign depending on the charge of the electron or positron involved in the scattering process. After correcting the measured ratio $R_{e^+e^-}$ for both charge-even (δ_{even}), and charge-odd (δ_{odd}) radiative terms, the ratio is expressed as:

$$R_{e^+e^-}(\varepsilon, Q^2) = \frac{1 - \delta_{2\gamma}}{1 + \delta_{2\gamma}} \approx 1 - 2\delta_{2\gamma}, \tag{2}$$

where $\delta_{2\gamma} = F(\varepsilon, Q^2)/\sigma_{\text{Born}}$ is the fractional TPE correction to σ_R. In this case, any deviation of $R_{e^+e^-}$ from unity is a clear signature of hard-TPE effect.

2 Method

Unlike several previous phenomenological extractions, which predicted the ratio $R_{e^+e^-}$ assuming proposed, model-independent parametrization of the function $F(\varepsilon, Q^2)$, I do not assume prior knowledge of the functional form of $F(\varepsilon, Q^2)$, and extract the ratio $R_{e^+e^-}$ directly using combined unpolarized, and polarized ep elastic scattering experimental data. The procedure, together with the constraints and assumption used to extract $R_{e^+e^-}$ is as follows:

I express $R_{e^+e^-}$ directly in terms of σ_R and σ_{Born} using Eqs. (1) and (2) as:

$$R_{e^+e^-} = [3 - 2\frac{\sigma_R(\varepsilon, Q^2)}{\sigma_{\text{Born}}(\varepsilon, Q^2)}]. \tag{3}$$

For a fixed Q^2 value, I use Eq. (3) to calculate $R_{e^+e^-}$ using σ_R measurements at each ε point, along with their uncertainties. I do not account for any residual uncertainty in the normalization between the different σ_R data subsets, and I use only the total uncertainties on σ_R as published. Here, $\sigma_{\text{Born}} = G_M^2[1 + \frac{\varepsilon}{\tau}R_p^2]$, and R_p is constrained to its value, along with its associated uncertainty, as given by Ref. [7]. The true G_M FF is extracted in unbiased approach relying on the fact that $F(\varepsilon = 1, Q^2) \to 0$ (Regge Limit) or: $\sigma_R(\varepsilon = 1, Q^2) = \left[G_M^2 + G_E^2/\tau\right]$. Because of the experimentally observed linearity of σ_R with ε, at a fixed Q^2 value, I fit the experimental σ_R linearly to ε using the form: $\sigma_R = [a_1(Q^2) + a_2(Q^2)\varepsilon]$, where $a_{(1,2)}(Q^2)$ are the parameters of the fit, and functions of Q^2 only. I solve for $G_M^2(Q^2)$ by equating the two expressions for $\sigma_R(\varepsilon = 1, Q^2)$:

$$G_M^2(Q^2) = \frac{[a_1(Q^2) + a_2(Q^2)]}{(1 + \frac{R_p^2}{\tau})}, \tag{4}$$

and the error on $R_{e^+e^-}$ is calculated by propagating the errors on σ_R, G_M, and R_p using Eq. (3). At each Q^2 point, the ratio $R_{e^+e^-}$ is fitted linearly to the form $R_{e^+e^-} = [1 -$

$\alpha_1(Q^2)(1-\varepsilon)]$, with the fit parameter $\alpha_1(Q^2)$ being the TPE coefficient, and represents the slope. I then parametrize the Q^2 dependence of $\alpha_1(Q^2)$, and for a fixed Q^2 value, I use this parametrization to extract the $R_{e^+e^-}$ as a function of ε. Finally, I compare my extractions of $R_{e^+e^-}$ to several previous phenomenological extractions, TPE hadronic calculations, and world's data with emphasis placed mainly on the kinematics range of the recent precise direct measurements from the OLYMPUS collaboration [24], VEPP-3 collaboration [23], and CLAS collaboration [22] experiments.

3 Results and Discussion

In this work, the ratio $R_{e^+e^-}$ is extracted, based on Eq. (3), using world's data on unpolarized σ_R taken from Refs. [1,19,25–30] for a total of 120 Q^2 points up to $Q^2 = 5.2$ (GeV/c)^2. The ratio $R_{e^+e^-}$ as a function of ε is shown in Fig. 1 for a sample of low- and high-Q^2 data points. The extracted ratio $R_{e^+e^-}$ for all Q^2 points shows clear linear ε dependence, and is fitted linearly to $R_{e^+e^-} = [1 - \alpha_1(Q^2)(1-\varepsilon)]$ (solid black line), where the TPE coefficient $\alpha_1(Q^2)$ is the parameter of fit. The extracted $\alpha_1(Q^2)$ for all Q^2 points is shown in Fig. 2. It is clearly that $\alpha_1(Q^2)$ is on the few percent level, and increases in magnitude with increasing Q^2 for most data sets. In addition, I compare the extracted TPE coefficient to different $\alpha_1(Q^2)$ parameterizations based

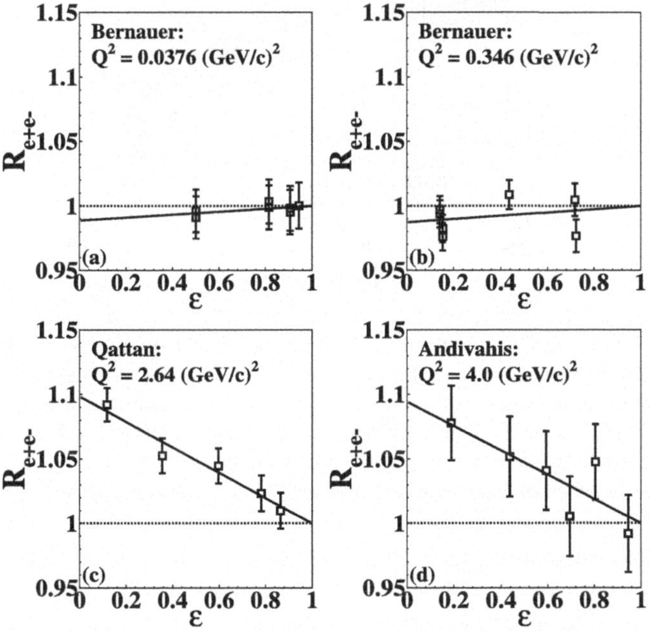

Fig. 1. (Color online) The ratio $R_{e^+e^-}$ as a function of ε as extracted in this work for a sample of low- and high-Q^2 data points from the data of Refs. [1,25,27]. The linear fit is shown in solid black line.

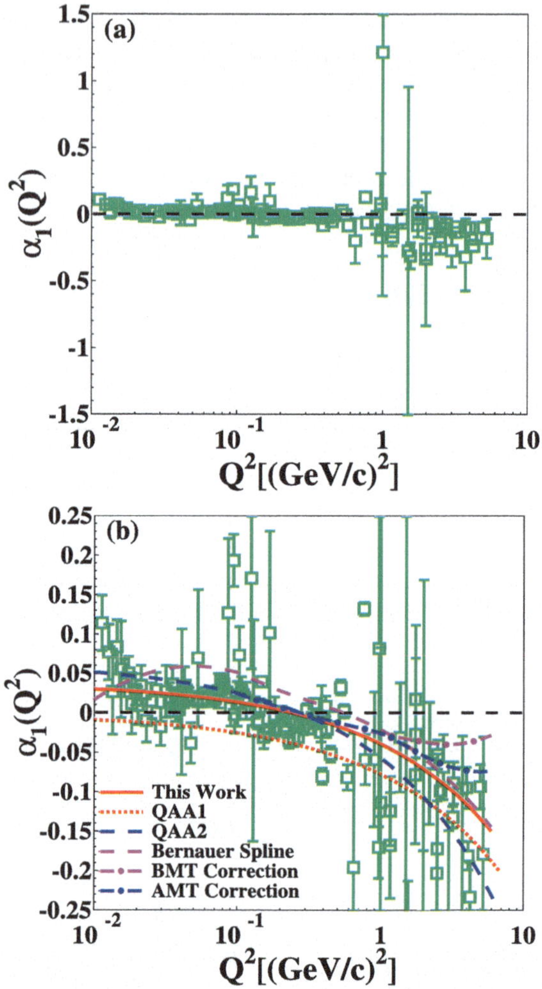

Fig. 2. (Color online) The TPE coefficient $\alpha_1(Q^2)$ as obtained from this work (open dark-green squares). The bottom plot is on a smaller vertical scale for clarity. Also shown are the simple parameterization from this work (solid red line), and $\alpha_1(Q^2)$ as determined using TPE-hadronic calculations: "BMT Correction" [31] (dashed-dotted magenta line), "AMT Correction" [32] (dashed-dotted blue line), and previous phenomenlogical analyzes "QAA1" [5] (dashed red line), "QAA2" [9] (long-dashed blue line), and "Bernauer Spline" [27] (large-dashed magenta line).

on TPE hadronic calculations from Ref. [31] "BMT-Correction" and Ref. [32] "AMT-Correction", which adds an additional phenomenological TPE contribution at higher Q^2, and from previous phenomenological extractions from Ref. [6] "QAA1", Ref. [9] "QAA2", and Ref. [27] "Bernauer Spline". The TPE coefficient $\alpha_1(Q^2)$ shows a change in sign at low Q^2, which is also seen in previous low-Q^2 TPE calculations [31,33,34]. In addition, the uncertainties I quote on the extracted $\alpha_1(Q^2)$ are an underestimate of the true uncertainties, as they are driven mainly by the uncertainties quoted on the Mainz

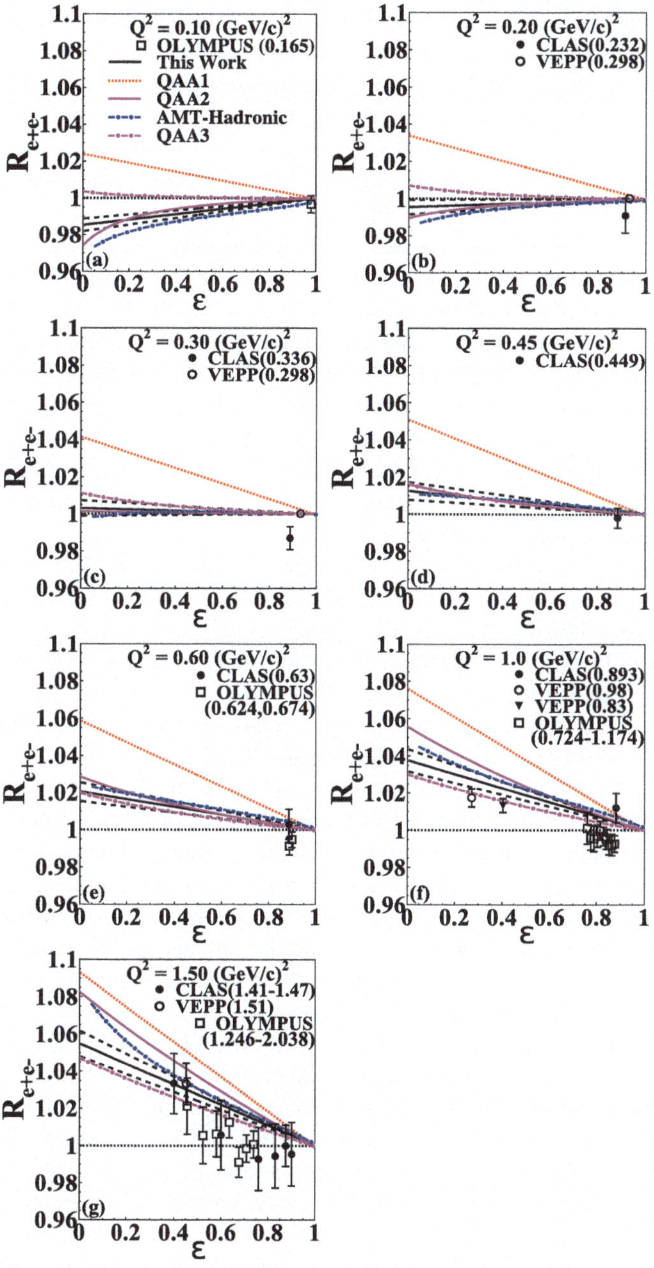

Fig. 3. (Color online) The ε dependence of the ratio $R_{e^+e^-}$ from this work (solid red line), and its associated error band (long-dashed dark-green line) at the Q^2 values listed in the figure. Also shown are TPE hadronic calculations "AMT-Hadronic" [32] (dashed-dotted black line), and phenomenological extractions: "QAA1" [5] (dashed red line), and "QAA2" [9] (long dashed blue line). The direct measurements of $R_{e^+e^-}$ from Refs. [22–24] are shown at the listed Q^2 value(s) in $(\text{GeV/c})^2$.

σ_R measurements [27], which do not include correlated systematic uncertainties, very significant, to the total uncertainty in their final FFs parameterization. It should be noted here that in the "QAA1" extraction [6], the σ_R measurements used in the fit were limited to $Q^2 \geq 0.39$, and the used R_p parametrization from Ref. [2] was not well constrained at low Q^2. As a result, $\alpha_1(Q^2)$ below $Q^2 \approx 1$ (GeV/c)^2 is driven mainly by the fitting function adapted, which is proportional to $\sqrt{Q^2}$. However, in this work, and that of "QAA2" [9], the low-Q^2 data from Mainz [27] were included, and a well-constrained R_p parametrization, along with its associated uncertainty, was used, which provided meaningful uncertainties for $\alpha_1(Q^2)$ below $Q^2 = 1$ (GeV/c)^2. I tried several functional forms to parametrize the Q^2 dependence of $\alpha_1(Q^2)$, and the best fit is obtained when I used a functional form similar to that used in the "QAA2" analysis with $\alpha_1(Q^2)$ best parametrized as: $\alpha_1(Q^2) = [(0.039 \pm 0.002) + (-0.076 \pm 0.004)\sqrt{Q^2}]$, with Q^2 is in (GeV/c)^2. I do not quote a reduced χ^2 value on the $\alpha_1(Q^2)$ fit as this would not be meaningful given that we do not have a complete evaluation of the uncertainties on $\alpha_1(Q^2)$ from the Mainz data. Overall, the $\alpha_1(Q^2)$ parametrization from this work is in generally good qualitative agreement with TPE hadronic calculations, and previous phenomenological extractions shown.

The ratio $R_{e^+e^-}$ as a function of ε extracted from this work using the new $\alpha_1(Q^2)$ parametrization is shown in Fig. 3 "This Work". The associated error band on $R_{e^+e^-}$ is computed using the covariance matrix of the $\alpha_1(Q^2)$ fit, and shown as (long-dashed dark-green line). In addition, I compare the results to TPE hadronic calculations "AMT-Hadronic" [32], phenomenological extractions: "QAA1" [6] and "QAA2" [9], and world's data from the recent precise direct measurements from the OLYMPUS collaboration [24], VEPP-3 collaboration [23], and CLAS collaboration [22]. At low Q^2, the ratio $R_{e^+e^-}$ from this work is below unity, and linear in ε. As Q^2 increases, the ratio increases, changes sign (above unity), and remains linear in ε. On the other hand, while the "QAA2" extraction is below unity at low Q^2, it shows strong non-linearity at low ε. However, the"QAA1" extractions predict a positive ratio (significantly above unity) for all Q^2 points. Overall, the $R_{e^+e^-}$ extractions from this work are in good qualitative agreement with the shown TPE hadronic calculations and phenomenological extractions, except with the "QAA1" extractions, which yield large value for $R_{e^+e^-}$ for all Q^2 range.

All recent direct measurements of $R_{e^+e^-}$ by the OLYMPUS collaboration [24], VEPP-3 collaboration [23], and CLAS collaboration [22] are taken for $Q^2 < 2.1$ (GeV/c)^2. This is below where the discrepancy on the form factors $\mu_p R_p$ is significant. The data taken by the CLAS and VEPP-3 collaborations are limited to the range (0.20–1.50) (GeV/c)^2, with precise measurements taken at $Q^2 \approx 1.0$ and 1.5 (GeV/c)^2. The measured ratio is larger than unity, with clear ε dependence mainly at low ε points. This clearly provides an evidence of a sizable TPE correction at larger Q^2 values. For the VEPP-3 data, a sharper Q^2 dependence is observed. However, such a Q^2 dependence seems to disappear when the data are compared to calculations that increase with Q^2 value. For the OLYMPUS experiment, $R_{e^+e^-}$ is measured in the range (0.165–2.038) (GeV/c)^2. At high ε, $R_{e^+e^-}$ is below unity, and then starts to change sign, and increase gradually with decreasing ε reaching approximately 2% at $\varepsilon = 0.46$.

The extracted $R_{e^+e^-}$ from this work is in generally good agreement with existing precise measurements at large ε, within the uncertainties of these measurements, and for all Q^2 points, with the exception of the high ε points taken by CLAS at $Q^2 = 0.336$ (GeV/c)^2, and OLYMPUS in the range $0.6 \leq Q^2 \leq 2.038$ (GeV/c)^2. At low ε points, the extracted $R_{e^+e^-}$ agrees very well, within uncertainty band, with precise measurements taken at $Q^2 = 1.0$ and 1.50 (GeV/c)^2, which provides clear evidence for hard-TPE correction.

4 Conclusions

Contrary to many phenomenological studies, which extracted $R_{e^+e^-}$ using proposed, model-independent parametrizations of the TPE correction to σ_R, I extracted $R_{e^+e^-}$ using combined unpolarized, and polarized ep elastic scattering measurements, and provided a simple parametrization of the TPE correction to σ_R. The ratio $R_{e^+e^-}$ extracted from this work is in generally good agreement with TPE hadronic calculations, previous phenomenological extractions, and recent precise direct measurements at large ε and for all the Q^2 range, within the uncertainties of these measurements. On the other hand, my extraction are below the high ε measurements taken by CLAS at $Q^2 = 0.336$ (GeV/c)^2, and OLYMPUS in the range $0.6 \leq Q^2 \leq 2.038$ (GeV/c)^2. At low ε, my extractions agree very well, within the error bands of the extractions, with precise measurements taken at $Q^2 = 1.0$ and 1.50 (GeV/c)^2. As all direct measurements are taken for $Q^2 < 2.1$ (GeV/c)^2, below where the discrepancy on R_p is significant, it is very early to claim that TPE corrections could account for such a discrepancy.

Acknowledgments. The author acknowledges the support provided by Khalifa University of Science and Technology.

References

1. Qattan, I.A., et al.: Phys. Rev. Lett. **94**, 142301 (2005)
2. Gayou, O., et al.: Phys. Rev. Lett. **88**, 092301 (2002)
3. Arrington, J., Blunden, P., Melnitchouk, W.: Prog. Part. Nucl. Phys. **66**, 782 (2011)
4. Carlson, C.E., Vanderhaeghen, M.: Annu. Rev. Nucl. Part. Sci. **57**, 171 (2007)
5. Qattan, I.A., Alsaad, A.: Phys. Rev. C **83**, 054307 (2011)
6. Qattan, I.A., Alsaad, A., Arrington, J.: Phys. Rev. C **84**, 054317 (2011)
7. Qattan, I.A., Arrington, J.: Phys. Rev. C **86**, 065210 (2012)
8. Qattan, I.A., Arrington, J.: Eur. Phys. J. WOC **66**, 06020 (2014)
9. Qattan, I.A., Arrington, J., Alsaad, A.: Phys. Rev. C **91**, 065203 (2015)
10. Qattan, I.A.: Phys. Rev. C **95**, 055205 (2017)
11. Qattan, I.A., Arrington, J.: J. Phys: Conf. Ser. **869**, 012053 (2017)
12. Qattan, I.A.: Phys. Rev. C **95**, 065208 (2017)
13. Qattan, I.A., Homouz, D., Riahi, M.K.: Phys. Rev. C **97**, 045201 (2018)
14. Qattan, I.A., Arrington, J.: J. Phys: Conf. Ser. **1258**, 012007 (2019)
15. Qattan, I.A., Patole, S.P., Alsaad, A.: Phys. Rev. C **101**, 055202 (2020)
16. Qattan, I.A.: J. Phys: Conf. Ser. **1643**, 012192 (2020)
17. Qattan, I.A., Alsaad, A., Ahmad, A.A.: Phys. Rev. C **103**, 055202 (2021)

18. Qattan, I.A., Alsaad, A., Ahmad, A.A.: Phys. Rev. C **103**, 045202 (2021)
19. Christy, M.E., et al.: Phys. Rev. C **70**, 015206 (2004)
20. Meziane, M., et al.: Phys. Rev. Lett. **106**, 132501 (2011)
21. Puckett, A.J.R., et al.: Phys. Rev. C **96**, 055203 (2017); **98**, 019907(E) (2018)
22. Adikaram, D., et al.: CLAS Collaboration. Phys. Rev. Lett. **114**, 062003 (2015)
23. Rachek, I.A., et al.: VEPP-3 Collaboration. Phys. Rev. Lett. **114**, 062005 (2015)
24. Henderson, B.S., et al.: (OLYMPUS Collaboration), Phys. Rev. Lett. **118**, 092501 (2017)
25. Andivahis., et al.: Phys. Rev. D **50**, 5491 (1994)
26. Walker, R.C., et al.: Phys. Rev. D **49**, 5671 (1994)
27. Bernauer, J.C., et al.: A1 Collaboration. Phys. Rev. C **90**, 015206 (2014)
28. Bartel, W., et al.: Nucl. Phys. B **58**, 429 (1973)
29. Litt, J., et al.: Phys. Lett. B **31**, 40 (1970)
30. Berger, C., et al.: Phys. Lett. B **35**, 87 (1971)
31. Blunden, P.G., Melnitchouk, W., Tjon, J.A.: Phys. Rev. C **72**, 034612 (2005)
32. Arrington, J., Melnitchouk, W., Tjon, J.A.: Phys. Rev. C **76**, 035205 (2007)
33. Borisyuk, D., Kobushkin, A.: Phys. Rev. C **75**, 038202 (2007)
34. Arrington, J.: J. Phys. G **40**, 115003 (2013)

Search for Majorana Neutrinos with 1 Ton-Year Exposure Data of KamLAND-Zen

Sei Ieki(✉) and on behalf of KamLAND-Zen collaboration

Research Center for Neutrino Science, Tohoku University, Sendi 980-8578, Japan
ieki@awa.tohoku.ac.jp

Abstract. KamLAND-Zen experiment is searching for neutrinoless double-beta decay of ^{136}Xe using a liquid scintillator detector at the Kamioka Observatory in Japan. The KamLAND-Zen detector contains 745kg of enriched xenon and we have been taking data since January 2nd, 2019. This paper reports the 1st result from the KamLAND-Zen experiment at 970 kg-yr exposure. Statistically significant excess was not observed in our region of interest (2.35$< E <$2.7 MeV), we set a limit on the neutrinoless double-beta decay half-life $T^{0\nu}_{1/2} > 2.3 \times 10^{26}$ yr at 90% C.L. Corresponding upper limits on the effective Majorana neutrino mass is $\langle m_{\beta\beta} \rangle < (36 - 156)$ meV using commonly adopted nuclear matrix element calculations.

Keywords: Neutrino · Double-beta decay · Liquid scintillator

1 Introduction

Neutrinoless double-beta ($0\nu\beta\beta$) decay is a lepton number violating process, only occur if neutrinos have Majorana nature. The decay rate is proportional to the square of the effective Majorana neutrino mass $\langle m_{\beta\beta} \rangle$. The discovery can be a proof to the absolute neutrino mass scale and their mass hierarchy. KamLAND-Zen experiment is searching for $0\nu\beta\beta$ decay of ^{136}Xe using a xenon-loaded liquid scintillator [1]. The experiment was upgraded to increase the target mass to 745 kg and it was almost doubled compared with our previous phase [2]. In the following sections, we introduce KamLAND detector and our first result of $0\nu\beta\beta$ decay search with almost 1 ton-year exposure data.

2 Detector

The Kamioka Liquid scintillator Anti-Neutrino Detector (KamLAND) is located in the Kamioka mine, Gifu, Japan. Figure 1 shows a schematic of KamLAND detector. It consists of an inner scintillation detector and an outer water Cherenkov detector. The inner detector contains a 18 m diameter stainless-steel spherical tank. A 13 m diameter nylon-EVOH balloon filled with 1 kton of LS (KamLAND-LS) is in the tank. The KamLAND-LS is a composition of dodecane, pseudocumene (PC), and PPO (2,5-diphyenyloxazole) as the fluor. The outside of the balloon is filled with non-scintillating buffer oil which is a mixture of 57% of isoparaffin and 43% of dodecane by volume. 1325 17-inch

H. J. Choi et al. (Eds.): APPC 2022, *Proceedings of the 15th Asia Pacific Physics Conference*, pp. 141–148, 2024.
https://doi.org/10.1007/978-981-96-0191-2_14

photomultiplier tubes (PMTs) and 554 20-inch PMTs are on the inner surface of the stainless-steel sphere and observe scintillation photons. The scintillation light yield is ~ 8000 photons/MeV.

On May 9th, 2018, a 3.8 m diameter inner balloon [3] was installed at the center of the nylon-EVOH balloon. It is filled with $30.5\pm0.3\,\mathrm{m}^3$ of xenon-loaded LS (Xe-LS) which is a composition of decane, PC and PPO. The Xe-LS contains 745 kg of xenon. The isotopic abundances in the enriched xenon gas is $(90.85 \pm 0.13)\%$ ^{136}Xe and $(8.82 \pm 0.01)\%$ ^{134}Xe. The inner-balloon is larger and clear than our previous phase, since the contamination during production process was suppressed by better mitigation measures. The scintillation light yield of KamLAND-LS is 10% higher than that of Xe-LS.

The outer detector is a shield for gamma-rays and fast neutrons from surrounding rock. It is a 20 m diameter and 20 m-high cylindrical cave filled with 3.2 kton of pure water. The Cherenkov light is detected by 140 20-inch PMTs to identify cosmic-ray muons.

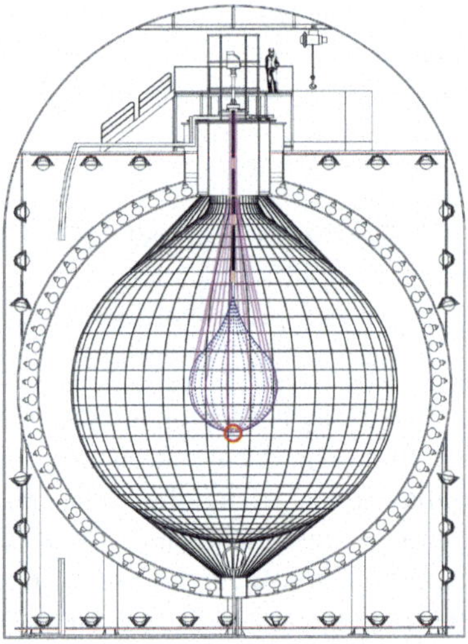

Fig. 1. (Color online) Schematic of KamLAND detector. A droplet like inner balloon contains Xe-LS. It is surrounded by KamLAND-LS filled in a 13 m diameter spherical balloon. A hot spot is indicated by a 0.7 m-diameter red circle whose center is at (x, y, z) = (0,0,-1.9 m).

3 Neutrinoless Double-Beta Decay Search

3.1 Calibration

The scintillation light yield in the Xe-LS is calibrated with ^{214}Bi $\beta^- + \gamma$ decays energy spectra, while its position dependence is corrected with monochromatic ^{214}Po α decays in ^{222}Rn series. A Geant4 based detector Monte Carlo simulation is tuned to reproduce the PMT hit timing and charge distributions. Energy and vertex resolutions in the Xe-LS are estimated to be $6.7\%/\sqrt{E(\mathrm{MeV})}$ and 13.7 cm/$\sqrt{E(\mathrm{MeV})}$, respectively. The position and time dependence of the energy scale in the Xe-LS is monitored using the 2.225 MeV monochromatic γ-ray from $^{1}\mathrm{H}(n,\gamma)^{2}\mathrm{H}$ where neutron is induced by cosmic-ray muon spallation. The energy scale fluctuation is less than 1%.

3.2 Event Selection

We used data obtained from February 5th, 2019 to May 8th, 2021. The data collected before that period was not used for a spectrum fit in Sect. 3.4 in order to avoid contamination from ^{222}Rn ($\tau = 5.5$ day). $0\nu\beta\beta$ decay candidates are selected by following first-level cuts:

- $r < 2.5$ m
  Reconstructed vertex should be within 2.5 m from the detector center. Also, the events at the bottom of the inner balloon are removed since backgrounds are concentrated (see Fig. 1).
- Cosmic-ray muon veto
  Comic-ray muons and events observed within 2 ms after muons are rejected.
- Rn veto and double-pulse veto
  Rn series such as ^{214}Bi-^{214}Po and ^{210}Bi-^{210}Po are vetoed by a delayed coincidence selection. This selection identifies sequential decays by requiring the time-space distance of prompt and delayed events to be less than 1.9 ms and 1.7 m. In addition, events with a double-pulse in an event time window are rejected.
- Reactor veto
  The inverse β decay induced by $\bar{\nu}_e$ from reactor is vetoed by a delayed coincidence selection.
- Badness cut
  Non-physical events such as electronic noise and pile-up events are eliminated by quality of vertex-time-charge fit.

Second-level cuts are applied to eliminate radioactive decays induced by cosmic-ray muon spallation.

- ^{12}B veto
  Events within 150 ms after muon crossing through the liquid scintillator are vetoed. It rejects 99.4% of ^{12}B β^- decay ($\tau = 29.1$ ms, Q=13.4 MeV).
- Carbon spallation veto
  Carbon spallation products accompanying with neutrons are identified by a triple coincidence of a muon, neutron captures and a β decay. The coincidence events are selected by the time difference between a cosmic-ray muon and the β decay $\Delta T < 180$ s, and the distance between the β decay and the nearest neutron capture $\Delta R < 1.6$ m. Hereafter, we use the same definition for ΔT and Δ R.
- Shower veto
  Spallation products are generated in muon-induced showers [4]. This selection utilizes that feature to remove the β decays isolated from neutrons ($\Delta R \geq 1.6$ m).

The spallation backgrounds are removed by a cut to a likelihood which uses shower charge along the muon track dQ/dl_{long}, transverse distance from the muon track Δl_{trans} and ΔT. The likelihood was profiled based on ^{12}B β^- decay.

- ^{137}Xe veto
  ^{137}Xe is produced by neutron capture on ^{136}Xe. The ^{137}Xe β^- decay (τ = 5.5 min, Q = 4.17 MeV) is rejected by a triple coincidence selection of a muon, high energy γ-rays from $^{136}\mathrm{Xe}(n,\gamma)^{137}$Xe ($E_\gamma$ = 4.03 MeV) and a ^{137}Xe β^- decay (ΔT <27 min). This selection rejects 74 ± 7% of ^{137}Xe β^- decay.
- Long-lived veto
  This selection is applied to reject xenon spallation products (long-lived isotopes) which typically have a few days or longer lifetime. Similar to carbon spallation veto, this is a triple coincidence selection while it is implemented as a cut to a likelihood and requires high neutron multiplicity. Parameters in the likelihood are ΔT, ΔR and the neutron multiplicity. The rejection efficiency is obtained to be 42.0 ± 8.8% as a fit parameter in Sect. 3.4. The events not vetoed by this selection is categorized as Singles data (SD), $0\nu\beta\beta$ decay candidates, while the other is categorized as Long-lived data (LD).

The total rejection efficiencies for ^{10}C β^+ decay and ^{6}He β^- decay are more than 99.3% and (97.6 ± 1.7)%, respectively.

3.3 Backgrounds

Sources of backgrounds are classified into four groups:

(1) Neutrino interactions
Neutrino interactions induce inevitable backgrounds. ^{136}Xe $2\nu\beta\beta$ decay can be discriminated only by measuring the total kinetic energy of two electrons. The background rate in ROI (2.35< E <2.70 MeV) is estimated by a fit to the energy spectrum. ^{8}B solar neutrino electron scattering (ES) deposit energy in liquid scintillator and make continuous spectrum. The event rate in the Xe-LS is estimated to be $(4.9 \pm 0.2) \times 10^{-3}$ (ton-day)$^{-1}$ using the number of electrons, ^{8}B solar neutrino flux and the cross-section.

(2) Radioactive impurities (RIs) in the Xe-LS
^{238}U series and ^{232}Th series contaminate the Xe-LS. The backgrounds are identified by the delayed coincidence, which selects ^{214}Bi-^{214}Po and ^{212}Bi-^{212}Po. If we assume the background rates in secular equilibrium, contamination of ^{238}U and ^{232}Th in the Xe-LS are measured to be $(1.5 \pm 0.4) \times 10^{-17}$ g/g$_{\mathrm{Xe-LS}}$ and $(3.0 \pm 0.4) \times 10^{-16}$ g/g$_{\mathrm{Xe-LS}}$, respectively.

(3) RIs from the inner balloon
The inner balloon film is also contaminated by ^{238}U and ^{232}Th. The contamination of ^{238}U and ^{232}Th is evaluated as $(3 \pm 1) \times 10^{-12}$ g/g$_{\mathrm{Film}}$ and $(3.8 \pm 0.2) \times 10^{-11}$ g/g$_{\mathrm{Film}}$, respectively. The amount of contamination was suppressed to almost one-tenth of previous inner balloon [2].

(4) Cosmic-ray muon spallation products
Cosmic-ray muons induce carbon spallation and xenon spallation in the liquid scintillator. The decays of the spallation isotopes are observed in the ROI. Each decay

rate was measured for the carbon spallation isotopes using a simultaneous likelihood fit to binned energy and ΔT. While the production rate of ^{137}Xe was estimated from the number of ^{136}Xe and the neutron capture cross section. The production rates of xenon spallation isotopes are not estimated separately due to limited statistics, a total contribution in the ROI is estimated from the spectrum fit in Sect. 3.4. Spallation products and their daughters are enumerated by FLUKA [5–7] and Geant4 [8–10] simulations. Also, the simulations are used to make long-lived likelihood PDFs, ΔT distribution and energy spectra for the fit.

3.4 Spectrum Fit

The event rates of $0\nu\beta\beta$ decay and backgrounds, and energy scale are estimated from the energy spectrum fit. After the second-level cuts, the candidates are divided into three time bins, 40 hemispherical-shell equal volume bins, 48 energy bins in 0.5 ∼ 4.8 MeV and two long-lived bins (SD/LD). We define a χ^2 as a sum of energy term χ_E^2 and penalty term χ_P^2:

$$\chi^2 = \chi_E^2 + \chi_P^2, \tag{1}$$

$$\chi_E^2 = \begin{cases} 2\sum_t \sum_v \sum_e \sum_l (\nu_{t,v,e,l} - n_{t,v,e,l}) & (\nu_{t,v,e,l} = 0), \\ 2\sum_t \sum_v \sum_e \sum_l \{(\nu_{t,v,e,l} - n_{t,v,e,l}) & \\ +n_{t,v,e,l} \log(n_{t,v,e,l}/\nu_{t,v,e,l})\} & (\nu_{t,v,e,l} \neq 0), \end{cases} \tag{2}$$

where $\nu_{t,v,e,l}(n_{t,v,e,l})$ is number of observed (expected) events, and t(ime), v(olume), e(nergy), and l(ong-lived) represent iteration over $3 \times 40 \times 48 \times 2$ bins. χ_P^2 constrains parameters which were estimated from other measurements. The $0\nu\beta\beta$ decay and the long-lived rates are estimated from the χ^2 scan by floating the rates freely. Not only SD but also LD were simultaneously fitted to take the most of statistics of the long-lived events. The SD events in the inner 10 volume bins ($r < 1.57$ m) and $2.35 < E < 2.70$ MeV give the best sensitivity since there are less backgrounds from the inner balloon. Combined energy spectra for the internal 10 volume bins and whole time bins are shown in Fig. 2. The detector livetime for SD and LD are 523.4 days and 49.3 days, respectively. The best fit event contributions in $r < 1.57$ m and $2.35 < E < 2.70$ MeV are summarized in Table 1.

Figure 3 shows the allowed region of the event rates. The best fit $0\nu\beta\beta$ decay rate is 0, while long-lived rate is 0.111 ± 0.019 events·(day· Xe-ton·ROI)$^{-1}$ which is about 1.5 σ discrepancy from the FLUKA prediction, 0.082 ± 0.006 events·(day· Xe-ton·ROI)$^{-1}$. Combined with our previous phase, no statistically significant excess is observed in the ROI, leading to a 90% C.L. limit on half-life $T_{1/2}^{0\nu} > 2.3 \times 10^{26}$ yr. This corresponds to $\langle m_{\beta\beta} \rangle < (36 - 156)$ meV using nuclear matrix elements of [11–23] and phase space factor of [24, 25]. We provided the most stringent constraint on the half-life and this is the first measurement exploring the inverted mass ordering region.

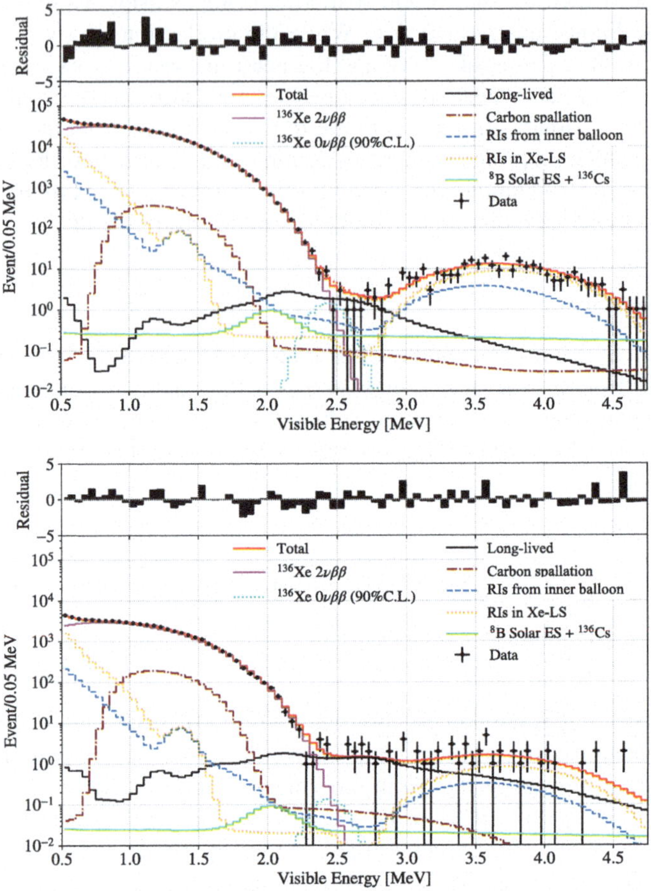

Fig. 2. (Color online) Energy spectra of SD (top) and LD (bottom) in $r < 1.57$ m, a total of 3 time bins.

Table 1. Number of events in 2.35< E <2.70 MeV and r <1.57 m are summarized for estimation and the best fit result from frequentist analysis.

Background	Estimated	Best fit
^{136}Xe $2\nu\beta\beta$	-	11.98
RIs in Xe-LS	-	0.98
RIs from inner balloon	-	3.06
^{8}B solar ν ES	1.65 ± 0.04	1.65
Carbon spallations	0.20 ± 0.14	0.22
Long-lived	7.75 ± 0.57	12.52
^{137}Xe	0.33 ± 0.28	0.34

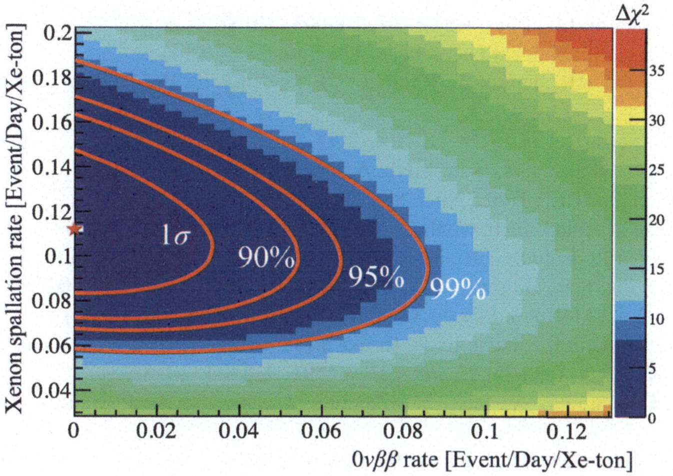

Fig. 3. (Color online) χ^2 is plotted as a function of $0\nu\beta\beta$ decay and long-lived rates in the ROI. The best fit is indicated by red star.

4 Conclusions and Future Prospects

We reported the first result of the $0\nu\beta\beta$ decay search from KamLAND-Zen800. The exposure of ^{136}Xe reached 970 kg-yr. No statistically significant excess was found in the ROI. We obtained the limit on the half-life $T^{0\nu}_{1/2} > 2.3 \times 10^{26}$ yr combining with our previous phase and this is the first search entering the inverted mass ordering region.

As a prospect to reach further sensitivity, rejection of the long-lived spallation backgrounds is one of the key issues. Our long-lived veto parameterize spacial correlation between radioactive decays and neutrons. It demonstrated ~42% rejection efficiency and we expect that will be improved by higher neutron detection efficiency. R&D of dead-time free electronics is in progress. Also, we plan to improve the energy resolution to reject the $2\nu\beta\beta$ decay.

Acknowledgments. The KamLAND-Zen experiment is supported by JSPS KAKENHI Grant Numbers 21000001, 26104002, and 19H05803; the U.S. National Science Foundation award numbers 2110720 and 2012964; the Heising-Simons Foundation; the Dutch Research Council (NWO); and under the U.S. Department of Energy (DOE) Grant No. DEAC02-05CH11231, as well as other DOE and NSF grants to individual institutions. The Kamioka Mining and Smelting Company has provided service for activities in the mine. We acknowledge the support of NII for SINET4.

References

1. Abe, S., et al.: KamLAND-Zen Collaboration. Phys. Rev. Lett. **130**, 051801 (2023)
2. Gando, A., et al.: KamLAND-Zen Collaboration. Phys. Rev. Lett. **117**, 082503 (2016)
3. Gando, Y., et al.: KamLAND-Zen Collaboration. JINST **16**, P08023 (2021)

4. Li, S.W., Beacom, J.F.: Phys. Rev. D **91**, 105005 (2015)
5. Bohlen, T.T., et al.: Nulcear Data Sheets **120**, 211–214 (2014)
6. Ferrari, A., Sala, P.R., Fasso, A., Ranft, J.: CERN-2005-10, INFN/TC_05/11, SLAC-R-773, (2005)
7. Felcini, M., Ferrari, A.: CERN-AB-Note-2006-01, EURISOL DS/Task 5/TN-06-01 (2006)
8. Agostinelli, S., et al.: Nucl. Instr. Meth. A **506**, 250–303 (2003)
9. Allison, J., et al.: IEEE Trans. Nucl. Sci. **53**(1), 270–278 (2006)
10. Allison, J., et al.: Nucl. Instr. Meth. A **835**, 186–225 (2016)
11. Vaquero, N.L., Rodríguez, T.R., Egido, J.L.: Phys. Rev. Lett. **111**, 142501 (2013)
12. Yao, J.M., Song, L.S., Hagino, K., Ring, P., Meng, J.: Phys. Rev. C **91**, 024316 (2015)
13. Rodríguez, T.R., Martínez-Pinedo, G.: Phys. Rev. Lett. **105**, 252503 (2010)
14. Deppisch, F.F., Graf, L., Iachello, F., Kotila, J.: Phys. Rev. D **102**, 095016 (2020)
15. Barea, J., Kotila, J., Iachello, F.: Phys. Rev. C **91**, 034304 (2015)
16. Coraggio, L., Gargano, A., Itaco, N., Mancino, R., Nowacki, F.: Phys. Rev. C **101**, 0443215 (2020)
17. Neacsu, A., Horoi, M.: Phys. Rev. C **91**, 024309 (2015)
18. Menedez, J., Poves, A., Caurier, E., Nowacki, F.: Nucl. Phys. A **818**, 139 (2009)
19. Terasaki, J.: Phys. Rev. C **102**, 044303 (2020)
20. Hyvärinen, J., Suhonen, J.: Phys. Rev. C **91**, 024613 (2015)
21. S̃imkovic, F., Rodin, V., Faessler, A., Vogel, P.: Phys. Rev. C **87**, 045501 (2013)
22. Mustonen, M.T., Engel, J.: Phys. Rev. C **87**, 064302 (2013)
23. Fang, D.L., Faessler, A., S̃imkovic, F.: Phys. Rev. C **97**, 045503 (2018)
24. Kotila, J., Iachello, F.: Phys. Rev. C **85**, 034316 (2012)
25. Stoica, S., Mirea, M.: Phys. Rev. C **88**, 037303 (2013); updated in arXiv:1411.5506v3
26. Abe, S., et al.: KamLAND Collaboration. Phys. Rev. C **81**, 025807 (2010)

Self-consistent Understanding of Nuclear Experiments and Astrophysical Observations Based on Relativistic Mean-Field Models with Isoscalar and Isovector-Meson Mixing

Tsuyoshi Miyatsu[1(✉)], Myung-Ki Cheoun[1], Kyungsik Kim[2], and Koichi Saito[3]

[1] Department of Physics and OMEG Institute, Soongsil University, Seoul 06978, Republic of Korea
tsuyoshi.miyatsu@ssu.ac.kr

[2] School of Liberal Arts and Sciences, Korea Aerospace University, Goyang 10540, Republic of Korea

[3] Department of Physics, Faculty of Science and Technology, Tokyo University of Science, Noda 278-8510, Japan

Abstract. Using the relativistic mean-field model with nonlinear couplings between the isoscalar and isovector mesons, we study the ground-state properties of finite, closed-shell nuclei as well as the characteristics of nuclear and neutron star matter. It is found that the isovector, scalar (δ) meson and the σ-δ mixing increase the neutron skin thickness, $R_{\rm skin}$, of ^{208}Pb. In contrast, the σ–δ mixing reduces the radius and tidal deformability of a neutron star. Thus, the present calculation can simultaneously account for the experimental data of $R_{\rm skin}$ of ^{208}Pb, reported by the PREX Collaboration, and the dimensionless tidal deformability of a canonical 1.4 $M_{\odot}$ neutron star, observed from the binary neutron star merger, GW170817.

Keywords: equation of state · neutron skin thickness · neutron stars · symmetry energy

1 Introduction

The properties of finite nuclei, nuclear matter, and neutron stars are generally governed by the nuclear equation of state (EoS), characterized by the relation between the energy density and pressure of the system. Not only the terrestrial experiments, such as heavy-ion collisions and electron scattering off nucleon and nuclei, but also the astrophysical observations of neutron stars are very useful to understand the isospin-asymmetric nuclear EoS [1].

Since the accurate determination of neutron skin thickness, $R_{\rm skin}$, of ^{208}Pb through parity-violation in electron scattering by the PREX collaboration [2], a critical discrepancy has been raised in nuclear physics. It is thus a very challenging task to understand the result from the PREX-2 experiment because the measured value, $R_{\rm skin} = 0.283 \pm 0.071$ fm, is remarkably larger than those expected theoretically. The $R_{\rm skin}$

H. J. Choi et al. (Eds.): APPC 2022, *Proceedings of the 15th Asia Pacific Physics Conference*, pp. 149–158, 2024.
https://doi.org/10.1007/978-981-96-0191-2_15

are determined by the nuclear symmetry energy, $E_{\rm sym}$, and its slope parameter, L, near the nuclear saturation density, ρ_0, mainly characterizing the properties of isospin-asymmetric nuclear matter and heavy nuclei. However, it is still hard to obtain the accurate value of $E_{\rm sym}$ and L [3].

On the other hand, the gravitational-wave (GW) signals from a binary neutron star merger play an important role in constructing the EoS for neutron star matter [4,5]. Especially, the observed tidal deformability of a neutron star has provided stringent restrictions on the mass–radius relation of neutron stars and hence the isospin-asymmetric nuclear EoS [6–8]. In Refs. [9–11], it has been insisted that the relatively small L is favored to satisfy the dimensionless tidal deformability of a canonical 1.4 $M_\odot$ neutron star, $\Lambda_{1.4} = 190^{+390}_{-120}$, observed from the binary neutron star merger, GW170817 [12–14].

In our previous study [15], we have developed the relativistic mean-field (RMF) model by introducing the isoscalar- and isovector-meson mixing, $\sigma^2\boldsymbol{\delta}^2$ and $\omega_\mu\omega^\mu\boldsymbol{\rho}_\nu\boldsymbol{\rho}^\nu$. It has been found that the quartic interaction due to the scalar mesons has a large influence on the radius and tidal deformability of a neutron star. In particular, the strong σ–δ mixing could reproduce the observational result of $\Lambda_{1.4}$ from GW170817.

In the present paper, we construct the effective interaction based on the RMF model with isoscalar- and isovector-meson mixing to explain both nuclear experiments and astrophysical observations of neutron stars. Particularly, we study the effects of the δ meson and σ-δ mixing on the ground-state properties of finite nuclei as well as the isospin-asymmetric nuclear EoS in order to account for the thick $R_{\rm skin}$ of ^{208}Pb measured by the PREX-2 experiment and the small $\Lambda_{1.4}$ observed from GW170817.

2 Formalism

The recently updated Lagrangian density in RMF approximation is employed to construct the EoS for isospin-asymmetric nuclear matter and to calculate the properties of finite nuclei [15]. We introduce the isoscalar (σ and ω) and isovector ($\boldsymbol{\delta}$ and $\boldsymbol{\rho}$) mesons as well as nucleons ($N = p, n$). The interacting Lagrangian density is then given by

$$\begin{aligned}\mathcal{L}_{\rm int} = &\sum_{N=p,n} \bar{\psi}_N \big[g_\sigma \sigma - g_\omega \gamma_\mu \omega^\mu + g_\delta \boldsymbol{\delta} \cdot \boldsymbol{\tau}_N \\ &- g_\rho \gamma_\mu \boldsymbol{\rho}^\mu \cdot \boldsymbol{\tau}_N \big] \psi_N - U_{\rm NL}(\sigma, \omega, \boldsymbol{\delta}, \boldsymbol{\rho}),\end{aligned} \tag{1}$$

where ψ_N is the nucleon field and $\boldsymbol{\tau}_N$ is its isospin matrix. The meson-nucleon coupling constants are respectively denoted by g_σ, g_ω, g_δ, and g_ρ. A nonlinear potential is here supplemented as

$$\begin{aligned}U_{\rm NL}(\sigma, \omega, \boldsymbol{\delta}, \boldsymbol{\rho}) = &\frac{1}{3} g_2 \sigma^3 + \frac{1}{4} g_3 \sigma^4 - \frac{1}{4} c_3 \left(\omega_\mu \omega^\mu\right)^2 \\ &- \Lambda_{\sigma\delta} \sigma^2 \boldsymbol{\delta}^2 - \Lambda_{\omega\rho} \left(\omega_\mu \omega^\mu\right) \left(\boldsymbol{\rho}_\nu \cdot \boldsymbol{\rho}^\nu\right),\end{aligned} \tag{2}$$

with three coupling constants (g_2, g_3, c_3) and two mixing parameters ($\Lambda_{\sigma\delta}$, $\Lambda_{\omega\rho}$) [16–19]. For describing the characteristics of finite nuclei, Coulomb interaction for photon is also taken into consideration [1].

Table 1. Results for the BigApple series.

BigApple series	$E_{\rm sym}(\rho_0)$ (MeV)	^{208}Pb	^{48}Ca
		$R_{\rm skin}$ (fm)	$R_{\rm skin}$ (fm)
BigApple	31.315	0.151	0.168
$\Lambda_{\sigma\delta} = 0$	33.092	0.197	0.186
$\Lambda_{\sigma\delta} = 60$	33.891	0.215	0.194
$\Lambda_{\sigma\delta} = 80$	34.221	0.222	0.196
$\Lambda_{\sigma\delta} = 85$	34.306	0.224	0.197

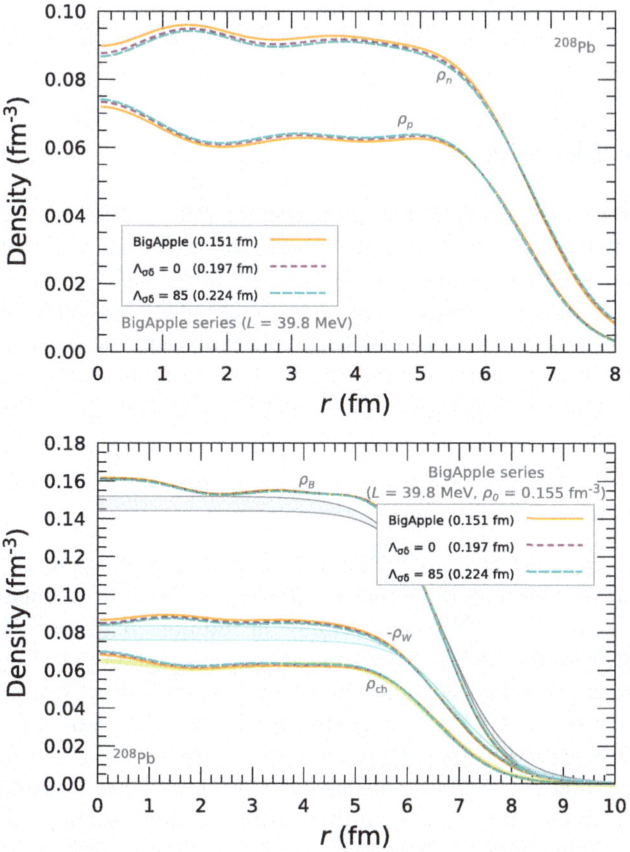

Fig. 1. (Color online) (Upper panel) Proton and neutron densities, ρ_p and ρ_n, in ^{208}Pb. (Lower panel) Baryon, charge, and weak charge densities, ρ_B, $\rho_{\rm ch}$, and ρ_W, in ^{208}Pb. The shaded bands represent the experimental data of ρ_B and ρ_W from the PREX-2 result [2], and the thick yellow line is the measured result of $\rho_{\rm ch}$ from the elastic electron scattering [24].

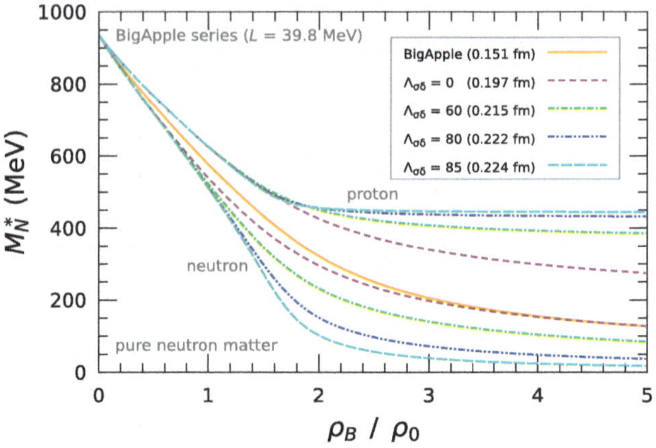

Fig. 2. (Color online) Effective nucleon mass, $M^*_{N=p,n}$, in pure neutron matter as a function of the baryon density ratio, ρ_B/ρ_0.

3 Numerical Results

In order to describe the ground-state properties of finite nuclei, we adopt the well-calibrated parameter set of the BigApple model [20], where the ω self-interaction and ω–ρ mixing are considered in Equation (2). Here, we additionally introduce the δ meson and σ–δ mixing by adjusting E_{sym} at ρ_0 so as to reproduce the same value of binding energy of ^{208}Pb to that of the original BigApple model. In the present study, we fix $g_\delta^2/4\pi = 3$, at which the curvature parameter of E_{sym} attains a minimum limit [15]. Note that the δ meson does not affect the properties of isospin-symmetric finite nuclei and nuclear matter.

3.1 Finite Nuclei

In Table 1, $E_{\text{sym}}(\rho_0)$ and R_{skin} of ^{208}Pb and ^{48}Ca are calculated using the BigApple parameter set with or without the δ meson (BigApple series). It is found that the δ-N coupling enlarges R_{skin} of ^{208}Pb considerably. In addition, R_{skin} of ^{208}Pb reaches the measured result from the PREX-2 experiment, $R_{\text{skin}} = 0.283 \pm 0.071$ fm [2], when the σ–δ mixing is taken into account. However, it is still difficult to simultaneously explain the data of R_{skin} of ^{48}Ca, reported by the CREX collaboration, $R_{\text{skin}} = 0.121 \pm 0.026(\text{exp}) \pm 0.024(\text{model})$ fm [21].

The density profiles of ^{208}Pb are displayed in Fig. 1. In the upper panel, the δ meson has an influence on the proton and neutron densities, ρ_p and ρ_n. In particular, the δ-N coupling affects both densities more than the σ–δ mixing does, which is the similar tendency to R_{skin} as shown in Table 1.

In the lower panel, we present the baryon, charge, and weak charge densities, ρ_B, ρ_{ch}, and ρ_W. Here, ρ_W is defined as

$$\rho_W(\boldsymbol{r}) \simeq Q_W^p \rho_{\text{ch}}(\boldsymbol{r}) + Q_W^n \int d\boldsymbol{r}'\, G_p^E(|\boldsymbol{r} - \boldsymbol{r}'|)\, \rho_n(\boldsymbol{r}), \tag{3}$$

with $Q_W^{p(n)}$ being the proton (neutron) weak charge and G_p^E being the proton electric form factor [22, 23]. Therefore, ρ_W is theoretically calculated by combining $\rho_{\rm ch}$ (hence ρ_p) and ρ_n. In contrast, the experimental data of $\rho_B (= \rho_p + \rho_n)$ is expressed as the sum of measured results of ρ_W from the PREX-2 experiment [2] and of $\rho_{\rm ch}$ from the elastic electron scattering [24]. It is found that, in the lower panel of Fig. 1, $-\rho_W$ slightly moves to the experimental data if the δ meson is taken into account. Meanwhile, it has almost no influence on ρ_B. Thus, we speculate that, as already explained in Ref. [25], it is preferable to choose the smaller ρ_0 than the current value, 0.155 fm^{-3}, in order to support the PREX-2 experiment.

3.2 Nuclear and Neutron Star Matter

The δ-meson effect can be clearly seen in the effective nucleon mass, M_N^*. In RMF approximation, M_N^* is given by

$$M^*_{N=\binom{p}{n}}(\bar{\sigma}, \bar{\delta}) = M_N - g_\sigma \bar{\sigma} \mp g_\delta \bar{\delta}, \tag{4}$$

where $\bar{\sigma}$ and $\bar{\delta}$ are the mean-field values of scalar mesons, and M_N is the nucleon mass in vacuum. Displayed in Fig. 2 is the density dependence of M_N^* in pure neutron matter. It is found that the δ meson is responsible for the mass splitting between protons and neutrons, where M_p^* is much heavier than M_n^* at high densities, which implies that the neutron distribution is more spread out than the proton one. Especially, the σ-δ mixing strongly affects the mass splitting as ρ_B increases. Finally, M_p^* and M_n^* respectively reach the almost constant values at high densities.

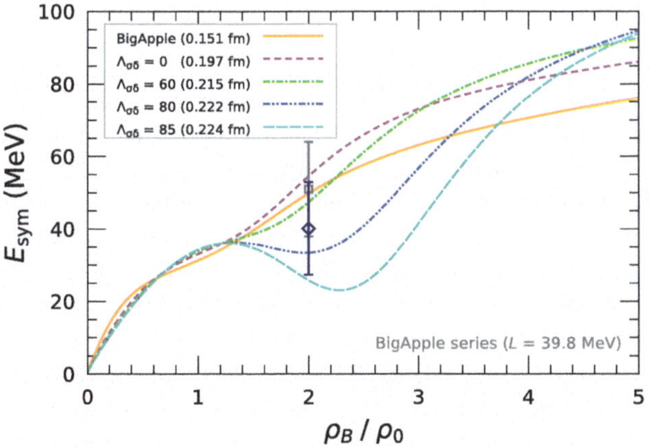

Fig. 3. (Color online) Nuclear symmetry energy, $E_{\rm sym}$, as a function of ρ_B/ρ_0. We show the constraints on the magnitude of $E_{\rm sym}$ at $2\rho_0$: $E_{\rm sym}(2\rho_0) \simeq 40.2 \pm 12.8$ MeV based on microscopic calculations with various energy density functionals [26], and $E_{\rm sym}(2\rho_0) \simeq 51 \pm 13$ MeV from neutron star observables since GW170817 [27].

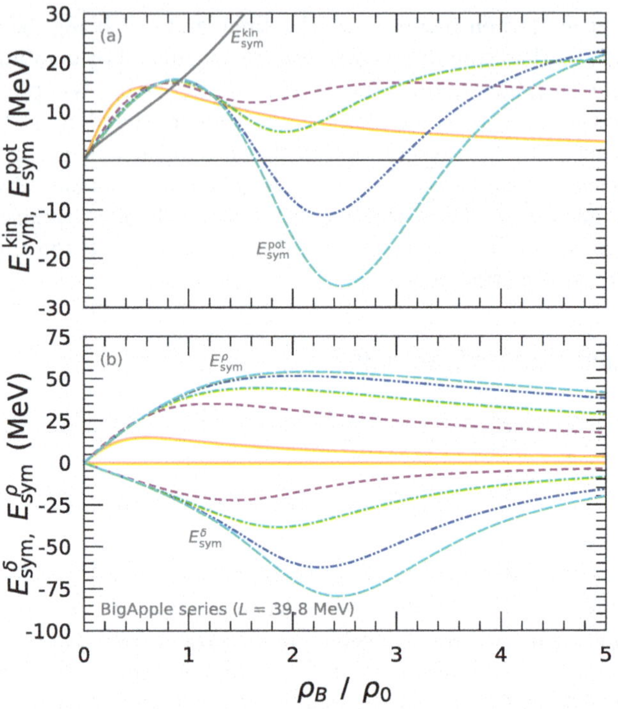

Fig. 4. (Color online) Contents of $E_{\rm sym}$ as a function of ρ_B/ρ_0. We present (a) the kinetic and potential terms, $E^{\rm kin}_{\rm sym}$ and $E^{\rm pot}_{\rm sym}$, and (b) the meson contributions, $E^{\delta}_{\rm sym}$ and $E^{\rho}_{\rm sym}$. The colored lines in both panels are the same as in the legend of Fig. 3.

The density dependence of nuclear symmetry energy, $E_{\rm sym}$, is depicted in Fig. 3. At low densities below $0.6\rho_0$, $E_{\rm sym}$ is slightly suppressed by the δ meson. In contrast, $E_{\rm sym}$ becomes stiff as ρ_B increases when the δ-N coupling is considered. Furthermore, the strong σ-δ mixing causes the temporal softness around $1.5\rho_0$–$2.5\rho_0$, and then $E_{\rm sym}$ increases rapidly at high densities. This behavior is similar to the cusp in $E_{\rm sym}$ using the skyrmion crystal approach [28,29] and to the results in the Skyrme Hartree-Fock calculations [30].

According to the Hugenholtz–Van Hove theorem [31,32], $E_{\rm sym}$ is generally divided into the kinetic and potential terms, $E^{\rm kin}_{\rm sym}$ and $E^{\rm pot}_{\rm sym}$, as $E_{\rm sym} = E^{\rm kin}_{\rm sym} + E^{\rm pot}_{\rm sym}$ [33]. We display the density dependence of $E^{\rm kin}_{\rm sym}$ and $E^{\rm pot}_{\rm sym}$ in Fig. 4(a). Because $E^{\rm kin}_{\rm sym}$ is calculated on the condition that $\rho_p = \rho_n$, it is common to all the cases. On the other hand, $E^{\rm pot}_{\rm sym}$ shows the unique behavior above ρ_0 due to the δ meson. When only the ρ meson and its quadratic interaction, $\omega_\mu\omega^\mu\boldsymbol{\rho}_\nu\boldsymbol{\rho}^{nu}$, are considered, $E^{\rm pot}_{\rm sym}$ shows the maximum point around $0.5\rho_0$, and then decreases monotonically as ρ_B increases. Once the δ meson is included, $E^{\rm pot}_{\rm sym}$ becomes larger at high densities. In particular, for $\Lambda_{\sigma\delta} \geq 60$, the rapid reduction occurs above ρ_0, and then $E^{\rm pot}_{\rm sym}$ turns to grow as ρ_B increases. Thus, the dip of $E_{\rm sym}$ around $2\rho_0$ in Fig. 3 is mainly caused by $E^{\rm pot}_{\rm sym}$.

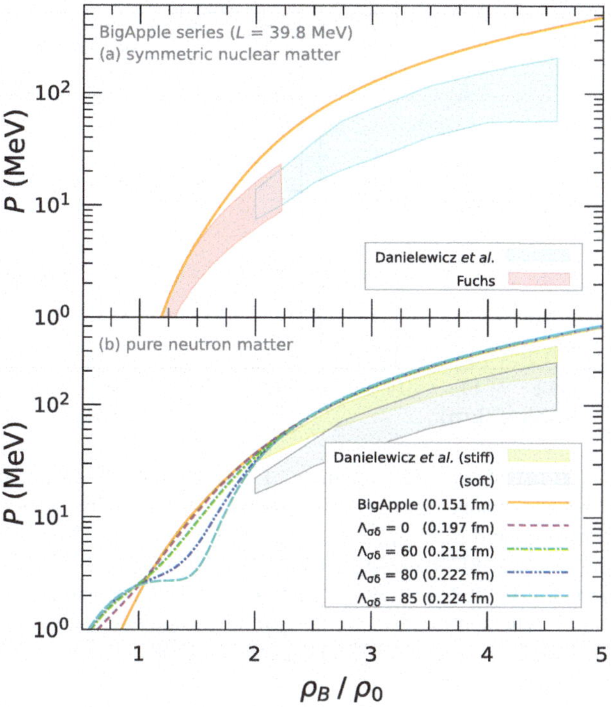

Fig. 5. (Color online) Pressure, P, as a function of ρ_B/ρ_0 in (a) symmetric nuclear matter and (b) pure neutron matter. The shaded areas represent the constraints from elliptical flow data [34] and kaon production data [35].

In RMF approximation, only the isovector mesons contribute to $E_{\rm sym}^{\rm pot}$ as $E_{\rm sym}^{\rm pot} = E_{\rm sym}^{\delta} + E_{\rm sym}^{\rho}$, where $E_{\rm sym}^{\delta}$ and $E_{\rm sym}^{\rho}$ are respectively the δ and ρ contributions [15], which are given in Fig. 4(b). It is found that $E_{\rm sym}^{\delta}$ contributes to $E_{\rm sym}^{\rm pot}$ negatively while $E_{\rm sym}^{\rho}$ behaves positively, which is similar to the general understanding of N-N interaction described by the nuclear attractive and repulsive forces. When the absolute value of $E_{\rm sym}^{\delta}$ is larger than that of $E_{\rm sym}^{\rho}$, $E_{\rm sym}^{\rm pot}$ has the rapid reduction in Fig. 4(a).

In Fig. 5, the density dependence of pressure, P, is displayed with the constraints on the nuclear EoS extracted from the analyses of particle flow data in heavy-ion collisions [34,35]. The δ meson does not affect P in symmetric nuclear matter. Nor does the nuclear EoS match the elliptical flow data, similar to the original BigApple model. In contrast, the δ meson has a large impact on P in pure neutron matter. The nuclear EoS is stiff below ρ_0 mainly due to the δ-N coupling. Meanwhile, the strong σ-δ mixing makes it soft extremely in the density region from ρ_0 to $2\rho_0$, around which the characteristics of a canonical $1.4\ M_\odot$ neutron star are generally determined. At high densities, the δ-meson effect practically disappears, and all the cases with the δ meson approach the BigApple nuclear EoS. Moreover, we can verify the matter stability against charge fluctuations, $dP/d\rho_B > 0$. However, when $\Lambda_{\sigma\delta}$ is larger than the current value, the

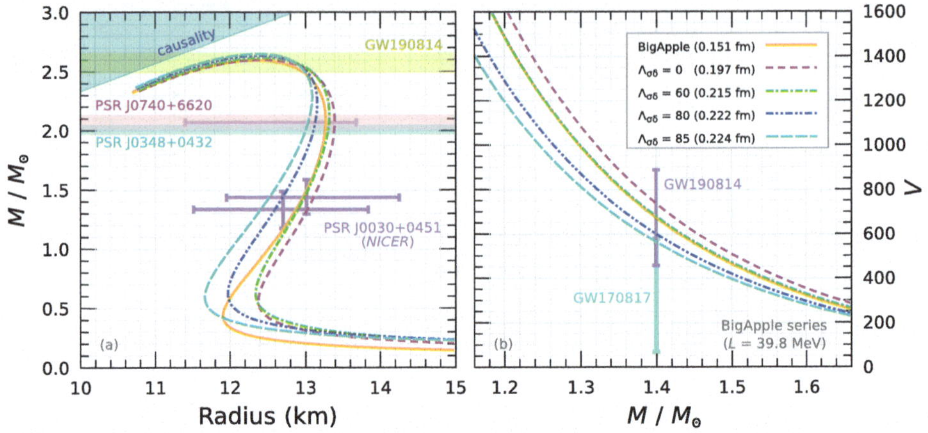

Fig. 6. (Color online) (a) Mass–radius relations of neutron stars. The observational data are supplemented by the constraints from PSR J0030+0451 by a NICER view ($1.34^{+0.15}_{-0.16}$ $M_\odot$ and $12.71^{+1.14}_{-1.19}$ km [37], and $1.44^{+0.15}_{-0.14}$ $M_\odot$ and $13.02^{+1.24}_{-1.06}$ km [38]), PSR J0348+0432 (2.01 ± 0.04 $M_\odot$) [39], PSR J0740+6620 ($2.072^{+0.067}_{-0.066}$ $M_\odot$ and $12.39^{+1.30}_{-0.98}$ km) [40–42], and the secondary object of GW190814 ($2.59^{+0.08}_{-0.09}$ $M_\odot$) [43]. (b) Dimensionless tidal deformability of a neutron star, Λ. The vertical bars are the astrophysical constraints on $\Lambda_{1.4}$ from binary merger events, GW170817 ($\Lambda_{1.4} = 190^{+390}_{-120}$) [13] and GW190814 ($\Lambda_{1.4} = 616^{+273}_{-158}$) [43].

nuclear matter becomes unstable at high densities and the phase transition should be considered [36].

The mass–radius relations of neutron stars are presented in Fig. 6(a). We here employ the EoS for nonuniform matter in the crust region, where nuclei are taken into account using the Thomas–Fermi calculation [44,45]. It is found that the δ meson have little impact on the neutron-star properties at the maximum-mass point. In contrast, the radius of a canonical 1.4 $M_\odot$ neutron star, $R_{1.4}$, is strongly affected by the δ meson. The δ-N coupling makes $R_{1.4}$ large for $\Lambda_{\sigma\delta} = 0$, while the σ-δ mixing reduces $R_{1.4}$ for $\Lambda_{\sigma\delta} > 0$. In particular, the strong σ-δ mixing makes $R_{1.4}$ smaller than that of the original BigApple model.

The dimensionless tidal deformability of a neutron star, Λ, is displayed in Fig. 6(b). We also present the recent constraints on $\Lambda_{1.4}$ based on the GW signals from the binary neutron star merger, GW170817 [13], and from the compact binary coalescence involving a 23 $M_\odot$ black hole and a 2.6 $M_\odot$ compact object, GW190814 [43]. When only the δ-N coupling is considered, $\Lambda_{1.4}$ becomes larger than that of the BigApple model. In contrast, the σ-δ mixing reduces $\Lambda_{1.4}$, and it is possible to explain both constraints for $\Lambda_{\sigma\delta} = 85$.

4 Summary and Conclusion

Using the RMF model with nonlinear couplings between the isoscalar and isovector mesons, we have studied the ground-state properties of finite nuclei as well as the characteristics of nuclear and neutron star matter. In particular, in order to explain both nuclear experiments and astrophysical observations of neutron stars, we have investigated the effects of the δ meson and σ-δ mixing on isospin asymmetry in magic nuclei and nuclear matter using the well-calibrated parameter set of the BigApple model.

It has found that the δ-N coupling predominantly increases the $R_{\rm skin}$ of ^{208}Pb. In addition, the proton and neutron density profiles are mainly affected by the δ-N coupling. By introducing both the δ-N coupling and σ-δ mixing, it is possible to account for the PREX-2 result. However, it is still difficult to explain the CREX data.

We have also found the important role of the δ meson in nuclear and neutron star matter. The δ meson is responsible for the proton-neutron mass splitting in matter, and the σ-δ mixing strongly affects the mass splitting at high densities. The $E_{\rm sym}^{\rm pot}$ also shows the rapid reduction above ρ_0 due to the σ-δ mixing, and thus the temporal softness of $E_{\rm sym}$ occurs around $1.5\rho_0$–$2.5\rho_0$. Furthermore, the σ-δ mixing extremely softens the EoS for pure neutron matter up to $2\rho_0$, and reduces the radius and tidal deformability of a neutron star.

In conclusion, we have found that the inclusion of the δ meson and the strong σ-δ mixing makes it possible to account for the thick $R_{\rm skin}$ of ^{208}Pb, reported by the PREX Collaboration, and the small $\Lambda_{1.4}$, observed from the binary neutron star merger, GW170817.

Acknowledgments. This work is supported by the National Research Foundation of Korea (Grant Nos. NRF-2021R1A6A1A03043957, NRF-2020K1A3A7A09080134, NRF-2020R1A2C3006177, and 2018R1A5A1025563).

References

1. Serot, B.D., Walecka, J.D.: Adv. Nucl. Phys. **16**, 1 (1986)
2. Adhikari, D., et al.: Phys. Rev. Lett. **126**, 172502 (2021)
3. Piekarewicz, J.: Phys. Rev. C **104**, 024329 (2021)
4. Hinderer, T.: Astrophys. J. **677**, 1216 (2008)
5. Hinderer, T., Lackey, B.D., Lang, R.N., Read, J.S.: Phys. Rev. D **81**, 123016 (2010)
6. Annala, E., Gorda, T., Kurkela, A., Vuorinen, A.: Phys. Rev. Lett. **120**, 172703 (2018)
7. Lim, Y., Holt, J.W.: Phys. Rev. Lett. **121**, 062701 (2018)
8. Raithel, C., Özel, F., Psaltis, D.: Astrophys. J. Lett. **857**, L23 (2018)
9. De, S., Finstad, D., Lattimer, J.M., Brown, D.A., Berger, E., Biwer, C.M.: Phys. Rev. Lett. **121**, 091102 (2018) [erratum: Phys. Rev. Lett. **121**, 259902 (2018)]
10. Hornick, N., Tolos, L., Zacchi, A., Christian, J.E., Schaffner-Bielich, J.: Phys. Rev. C **98**, 065804 (2018) [erratum: Phys. Rev. C **103**, 039902 (2021)]
11. Choi, S., Miyatsu, T., Cheoun, M.K., Saito, K.: Astrophys. J. **909**, 156 (2021)
12. Abbott, B.P., et al.: Phys. Rev. Lett. **119**, 161101 (2017)
13. Abbott, B.P., et al.: Phys. Rev. Lett. **121**, 161101 (2018)
14. Abbott, B.P., et al.: Phys. Rev. X **9**, 011001 (2019)
15. Miyatsu, T., Cheoun, M.K., Saito, K.: Astrophys. J. **929**, 82 (2022)
16. Boguta, J., Bodmer, A.R.: Nucl. Phys. A **292**, 413 (1977)
17. Todd-Rutel, B.G., Piekarewicz, J.: Phys. Rev. Lett. **95**, 122501 (2005)
18. Miyatsu, T., Cheoun, M.K., Saito, K.: Phys. Rev. C **88**, 015802 (2013)
19. Zabari, N., Kubis, S., Wójcik, W.: Phys. Rev. C **99**, 035209 (2019)
20. Fattoyev, F.J., Horowitz, C.J., Piekarewicz, J., Reed, B.: Phys. Rev. C **102**, 065805 (2020)

21. Adhikari, D., et al.: Phys. Rev. Lett. **129**, 042501 (2022)
22. Horowitz, C.J., Pollock, S.J., Souder, P.A., Michaels, R.: Phys. Rev. C **63**, 025501 (2001)
23. Horowitz, C.J., et al.: Phys. Rev. C **85**, 032501 (2012)
24. De Vries, H., De Jager, C.W., De Vries, C.: Atom. Data Nucl. Data Tabl. **36**, 495 (1987)
25. Horowitz, C.J., Piekarewicz, J., Reed, B.: Phys. Rev. C **102**, 044321 (2020)
26. Chen, L.W.: EPJ Web Conf. **88**, 00017 (2015)
27. Li, B.A., Cai, B.J., Xie, W.J., Zhang, N.B.: Universe **7**, 182 (2021)
28. Ma, Y.L., Rho, M.: AAPPS Bull. **31**, 16 (2021)
29. Lee, H.K., Ma, Y.L., Paeng, W.G., Rho, M.: Mod. Phys. Lett. A **37**, 2230003 (2022)
30. Chen, L.W., Ko, C.M., Li, B.A.: Phys. Rev. C **72**, 064309 (2005)
31. Czerski, P., De Pace, A., Molinari, A.: Phys. Rev. C **65**, 044317 (2002)
32. Cai, B.J., Chen, L.W.: Phys. Lett. B **711**, 104 (2012)
33. Miyatsu, T., Cheoun, M.K., Ishizuka, C., Kim, K.S., Maruyama, T., Saito, K.: Phys. Lett. B **803**, 135282 (2020)
34. Danielewicz, P., Lacey, R., Lynch, W.G.: Science **298**, 1592 (2002)
35. Fuchs, C.: Prog. Part. Nucl. Phys. **56**, 1 (2006)
36. Kubis, S., Wójcik, W., Zabari, N.: Phys. Rev. C **102**, 065803 (2020)
37. Riley, T.E., et al.: Astrophys. J. Lett. **887**, L21 (2019)
38. Miller, M.C., et al.: Astrophys. J. Lett. **887**, L24 (2019)
39. Antoniadis, J., et al.: Science **340**, 6131 (2013)
40. Cromartie, H.T., et al.: Nature Astron. **4**, 72 (2019)
41. Fonseca, E., et al.: Astrophys. J. Lett. **915**, L12 (2021)
42. Riley, T.E., et al.: Astrophys. J. Lett. **918**, L27 (2021)
43. Abbott, R., et al.: Astrophys. J. Lett. **896**, L44 (2020)
44. Miyatsu, T., Yamamuro, S., Nakazato, K.: Astrophys. J. **777**, 4 (2013)
45. Miyatsu, T., Cheoun, M.K., Saito, K.: Astrophys. J. **813**, 135 (2015)

The SABRE South Experiment at the Stawell Underground Physics Laboratory

I. Bolognino[1,2(✉)], for the SABRE South collaboration, E. Barberio[2,3], T. Baroncelli[2,3], J. L. Bignell[2,4], I. Bolognino[1,2], G. Brooks[2,5], F. Dastgiri[2,4], A. R. Duffy[2,6], M. Froehlich[2,4], G. Fu[2,3], M. S. M. Gerathy[2,3], G. C. Hill[1,2], S. Krishnan[2,5], G. J. Lane[2,4], G. Lawrence[2,6], K. T. Leaver[1,2], I. Mahmood[2,3], P. McGee[1,2], L. J. McKie[2,4], P. C. McNamara[2,4], M. Mews[2,3], W. J. D. Melbourne[2,3], G. Milana[2,5], L. J. Milligan[2,3], J. Mould[2,6], F. Nuti[2,3], F. Scutti[2,6], Z. Slavkovská[2,4], N. J. Spinks[2,4], O. Stanley[2,3], A. E. Stuchbery[2,4], G. N. Taylor[2,3], P. Urquijo[2,3], A. G. Williams[1,2], Y. Y. Zhong[2,4], and M. J. Zurowski[2,3]

[1] Department of Physics, The University of Adelaide, Adelaide, SA 5005, Australia
[2] ARC Centre of Excellence for Dark Matter Particle Physics, Adelaide, Australia
irene.bolognino@adelaide.edu.au
[3] School of Physics, The University of Melbourne, Melbourne, VIC 3010, Australia
[4] Department of Nuclear Physics and Accelerator Applications, The Australian National University, Canberra, ACT 2601, Australia
[5] School of Engineering, Swinburne University of Technology, Hawthorn, VIC 3122, Australia
[6] Centre for Astrophysics and Supercomputing, Swinburne University of Technology, Hawthorn, VIC 3122, Australia

Abstract. The SABRE (Sodium iodide with Active Background REjection) experiment aims to detect an annual rate modulation from dark matter interactions in ultra-high purity NaI(Tl) crystals in order to provide a model independent test of the signal observed by DAMA/LIBRA. It is made up of two separate detectors; SABRE South located at the Stawell Underground Physics Laboratory (SUPL), in regional Victoria, Australia, and SABRE North at the Laboratori Nazionali del Gran Sasso (LNGS). SABRE South is designed to disentangle seasonal or site-related effects from the dark matter-like modulated signal by using an active veto and muon detection system. Ultra-high purity NaI(Tl) crystals are immersed in a linear alkyl benzene (LAB) based liquid scintillator veto, further surrounded by passive steel and polyethylene shielding and a plastic scintillator muon veto. Significant work has been undertaken to understand and mitigate the background processes, that take into account radiation from the detector materials, from both intrinsic and cosmogenic activated processes, and to understand the performance of both the crystal and veto systems. SUPL is a newly built facility located 1024 m underground (∼2900 m water equivalent) within the Stawell Gold Mine and its construction has been completed in mid-2022. The laboratory will house rare event physics searches, including the upcoming SABRE dark matter experiment, as well as measurement facilities to support low background physics experiments and applications such as radiobiology and quantum computing. The SABRE South commissioning is expected to occur in 2023. This paper describes the setup and projections for the experiment, and the description of the underground laboratory.

H. J. Choi et al. (Eds.): APPC 2022, *Proceedings of the 15th Asia Pacific Physics Conference*, pp. 159–166, 2024.
https://doi.org/10.1007/978-981-96-0191-2_16

Keywords: dark matter · model-independent annual modulation signature · low-background NaI(Tl) scintillators · SUPL

1 Introduction

Dark Matter (DM) represents one of the most important open problems in modern physics [1]. One search technique is direct detection which attempts to observe the recoil of a target after scattering with a DM particle, hypothesised as a WIMP (Weakly Interacting Massive Particles) with masses on the Gev/c^2-TeV/c^2 and cross sections on the weak scale. This interaction rate should modulate annually, with a sinusoidal trend, due to the rotation of the Earth around the Sun as it moves through the galactic dark matter halo. To date, only one direct detection experiment, DAMA/LIBRA, has claimed to observe a modulation compatible with that expected from DM [2]. However, this signal is in tension with every other direct detection experiment; all of which report null results for a standard WIMP [3]. These are possible explanations: the majority of experiments do not use the same target material as DAMA/LIBRA and the current experiments with sodium iodide crystals do not achieve the same sensitivity [4,5]. Further, it can be argued that the modulation observed by DAMA/LIBRA is due to seasonal effects. Indeed, muons also present a modulation with a maximum in summer and a minimum in winter, as they are affected by atmospheric pressure and temperature. [6]. The SABRE (Sodium iodide with Active Background REjection) Collaboration was formed to provide a model independent test of this DAMA/LIBRA signal. It foresees two detectors located in two different hemispheres: SABRE North at Gran Sasso National Laboratory (LNGS), in Italy, and SABRE South at the Stawell Underground Physics Laboratory (SUPL), in Australia. The two detectors have a number of common features as they are centred around the same detector module concept, and use common simulation, DAQ, and software frameworks. The two detectors differ in their shielding designs - where SABRE South will utilise a liquid scintillator system for in-situ evaluation and validation of the background to provide background rejection and particle identification. SABRE North has opted for a fully passive shielding since the organic scintillator has been phased out by LNGS. The ultra-low background components will achieve the same sensitivity as DAMA/LIBRA; moreover, the site in the Southern Hemisphere will allow for the exclusion of systematic seasonal effects.

2 The SABRE South Setup

SABRE South is made up of three different subdetector systems: the NaI(Tl) crystal detector system, the liquid scintillator veto system, and the muon paddle detectors. The crystals and the liquid scintillator veto system are further shielded by steel and polyethelyne walls. The full experimental setup is shown in Fig. 1. The experiment can host seven NaI(Tl) cylindrical crystals 25 cm long and 5 cm in radius resulting in a mass

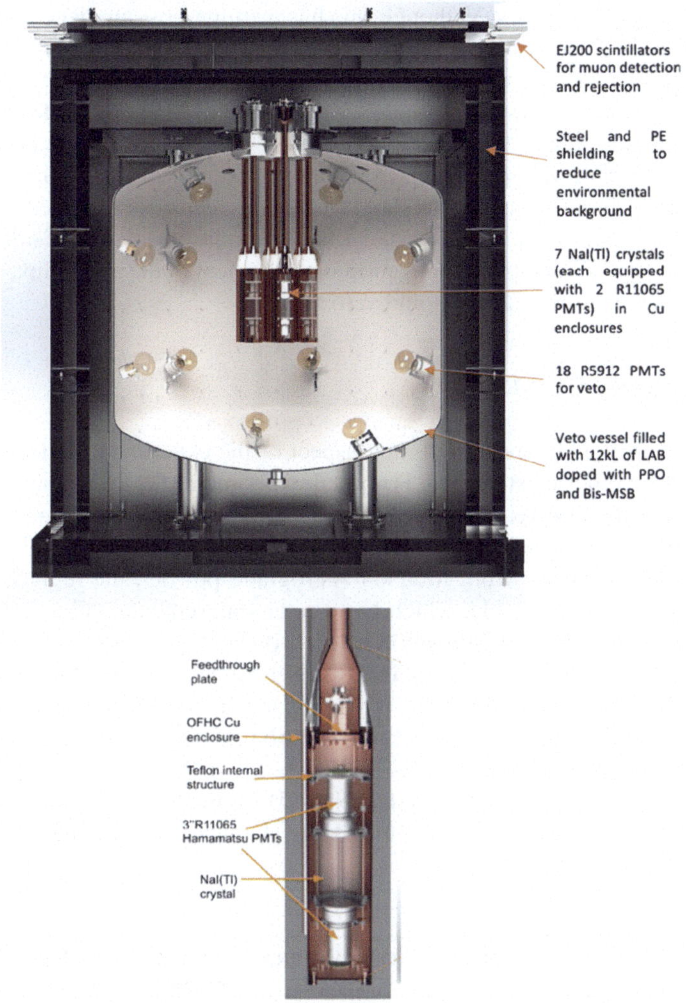

Fig. 1. Rendering of the SABRE South detector (top). Detail of the crystal enclosure (bottom).

of 7.2 kg per crystal (50.4 kg in total). Each crystal is wrapped in PTFE foil, mounted on by PTFE crystal holders, and coupled to a Hamamatsu R11065 PMT (7.6 cm diameter). Crystals are then encapsulated in cylindical oxygen-free high-thermal-conductivity copper enclosures flushed with nitrogen.

Enclosures are submersed into a liquid scintillator which is primarily used to tag and remove high energy decay products such as those from ^{40}K. The system has a 4π coverage, made up of 12 kL of linear alkyl benzene (sourced from JUNO) doped with PPO and Bis-MSB. Around the sides of the steel vessel this is encased in eighteen 20.4 cm Hamamatsu R5912 PMTs that are sampled at a rate of 500 MS/s. Based on preliminary optical simulations in Geant4, this system is expected to have a light yield of approximately 0.12 photoelectrons/keV, though this has a strong position dependence. With a threshold of 50 keV the inclusion of this system is able to reduce the total background by 27%, providing a background of less than 1 cpd/kg/keV [7].

The muon detection system is made up of eight 3 m long EJ200 detector paddles and has a total coverage of 9.6 m^2 above the main vessel. Each paddle is coupled to a Hamamatsu R13089 PMT on each end and sampled at a rate of 3.2 GS/s. Calibrations are still ongoing, but the energy threshold of these detectors is expected to be around 1 MeV. The detectors have approximately 400 ps timing resolution resulting in 5 cm position resolution, allowing for a long term measurement of the muon flux, and particle identification when used with the liquid veto system.

3 High-Purity Crystals

Crystal radiopurity is the most important aspect of the experiment. Indeed, as already demonstrated by other existing NaI(Tl)-based experiments,[2,4,5], a large fraction of the background for DM search comes from residual radioactive contaminants in the crystal themselves, especially ^{40}K and ^{210}Pb; the latter is a daughter of the uranium decay. Crystals are grown from Merck's Astrograde powder, the highest purity NaI powder commercially available, which has a potassium contamination below 10 ppb, and uranium and thorium contamination below 1 ppt.

The SABRE collaboration, through an R&D effort based out of Princeton in collaboration with Radiation Monitoring Devices company, has developed some of the lowest background crystals in the world with only 4.6 ppb of potassium (Fig. 2). The analysed crystal mass is equal to 4.3 kg.

RMD recently grew a NaI(Tl) crystal (NaI-35) for the SABRE South collaboration of a final mass of 3.3 kg which is currently being characterized at LNGS. Its light yield is ∼ 11.6 Photoelectrons/keV.

	NaI-33	DAMA/LIBRA crystals	ANAIS crystals	COSINE crystals
LY [phe/keV]	12.1 ± 0.2	6-10	15	15
FWHM/E @59.5 keV	13%	16%	11%	12%
^{238}U [ppt]	< 0.5	0.7-10	0.2-0.8	< 0.02-0.12
^{232}Th [ppt]	< 0.5	0.5-7.5	0.1-1	0.3-2.4
Alpha rate [mBq/kg]	0.54 ± 0.01	0.08-0.12	0.7-3.15	0.74-3.20
^{nat}K [ppb] (from ICP-MS)	4.6 ± 0.2	< 20	17-43	17-82

Fig. 2. NaI-33 crystal measurements [8] in red compared the ones from DAMA/LIBRA [9], ANAIS [10] and COSINE [11]. Potassium, Thorium and Uranium concentrations are very low respect to the other NaI(Tl) experiments.

4 Background Model and Sensitivity

A full detector simulation was conducted to provide a background model for the radioactive contamination [7]. The total radioactive background is expected to be less than 0.72 cpd/kg/keV (Table 1), dominated by NaI background (radiogenic and cosmogenic) with ^{40}K effectively suppressed with ∼80% efficiency [7]. This meets SABRE South's goal of having less than 10% of background from non-crystal sources. Simulated external backgrounds, shielding, LS bulk, and veto PMTs contribute less than 10^{-3} cpd/kg/keV.

These simulations also indicated that as well as providing shielding and vetoing, the liquid scintillator system also allows for in-situ measurements of background contamination to help inform fits due to well defined correlations in the liquid scintillator and crystals, an example of which is shown in Fig. 3.

The projected sensitivity of SABRE South assuming this background model from simulated radioactivity and a total crystal mass of 50 kg is shown in Figs. 4 and 5 assuming a standard spin independent WIMP and showing the evolution of discovery and exclusion power for the DAMA/LIBRA modulation. Based on these results SABRE South will be able to refute the interpretation of the DAMA/LIBRA modulation as a dark matter signal with 3σ CL. In the event of observation of the annual modulation, this signal would reach a significance of 5σ CL with two full years of data.

5 The Stawell Underground Physics Laboratory

SABRE South is to be placed at the recently completed Stawell Underground Physics Laboratory (SUPL). This is located in Western Victoria, Australia, 240 km from Melbourne. The laboratory is 1025 m below ground with a flat over burden of basalt, providing almost 3 km of water equivalent shielding, and a muon flux similar to that of Boulby,

Table 1. Background rate in the dark matter region for the SABRE South components, and the corresponding veto efficiency.

	Rate [cpd/kg/kev]	Veto Efficiency [%]
Crystal radiogenic	$5.2 \cdot 10^{-1}$	13
Crystal cosmogenic	$1.6 \cdot 10^{-1}$	40
Crystal PMTs	$3.8 \cdot 10^{-2}$	60
PTFE wrap	$4.5 \cdot 10^{-3}$	13
Enclosures	$3.2 \cdot 10^{-3}$	85
Conduits	$1.9 \cdot 10^{-5}$	96
Liquid scintillator	$4.9 \cdot 10^{-8}$	> 99
Steel vessel	$1.4 \cdot 10^{-5}$	> 99
Veto PMTs	$1.9 \cdot 10^{-5}$	> 99
Shielding	$3.9 \cdot 10^{-6}$	> 99
External	$O(10^{-4})$	> 99
Total	$7.2 \cdot 10^{-1}$	27

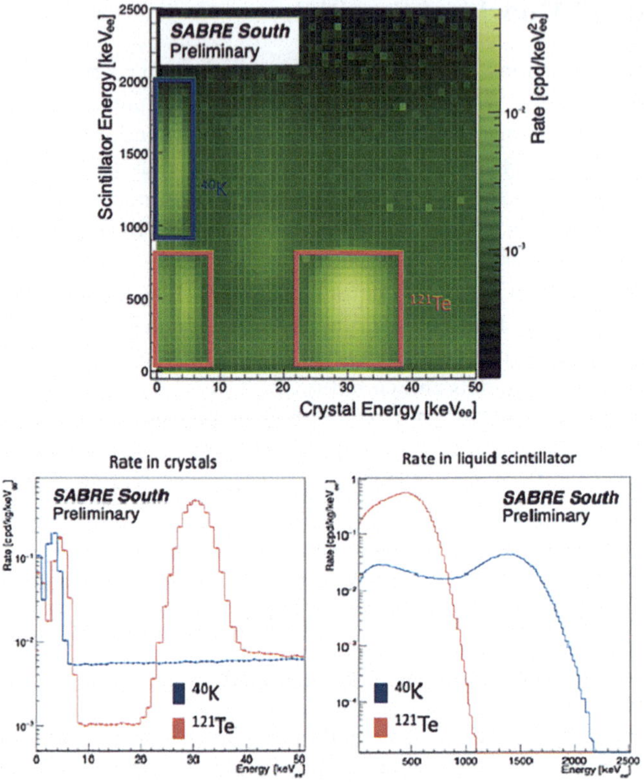

Fig. 3. Top: energy depositions in both the crystal and liquid scintillator. Indicated in blue and pink are events due to ^{40}K and ^{121}Te. Bottom left: rate recorded in crystals from ^{40}K and ^{121}Te. Bottom right: rate recorded in liquid scintillator from ^{40}K and ^{121}Te.

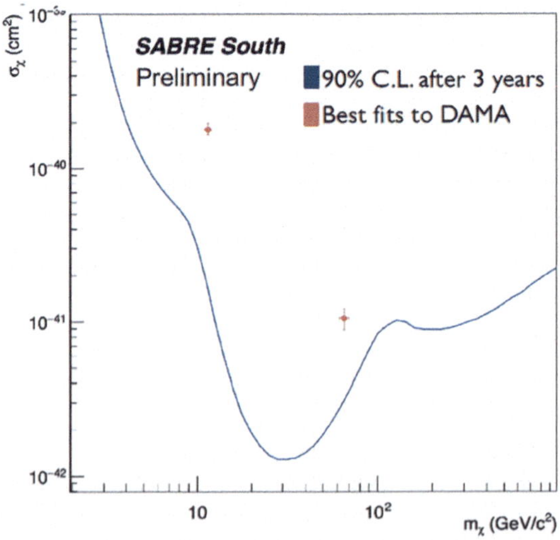

Fig. 4. 90% C.L. assuming a standard spin independent WIMP, a SABRE South total crystal mass of 50 kg and background of 0.72 cpd/kg/keV.

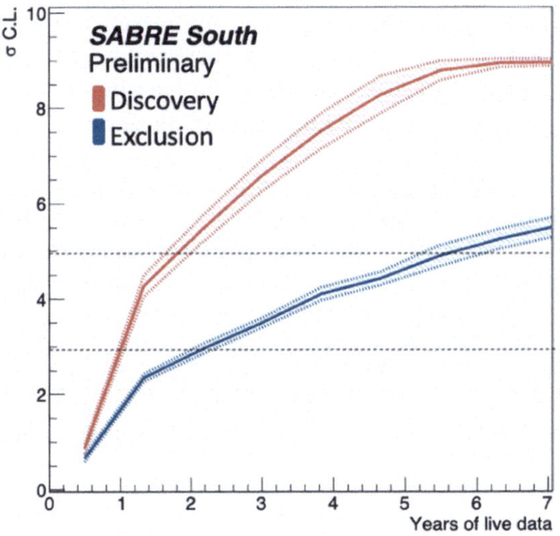

Fig. 5. Exclusion and discovery power of SABRE South for the DAMA/LIBRA modulation.

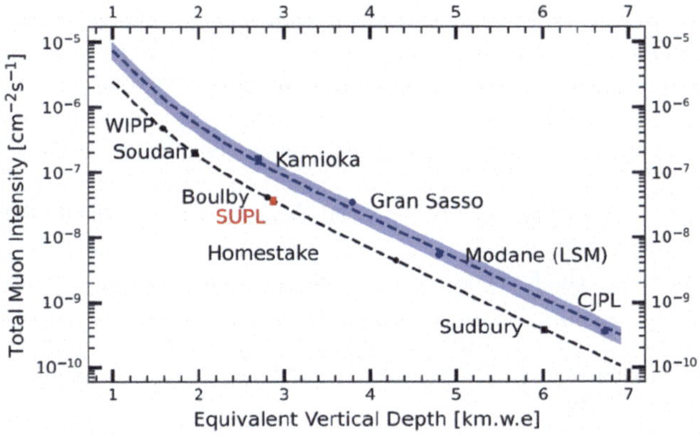

Fig. 6. Muon flux at a number of underground laboratories.

Fig. 6. Internal walls are pinned with steel, sprayed with low radioactivity shotcrete and coated with Tekflex to keep the radioactivity level low. Construction was completed in mid-2022; background muon, gamma, and neutron measurements are planned to start in late 2022.

6 Conclusions

SABRE South is part of the SABRE Collaboration designed to test DAMA/LIBRA modulation. SABRE South is the first dark matter direct-detection experiment in the Southern Hemisphere and will be located inside the new SUPL underground laboratory.

High purity crystals and large active veto give ultra-low background of 0.72 cpd/kg/keV. SABRE South will have 5σ discovery (3σ exclusion) power to a DAMA-like signal with 2 years of data taking. Construction will begin later this year with commissioning expected mid-2023, after which data taking will commence.

Acknowledgments. The SABRE South program is supported by the Australian Government through the Australian Research Council (Grants: CE200100008, LE190100196, LE170100162, LE160100080, DP190103123, DP170101675, LP150100705). This research was partially supported by Australian Government Research Training Program Scholarships and Melbourne Research Scholarships. This research was supported by The University of Melbourne's Research Computing Services and the Petascale Campus Initiative. We thank the SABRE North collaboration for their contribution to the SABRE South experiment design and to the simulation framework. We also thank the Australian Nuclear Science and Technology Organisation for the assistance with the material screening and the measurement of background radiation at SUPL.

References

1. Bertone, G., Hooper, D.: History of dark matter. Rev. Mod. Phys. **90**, 045002 (2018)
2. Bernabei, R., et al.: The dark matter: DAMA/LIBRA and its perspectives. arXiv:2110.04734
3. Workman, R.L., et al.: (Particle Data Group), Prog. Theor. Exp. Phys. 2022, 083C01 (2022)
4. Adhikari, G., et al.: Three-year annual modulation search with COSINE-100, arXiv:2111.08863
5. Amaré, J., et al.: Annual modulation results from three-year exposure of ANAIS-112. Phys. Rev. D **103**, 102005 (2021)
6. Agostini, M., et al.: Modulations of the cosmic muon signal in ten years of Borexino data, JCAP02, 046 (2019)
7. Barberio, E., et al.: Simulation and background characterisation of the SABRE South experiment. arXiv:2205.13849
8. Characterization of SABRE crystal NaI-33 with direct underground counting. Eur. Phys. J. C **81**, 299 (2021)
9. Bernabei, R., et al.: The DAMA/LIBRA apparatus, NIMA 592(3) (2008)
10. Amaré, J., et al.: Analysis of backgrounds for the ANAIS-112 dark matter experiment. Eur. Phys. J. C **79**, 412 (2019)
11. Adhikari, G., et al.: Search for a Dark Matter-Induced Annual Modulation Signal in NaI(Tl) with the COSINE-100 Experiment. Phys. Rev. Lett. **123**, 031302 (2019)

Particles and Fields

Explore the Origin of SSB from Adaptive Perturbation Method

Chen-Te Ma(✉)

Asia Pacific Center for Theoretical Physics, Pohang University of Science and Technology, Pohang, Gyeongsangbuk-do 37673, South Korea
yefgst@gmail.com

Abstract. This paper is a review of arXiv: 2205.00414. The spontaneous symmetry breaking is crucial for inducing a non-trivial vacuum. We focus on the (3+1)d ϕ^4 theory to explore the origin. The study needs efficient analysis in the strong-coupling regime. We quantize the interacting scalar field, which introduces a renormalized mass. This approach is equivalent to resuming the Feynman diagrams. The perturbation method matches the lattice solutions in the strongly coupled regime. We show the phase boundary and analyze the transition from the RG flow. This result suggests that the ϕ^4 theory alone does not experience SSB. Hence we expect that introducing other matters is necessary.

Keywords: Vacuum · ϕ^4 Theory · Strong-Coupling Regime · Effective Mass · Resummation

1 Introduction

The gauge boson masses in the Standard Model originate from the coupling between the Higgs field and the gauge bosons [1]. The mechanism induces a non-zero vacuum expectation value (VEV) to the (charged) Higgs field. The observed non-zero VEV in the Higgs mechanism breaks the global symmetry. This phenomenon is, in general, *spontaneous symmetry breaking* (SSB).

The one-loop effective action in (3+1)d $\lambda\phi^4$ theory generates the SSB when the scalar mass vanishes [2]. However, the weak-coupling expansion is not realizable for the massless regime [2]. The generation of a non-trivial vacuum from the quantum correction becomes subtle. This issue arsed due to the lack of a feasible computation method in a strongly coupled theory. In Quantum Mechanics (QM), one developed the adaptive perturbation method beyond the weak coupling regime [3,4]. This computation method only has the Hamiltonian formalism. The computation is troublesome, and the formulation does not have the manifest Lorentz symmetry. Here we propose a Lagrangian formalism with the manifest Lorentz symmetry. The central question that we would like to address in this letter is the following: *Uncovering the Origin of SSB from Adaptive Perturbation Method?*

In this review paper [5], we first minimize the vacuum energy for the renormalized mass $\gamma(\vec{k})$. Then the scalar field acting on the vacuum state allows the plane-wave solution [5]. The weak-coupling solution for $\gamma(\vec{k})$ shows the bubble resummation

H. J. Choi et al. (Eds.): APPC 2022, *Proceedings of the 15th Asia Pacific Physics Conference*, pp. 169–173, 2024.
https://doi.org/10.1007/978-981-96-0191-2_17

[5]. Our perturbation result matches the lattice simulation even when the coupling is strong in Ref. [5]. Therefore, the analytical result from the perturbation is reliable. We showed the condition of phase boundary (order parameter is the condensation) [5]. In Ref. [5], the result showed no SSB from the Renormalization Group (RG) flow. We should include other matters to induce the non-trivial vacuum [5].

2 Dynamics

We begin from the Hamiltonian of $\lambda\phi^4$,

$$\mathrm{H} = \int d^3x\, \left(\frac{1}{2}\Pi^2(x) + \frac{1}{2}\partial_j\phi(x)\partial_j\phi(x) + \frac{m^2}{2}\phi^2(x) + \frac{\lambda}{4!}\phi^4(x)\right), \tag{1}$$

where the index $j = 1, 2, 3$ denotes the space dimensions. The parameters m and λ are the bare mass and the (positive) coupling constant, respectively, and Π is the conjugate momentum. The ∂_0 and ∂_j are the time and spatial derivatives, respectively. The commutation relation between ϕ and Π is:

$$\begin{aligned} [\phi(\vec{x}), \Pi(\vec{y})] &= i\delta^{(3)}(\vec{x} - \vec{y}), \\ [\phi(\vec{x}), \phi(\vec{y})] &= [\Pi(\vec{x}), \Pi(\vec{y})] = 0. \end{aligned} \tag{2}$$

In the standard quantization, the solution of a free scalar field is a superposition of plane wave

$$\phi(x) = \int \frac{d^3\vec{k}}{(2\pi)^3} \frac{1}{\sqrt{2E_{\vec{k}}}} (a_{\vec{k}} e^{ik\cdot x} + a^\dagger_{\vec{k}} e^{-ik\cdot x}) \,, \tag{3}$$

with

$$E_{\vec{k}} \equiv \vec{k}^2 + m^2, \; k \cdot x \equiv -E_{\vec{k}} x^0 + \vec{k} \cdot \vec{x}. \tag{4}$$

The $a_{\vec{k}}$ and $a^\dagger_{\vec{k}}$ are annihilation and creation operators for each momentum $\vec{k}$, respectively.

We will show that the interacting scalar field allows the plane wave solution. The scaling transformation acting on ϕ does not change the commutation relation [3,4]. Therefore, we can introduce a renormalized mass $\gamma(\vec{k})$ (an even function for $\vec{k}$ and non-negative) to the scalar and momentum fields:

$$\begin{aligned} \phi(x)|0\rangle &\equiv \left(\int \frac{d^3\vec{k}}{(2\pi)^3} \frac{1}{\sqrt{2\gamma(\vec{k})}} (a_{\vec{k}} e^{ikx} + a^\dagger_{\vec{k}} e^{-ikx})\right)\Big|0\Big\rangle, \\ \Pi(x)|0\rangle &\equiv \partial_0\phi(x)|0\rangle. \end{aligned} \tag{5}$$

The $|0\rangle$ is the associated vacuum annihilated by $a_{\vec{k}}$. In general, the interacting fields cannot have the plane wave solution. We will determine $\gamma(\vec{k})$ to allow the solution when the scalar field acts on the vacuum state. Two solutions of ϕ are given by the Bogoliubov transformation $U(\theta)$ [3,4]:

$$U(\theta) a_{\vec{k}} U^{-1}(\theta) \to a_{\vec{k}}; \; U(\theta)|0\rangle \to |0\rangle. \tag{6}$$

Although the transformation changes the operators, the expectation value of all observables is invariant.

We choose the renormalized mass by minimizing the vacuum energy $\langle 0|H|0\rangle$. The saddle point equation from $\gamma(\vec{k})$ is:

$$\gamma^2(\vec{k}) = \vec{k}\cdot\vec{k} + \gamma^2(0),\ \gamma^2(0) = m^2 + \frac{\lambda}{4}\int \frac{d^3\vec{k}}{(2\pi)^3}\frac{1}{\gamma(\vec{k})}\,. \tag{7}$$

Because the $\gamma(\vec{k})$ is the adaptive parameter for minimizing energy, the name of this method is the adaptive perturbation method [3,4].

In QM, the unperturbed Hamiltonian H_0 selects terms with an equal number of $a^\dagger$ and a. This choice shows solvability (although it contains the coupling terms). We also consider an equal number of creation and annihilation operators in H_0 for QFT. The commutator of H_0 and $a^\dagger_{\vec{p}}$ shows a non-linear term. This term shows that the scalar field does not allow the plane wave solution. However, we are only interested in the expectation value. When the commutator acts on the vacuum state, the non-linear term vanishes. Hence when computing the observable, the plane wave solution is applicable to the interacting scalar field. The time evolution of fields is the same as in quantizing a non-interacting theory:

$$[H_0, a^\dagger_{\vec{p}}]|0\rangle = \gamma_{\vec{p}}a^\dagger_{\vec{p}}|0\rangle \quad\Rightarrow\quad e^{iH_0t}a^\dagger_{\vec{p}}e^{-iH_0t}|0\rangle = e^{i\gamma_{\vec{p}}t}a^\dagger_{\vec{p}}|0\rangle\,. \tag{8}$$

We use Eq. (7) to reach this result. Therefore, this result does not hold for an arbitrary $\gamma_{\vec{k}}$. The adaptive parameter plays a non-trivial role in obtaining the interacting vacuum.

The unperturbed spectrum is

$$\gamma_{\vec{k}} = \sqrt{\vec{k}\cdot\vec{k} + \gamma^2(0)}. \tag{9}$$

One can organize the computation from the following Lagrangian

$$\begin{aligned}\mathcal{L}_0 &= -\frac{1}{2}(\partial_\mu\phi\partial^\mu\phi + \gamma^2(0)\phi^2),\\ \mathcal{L}_I &= \frac{\lambda}{4}\left(\int\frac{d^4q_E}{(2\pi)^4}\frac{1}{q_E^2+\gamma^2(0)}\right)\phi^2 + \frac{\lambda}{4!}\phi^4,\end{aligned} \tag{10}$$

where $\mathcal{L}_0$ and $\mathcal{L}_I$ are the unperturbed and perturbed parts, respectively. The Euclidean spacetime index is given by $\mu = 0, 1, 2, 3$. The momentum in the Euclidean signature is q_E. In general, the decomposition is different between Hamiltonian and Lagrangian formulations. Because the Lagrangian formulation has the manifest Lorentz symmetry, the result should be better.

We can solve $\gamma^2(0)$ from the weak-coupling expansion, and the solution is given by

$$\begin{aligned}\gamma^2(0) &= m^2 - \frac{\lambda}{2}\int\frac{d^4k}{(2\pi)^4}\frac{i}{k^2+m^2}\\ &-\frac{\lambda^2}{4}\int\frac{d^4k}{(2\pi)^4}\frac{i}{(k^2+m^2)^2}\int\frac{d^4q}{(2\pi)^4}\frac{i}{q^2+m^2} + \cdots\,.\end{aligned} \tag{11}$$

When we substitute this solution with $\gamma^2(0)$, the tree-level propagator is equivalent to resuming bubble diagrams. Therefore, the number of relevant diagrams decreases significantly. Hence adaptive perturbation method is an efficient method for organizing Feynman diagrams.

We can use a similar way to include the case of non-trivial condensation [5]. The renormalized mass and the constant condensation satisfy that [5]:

$$\Gamma^2(k_E) = m^2 + \frac{\lambda}{2}\phi_0^2 + \frac{\lambda}{2}\int \frac{d^4q_E}{(2\pi)^4}\frac{1}{q_E^2+\Gamma^2(q_E)} - \frac{\lambda^2\phi_0^2}{2}\int \frac{d^4q_E}{(2\pi)^4}\frac{1}{q_E^2+\Gamma^2(q_E)}\frac{1}{(k_E-q_E)^2+\Gamma^2(k_E-q_E)}, \quad (12)$$

$$\phi_0\left(m^2 + \frac{\lambda}{3!}\phi_0^2 + \frac{\lambda}{2}\int \frac{d^4k_E}{(2\pi)^4}\frac{1}{k_E^2+\Gamma^2(k_E)}\right) = 0. \quad (13)$$

Hence this method is general enough to study SSB. The non-trivial condensation $\phi_0 \neq 0$ introduces the momentum dependence to the renormalized mass $\Gamma(k_E)$. The coupled equations for momenta are complicated but computable. We can solve them and do the perturbation to compute the correlation functions. To compare with lattice simulations, we replace the propagator with the lattice propagator in the perturbation formula

$$\int \frac{d^4k_E}{(2\pi)^4}\frac{1}{k_E^2+\gamma^2(0)} \rightarrow \int \frac{d^4k_E}{(2\pi)^4}\frac{1}{\gamma^2(0)+\sum_{\mu=1}^{4}(2--2\cos(k_{E,\mu}))}. \quad (14)$$

Here we choose the unit lattice spacing. The perturbation results precisely match the lattice simulation up to $\lambda = 16$ [5]. Therefore, the perturbation analysis does not lose reliability in the strongly coupled regime.

3 Phase Boundary

The renormalized mass is continuous across the two phases ($\langle\phi\rangle = 0$ and $\langle\phi\rangle \neq 0$). Therefore, we can show the phase boundary by solving the $\Gamma^2(k_E)$ using Eqs. (12) and (13) with $\phi_0 \rightarrow 0$. The continuity implies that the parameters of the phase boundary satisfy the equation

$$m^2 = -\frac{\lambda}{2}\int \frac{d^4k_E}{(2\pi)^4}\frac{1}{k_E^2}. \quad (15)$$

This equation also implies that the phase transition happens when $\gamma^2(0) = 0$. The parameters of the phase boundary in the lattice theory satisfy that

$$m^2 = -\frac{\lambda}{2}\int \frac{d^4k_E}{(2\pi)^4}\frac{1}{\sum_{\mu=1}^{4}(2--2\cos(k_{E,\mu}))}. \quad (16)$$

One solved the integration equation to obtain the phase boundary on the lattice in Ref. [5]. We will study the RG flow to investigate whether the SSB happens in a physical region.

4 RG Flow

We begin from the symmetric phase $\phi_0 = 0$. The relationship between the physical mass and the renormalized mass is [5]

$$m_p^2 = \gamma^2(0) - \frac{\lambda_p^2}{3!}\frac{1}{(16\pi^2)^2}\int_0^1 dx_1 \int_0^{1-x_1} dx_2 \left[\frac{\Lambda^4}{\alpha_1\alpha_2\big((1-\alpha_3)m_p^2 + (\alpha_1+\alpha_2)\Lambda^2\big)} + \frac{(1-\alpha_3)m_p^2}{\alpha_1^2\alpha_2^2}\ln\left(\frac{(1-\alpha_3)m_p^2\big((1-\alpha_3)m_p^2 + (\alpha_1+\alpha_2)\Lambda^2\big)}{\big((1-\alpha_3)m_p^2+\alpha_1\Lambda^2\big)\big((1-\alpha_3)m_p^2+\alpha_2\Lambda^2\big)}\right)\right], \tag{17}$$

where

$$\begin{aligned} \alpha_1 &= x_1 + x_3, \\ \alpha_2 &= \frac{x_1x_2 + x_2x_3 + x_3x_1}{x_1+x_3}, \\ \alpha_3 &= \frac{x_1x_2x_3}{x_1x_2 + x_2x_3 + x_3x_1}, \\ x_1 + x_2 + x_3 &= 1. \end{aligned} \tag{18}$$

We regularize the integration from the momentum cut-off Λ is a momentum cut-off. The λ_p is the physical coupling constant. This equation implies that $\gamma^2(0)$ is not negative when $m_p^2 \geq 0$ [5]. The condition of the phase boundary $\gamma^2(0) = 0$ is not compatible with $m_p^2 \geq 0$ when $\lambda_p \neq 0$ [5]. Hence the phase boundary corresponds to the Gaussian fixed-point as expected [5]. The result of renormalizable group flow cannot generate non-trivial condensation [5]. Therefore, the self-interaction cannot induce SSB. It is necessary to include other matters for obtaining SSB.

Acknowledgments. The author acknowledges the YST Program of the APCT and would like to thank Nan-Peng Ma for his encouragement.

References

1. Goldstone, J.: Nuovo Cim. **19**, 154–164 (1961)
2. Coleman, S.R., Weinberg, E.J.: Phys. Rev. D **7**, 1888–1910 (1973)
3. Weinstein, M.: Adaptive perturbation theory. I. Quantum mechanics. [arXiv:hep-th/0510159 [hep-th]]
4. Weinstein, M.: Nucl. Phys. B Proc. Suppl. **161**, 238-247 (2006) [arXiv:hep-th/0510160 [hep-th]]
5. Ma, C.T., Pan, Y., Zhang, H.: Explore the Origin of SSB from Adaptive Perturbation Method. [arXiv:2205.00414 [hep-th]]

Institutionalization of High-Energy Physics in Japan

Yoshinobu Takaiwa[1(✉)] and Kohji Hirata[2]

[1] Archives Office, Tsukuba, Japan
yoshinobu.takaiwa@gmail.com

[2] Accelerator Laboratory, High Energy Accelerator Research Organization (KEK), Tsukuba, Ibaraki, Japan

Abstract. The high-energy physics (HEP) in Japan started around 1960s by performing experiments using ES at INS and making the "Future Plan of Nuclear Physics Research" to build a high energy accelerator, and the community of the HEP physicists was also created by performing the future plan. This process completed when the National Laboratory for High Energy Physics (KEK) was established as an Inter-University Research Institute, which was a research facility of a new type. In this article the history of how they were established and some consideration on its historical background.

Keywords: High Energy Physics · Future Plan · Researchers' Organization · Inter-University Research Institute

1 Introduction

The High Energy Physics (HEP) is a typical big science in the sense that it requires large high energy accelerators.

In Japan, during the period of the Occupation after the War, the nuclear physicists spent hard times, and some ten years after they hoped to do HEP. For this purpose they demanded the Government or social establishment to build a high energy accelerator for particle physics experiments. In this paper the term high energy accelerator represents the accelerators with newer technologies than the cyclotrons which have to take relativistic motion of high energy beam particles, typically higher than 1 GeV, into account and the synchrotron is one of such machines. Although Japanese nuclear physicists had experiences of building cyclotrons, the synchrotron was quite new to them. It was necessary not only to learn it but also to create a group of physicists who could work together for the project.

Since by the defeat in the War the social structures changed very much, the academic system of Japan was modified; the Science Council of Japan (日本学術会議. SCJ, in short) was established in 1949. The SCJ was to make advices to the Government on the issues related to the academic system and academic activities. The members of the SCJ were elected from among all the scientists in Japan. Among other things, its

H. J. Choi et al. (Eds.): APPC 2022, *Proceedings of the 15th Asia Pacific Physics Conference*, pp. 174–180, 2024.
https://doi.org/10.1007/978-981-96-0191-2_18

important job was to give recommendations to the Government on some specific matters, according to the requests of members who were representing the all academicians of the country. The SCJ was sometimes called as the Diet of academicians.

In this new system, having organizations of researchers had essential meaning in creating and promoting new projects or facilities. The organization of HEP in Japan had not been created, yet, and whether the new project they proposed went well or not might depend on how its formation was made. The term institutionalization of HEP in this context means the formation of organization of researchers' community of HEP which is capable of promoting its activities, administrating the community, and, on behalf of it, interacting with external social structures in which the research activities are embedded, typically the SCJ and the Government.

The interactions between academic activities and social systems are always interesting subjects from view points of science history and sociology. This paper describes how such institutionalization of HEP occurred in Japan as an example of historical studies of such interactions.

The authors previously published two papers related to the current subject [1,2] (in Japanese), and in this paper the essence of them is summarized with supplemental discussion of historical background and the some implications to the HEP researches after it.

2 High Energy Accelerators

The idea of synchrotron was first conceived in US and Europe around 1945 and it grew up rapidly to make it into real facilities with proton beams such as Cosmotron at Brookhaven (1953) and Bevatron at Berkeley (1954) [3]. Knowing this, ambitious Japanese physicists, some of whom had experiences of building cyclotrons, wanted to chase them when the situation allowed.

2.1 Electron Synchrotron

The studies of synchrotrons by Japanese physicists started in late 1950s with the electron synchrotrons (ES) at a few laboratories in Japan, Phys. Department of Tohoku University and Phys. Department of University of Tokyo, and the Institute for Nuclear Study (INS) of University of Tokyo. The former two were built for accelerator studies, and the last one was also built as the study machine but it had been used for real experiments fairly long time from its completion in 1962 until its shut-down in 1997 as a matter of fact.

2.2 The Future Plan with Proton Synchrotron

However, they were not powerful enough for particle physics and the proton accelerator with higher energy was desired. Thus, the proposal to construct a large proton synchrotron (PS) followed, which was included in "The Future Plan of Nuclear Physics Research" recommended to the Government (the Prime Minister) by the SCJ in 1962.

The recommendation did not show detailed parameters for the accelerator but it simply stated that the beam current would exceed 0.1 μA and beam energy, higher than 12 GeV, although much aggressive design parameters were suggested in previously presented reports of the study group [4].

It must be noted that the Future Plan was also proposed to realize the establishment of research institution managed by the autonomy of the researchers; this was an ideal way of realizing "democratic" administration of research activities in the meaning of its historical time.

2.3 Debates on the Machine Type Selection

The Future Plan was not approved immediately, probably due to its large budget which was almost same order of magnitude as the total budget of whole domestic academic activities at that time. Until the recommendation got approved, intense debates on the parameters of the PS were going on. The major point was whether the agressive parameters for the beam intensity proposed by the study group could be realized technically, and the discussion concluded to reject the aggressive designs and adopted modest one from viewpoint of accelerator technologies but raised the beam energy to 40 GeV.

3 Creating New Discipline and Laboratory

The Future Plan eventually would reach at the first stage goal in 1971 by the establishment of National Laboratory for High Energy Physics (KEK) in Tsukuba, and the proton synchrotron was built with the beam energy 8 GeV for the beginning and increased to 12 GeV soon after. The KEK was the first of Inter-University Research Institutes (IURI) in Japan. The IURI is a unique Japanese system for research institutes; the facilities or equipments of the institute are provided to the scientists who want to use them (called users), and representatives of the users may participate the administration of the institutes in a sense.

Forming organization for community of user scientists had special meaning. The KEK, as an IURI, had a committee to discuss its administration issues, and the committee accepted the representatives of the users' organization as its members. The selection of the members of the committee was done by a democratic manner, election by the users community members. The word "democracy" had special and strong meanings in those days that this paper is describing, as described above; the self-governance of the institute by researchers or the community of them was desired. Without knowing this as the background, the original idea of the IURI system may be difficult to understand, although it has become obscured recent days.

3.1 Scientists Communities of Nuclear Physics

Soon after the War there formed three groups or communities of *Nuclear Physics* in a broader sense; the Elementary Particle Theory Group called "Soryushiron-Gurupu" (SG, in short), Experimental Nuclear Physics Group "Genshikaku-Danwakai" (Kaku-Dan), and Cosmic-ray Researchers Committee "Uchusen-Kenkyusha-Kaigi" (CRC).

These three groups in turn constituted a larger organization which had special function in the SCJ as its committee, "Special Committee for Nuclear Research" (原子核特別委員会. Kakutokui, for short). By the fact that this organization was defined as the special committee of the SCJ, its suggestions to make recommendations to the Prime Minister (the Goverment) had special value.

3.2 Joint Research Laboratories Attached to Universities

Before the IURI was introduced, as the result of such recommendations, a new type of research institutes, Joint Research Institute (大学共同利用研究所, JRI), was legally established and it was designed to allow its research activities partially managed under control of the "users", researchers, or the organized group of them, who would perform the research at the laboratory. Each JRI was attached to a national university.

This system of the JRI was first adopted by Research Institute for Fundamental Physics (RIFP, Yukawa Memorial Hall) of Kyoto University for SG and Cosmic Ray Observatory of University of Tokyo for CRC in 1953 (August), followed by the Institute for Nuclear Study (INS) of University of Tokyo in 1955 (July). Although the Kakudan was most benefited by the INS because it had a cyclotron as its main facility, it was virtually for all of three groups.

As the INS was established by the recommendation of the SCJ and the Kakutokui acknowledged itself as the organization of users of it, the Kakutokui made "Subcommittee for INS" (原子核研究所小委員会. Kakushoi, for short) and this acted as the virtual administrative committee of the INS from its beginning 1955 until 1975, as the consequence the SCJ behaved as if it was administrating the INS, which is formally an attached laboratory to the University of Tokyo.

It is apparent that the JRI is the forerunner of the IURI.

3.3 Formation of HEP Community and Accelerator Design

The Future Plan contained programs for all three groups and two new laboratories were recommended; one for nuclear physics in a narrower sense (low energy nuclear physics) for Kakudan and the other for particle physics having a high energy accelerator for "HEP" and the division of ultra-high energy physics for CRC. The program of building the accelerator was promoted by the group of people consisted of interested members of all three groups and there was no organized group yet for HEP at this moment. At this stage those who wanted to build and use the accelerator tried to make informal group for information exchange at least, hoping that it would acquire some political power promoting their plan. This is the preliminary group of the HEP researchers, and it was called "High Energy Physics Club (高エネルギー物理学同好会. HEP-Club)".

When the HEP-Club started the ES of INS had just begun its operartion for experiments and the HEP-Club contributed to make and arrange the schedule of those experiments, so that it worked not only for information exchange but contributed to administration work of the INS to some extent. Thus, members of HEP-Club could accumulate the experiences of doing HEP works.

In 1964, the budget for preliminary studies of the Future Plan, which is mainly for accelerator design, was approved, and the Kakutokui created its subcommittee for acomplishment of the Future Plan, "Soryushi-kenkyujo Junbi-Chousa Iinkai" (素粒子研究所準備調査委員会. SJC, for short), because the new laboratory would be the Elementary Partcle Research Laboratory "Soryushi-Kenkyujo" (素粒子研究所, Soken, for short) and serious work of selection for the type of the accelerator restarted as mentioned in the Sect. 2.

After the accelerator parameters were decided, the leader of the design team was replaced and the team, previously established as the semi-official division of the INS called Soken Junbi Shitu (SJS), was re-organized to complete the design for the realistic and technically achievable accelerator. By this, the team SJS became recognized as the core of the HEP group and the Kakutokui admitted the HEP as independent group besides SG, Kakudan and CRC in 1967. The High Energy Physics Club became a formal constituent sub-group of the Kakutokui of SCJ. This is the big step towards the institutionalization of HEP.

The Kakutokui assumed that the SJC would become a committee of administration of new laboratory, Soken, when it would get established, just as the Kakushoi was for the INS, and the Soken would be also administrated under the Kakutokui of the SCJ.

Now, HEP had its own organization and it might stand by itself, so that, when the Kakutokui could not make proper reaction against the Goverment decision to reduce the budget and, consequently, the beam energy of the accelerator, and to propose different ideas for the administration strategy of the new laboratory, the HEP group admitted the decision and made the creation of KEK possible in spite of the reluctance of the Kakutokui.

3.4 Inter-University Research Institutes

The original idea of the structure to govern the new laboratory and its academic acitivies in the Future Plan, was rather radical in the sense that it must allow self-governance by the researchers avoiding unacceptable control by the Government. This idea emerged from general discussion in the SCJ on proposing the Basic Law of Science-and-Technology Policies, and the experience of INS administration by the Kakushoi.

Whole set of the ideas was not accepted by the Government but the part of it was realized in the law to establish a new laboratory, compromizing with the rest of the legal system. The IURI is the new system of the laboratory, and it is established as the Governmental institution—independent from universities, and the users or representatives of the organized group may participate in a committee for management. The committee oversees and gives advices about the matters restricted to its research activities. In this sense, KEK was, when it started (1971, April), regarded as a shared facility of the HEP physicists belonging to any universities in Japan. The users organization, the HEP-Club consisting of researchers of such universities, created the HEP Committee to represent its opinions to the external world, and it may be able to control research activities of the institute more or less. This is the completed HEP institutionalization.

It should be noted that there was no place for Kakutokui or SCJ formally in the administration system of the IURI. The formal external participation in the administration of the institute were through its committee members nominated by the users organization. The Kakutokui ceased to exist 1972 in the SCJ and INS changed the system from Kakutokui based administration to the system University of Tokyo admits to its attached laboratories.

4 Conclusion

When the plan of HEP research in Japan was proposed, no real acitivities of HEP were seen, except older nuclear physics activities like theoretical physics, low energy nuclear physics experiments using cyclotrons, and cosmic-rays observation. Each of them had its own organization at that time, and they collectively supported the Special Committee for Nuclear Research, Kakutokui, of the SCJ, which discussed the nuclear physics research promotion and it could be regarded as an organization of nuclear physics as a whole.

The Kakutokui proposed the Future Plan of nuclear physics which included a HEP accelerator as its important component. At the same time, the ES of INS which was made as a study machine for new technology, synchrotron, started its operation for experiments. These two backgrounds caused the emergence of the HEP community and prepared the formation of its organization.

The formation progressed through real works of them; contribution to the coordination and scheduling of experiments at ES of INS and the practical and concrete design of the accelerator for the Future Plan, and finally the efforts to establish a new laboratory system by defining the roles of researchers and their organization which contributed to its management to some extent. Thus, the establishment of the new laboratory marked the completion of the institutionalization of HEP.

It may be difficult to generalize this consideration to historically and regionally different conditions, it is hoped that it helps to understand the structure of science and social systems and their interactions.

Acknowledgments. For this work the documents in the Archives Office of KEK, the Sakata Memorial Archival Library of Nagoya University and others are referred and the authors thank these libraries and archives [5,6].

This work was supported by JSPS KAKENHI Grant Numbers JP19K00288 and JP22K00286.

References

1. Hirata, K., Takaiwa, Y.: A comprehensive research group based on the grants-in-aid for scientific research (KAKENHI) and the formation of high energy physics community in Japan. Kagakushi Kenkyu「科学史研究」**56** (No. 281), 12 (2017)
2. Hirata, K., Takaiwa, Y.: Inter-university research institutes and the conflict of two types of autonomy, Kagakushi Kenkyu「科学史研究」**59** (No. 293), 38 (2020)
3. McMillan, E.M.: A history of the synchrotron. Phys. Today **37**, 31 (1984). According to this article, the idea of synchrotron was independently proposed by three people, V. I. Veksler (USSR), M. Oliphant (UK), and E. M. McMillan (USA)
4. Kitagaki, T.: A Proposed 12 BeV Proton Synchrotron. In: Proceedings of International Conference on High Energy Accelerators, Brookehaven, pp. 64–69 (1961)

5. KEK Archives Office: Collection of documents from various sources. (in Japanese) Many documnets are referred and the following list is only samples of them: Minutes of Kakuken Shoiinkai, Minutes of Kakuken Kyojukai, GenshikakuTokubetsuIinkai Sogojimukyokuho, Ko-Enerugi-Dokokai Tsushinn, and Reports of various study groups (1955–)
6. Sakata Memorial Archive Library: Sakata Memorial Aechival Library Catalogue of Materials Vol.1 (rev. 2013)/ Vol.2 (2014). (in Japanese). This library has comprehensive documents of the SCJ (1949—1970)

The Overview and Status of the GBAR Experiment

Bongho Kim(✉)

The Center for Underground Physics, Institute for the Basic Science, 55, Expo-ro, Yuseong-gu, Daejeon 34126, South Korea
bhokim85@ibs.re.kr

Abstract. The GBAR (Gravitational Behaviour of Anti-hydrogen at Rest) experiment aims to perform a free-fall experiment of an ultra-cold anti-hydrogen at the terrestrial gravitational field. An anti-hydrogen ion which is trapped and cooled is neutralized and kicked by photo-detachment laser and free-falls in a vacuum chamber. By measuring secondary particles from anti-hydrogen annihilation after the free-fall motion, the gravitational acceleration of the anti-hydrogen atom can be measured within 1% precision. The GBAR experiment plans to produce anti-hydrogen atom and ion as a first goal. In here, the experimental technique with some detail and the current status of the experiment will be shown.

Keywords: Antimatter · Gravity · Antiproton trap · Antihydrogen · Antihydrogen ion

1 Introduction

After the discovery of anti-particles (positron and anti-proton), the question about anti-gravity arouse. From arising of several theoretical arguments about the anti-gravity since 1960, there have been continuous interests about anti-matter's gravitational interaction [1] and possible cosmological models [2–4]. The gravitational interaction between matter and anti-matter which is a fundamental property of the nature has never been measured precisely. Not only the repulsive interaction, the gravitational interaction between matter and anti-matter which can be slightly different compared with interaction between matters is an intriguing topic.

The gravitational interaction measurement of anti-matter is a confirmation of Weak Equivalence Principle (WEP) for anti-matter. Recently, there's a test of universality of clock, i.e., clock Weak Equivalence Principle (WEPc) violation for anti-matter by measuring anti-proton's cyclotron frequency from the BASE experiment [6]. The measurement constrains hypothetical interactions mediating violations of WEPc for anti-matter to less than 1.8×10^{-7}. As another way to test WEP, a test of universality of free-fall, i.e., free-fall Weak Equivalence Principle (WEPff) test [7] which is a direct measurement of the anti-matter's gravitational

H. J. Choi et al. (Eds.): APPC 2022, *Proceedings of the 15th Asia Pacific Physics Conference*, pp. 181–188, 2024.
https://doi.org/10.1007/978-981-96-0191-2_19

interaction based on free-fall experiment has been studied. The gravitational interaction which is relatively weaker than any other interaction makes it hard to measure by other effects like patch effect, etc. With development of technique about antimatter cooling, it becomes possible to test WEPff by neutral bound state of matter and anti-matter mainly using anti-hydrogen atoms. The ALPHA experiment excludes initial limit for parameter F which is a ratio of gravitational to inertial mass of anti-hydrogen to $-65 < F > 110$ at 95% significance level. There are three experiments AEGIS, ALPHA-g and GBAR which plan to measure anti-hydrogen atom's gravitational acceleration in the terrestrial gravitational field to higher precision below normal gravity [8,9]. Also, there have been some experimental efforts to measure the anti-matter's gravitational interaction by using Muonium [10] and Positronium (Ps) [11].

The GBAR (AD-7) experiment plans to drop a ultra-cold anti-hydrogen as like Galileo's Pisa experiment to measure the gravitational acceleration of anti-matter in the terrestrial gravitational field. For this, firstly, GBAR experiment will produce anti-hydrogen ion by double charge exchange process between an anti-proton and two Pss. The produced anti-hydrogen ion can be cooled down to ~uK temperature by sympathetic cooling with laser cooled ions. By photo-detachment laser to the ultra-cold anti-hydrogen ion, the neutralized anti-hydrogen will be dropped from a final trap and it will free-fall by the gravitational interaction. By strong interaction between anti-hydrogen and matter at the destination of the anti-hydrogen atom, the atom will be annihilated and the secondary particles from the annihilation will give signals to detectors which will be used to get the information of destination point and arrival time. By about 1500 anti-hydrogen free-falling test, the gravitational acceleration within 1% precision is expected to be measured.

2 Experiment Detail

Since the approval by CERN in 2012 [12], the GBAR experiment has been developed. The schematic view of the GBAR experiment is described in the Fig. 1. This is changed from previous summary paper [13] with adding new apparatus. The detail of each apparatus will be explained in the corresponding sections. The GBAR experiment has two main beam lines which are anti-proton beam line and positron beam line. The anti-proton beam line is connected from ELENA ring at CERN which provides 100 keV anti-proton pulsed beam for every 110 s duration. The positron beam line is started from electron linac in the GBAR experimental zone and provides 4 keV positron pulsed beam to positronium target. Two streams of beam line meet in the reaction point for the production of anti-hydrogen ions. The created anti-hydrogen ion by the interaction between anti-proton and Ps is guided to the free-fall chamber (FFC) which has ion traps inside. The detection system surrounding the FFC measures the secondary particles from annihilation.

The picture of the current apparatus is shown in Fig. 2. The apparatus to detect generated anti-hydrogen atom and anti-hydrogen ion has been prepared.

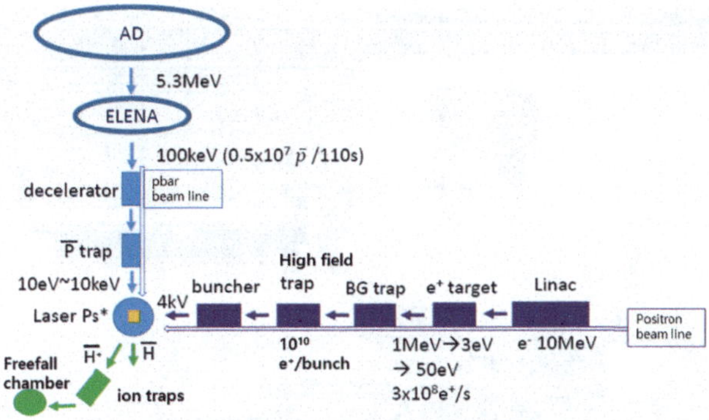

Fig. 1. The schematic view of the GBAR experiment : (sky-blue line) anti-proton beam line started from upside, (violet line) positron beam line started from right side, (green line) anti-hydrogen beam line to free-fall chamber ended at the left bottom

At the anti-proton beam line, the anti-proton decelerator and beam optics for guiding and focusing are prepared. At the positron beam line, the electron linac, positron target&moderator, buffer-gas trap, high field trap and positron beam accelerator are placed. In the reaction chamber, few types of Ps target are prepared for the anti-hydrogen atom and ion production. After the reaction chamber, the Lamb-shift measurement setup [17] and the switchyard are place. The anti-proton trap is mounted at the end of the anti-hydrogen ion beam line for test before installation at the final position in the anti-proton beam line. The current milestone is production of anti-hydrogen atom and anti-hydrogen ion as a first step.

2.1 Positron Beam Line

To produce high positronium density, linac based positron beam line has been developed for enough positron beam intensity. The linac designed for a kinetic energy of 9 MeV in a pulse width duration of 2.85 μs and a repetition rate of up to 300 Hz is installed inside of the concrete bunker [18]. A tungsten target and moderator with biased potential as a mono-energetic positron emitter is mounted inside. The goal of positron beam from the bunker is 3 $\times 10^8 e^+/s$ and has been commissioned for stable operation with full power.

The positron beam is guided to out of linac by transport system and initially trapped and accumulated at the buffer-gas trap. For the higher accumulation, the accumulated positrons are moved to the high-field trap. At the high field trap which is a penning-malmberg trap with 5T superconducting magnet, positron is accumulated up to $\sim 10^{10} e^+/bunch$ and the current reported record in the GBAR experiment is 1.4×10^9 $e^+/bunch$. The accumulated positron is accelerated to a kinetic energy of 4 keV by the positron accelerator after the high-field trap which is a uni-potential lens system with fast switching. After the acceleration, a lens system is used to guide positron bunch into the Ps target cavity.

Fig. 2. The picture of current GBAR experimental zone. The alphabet letters are inserted in the picture for below explanation. The linac is installed at the right side concrete bunker. The positron beam line (b) as marked by violet arrow is started from right side and ended in the sky-blue box. The ELENA ring is hided upside left and the anti-proton beam line (a) as marked by sky-blue arrow is started from upside and ended at the sky-blue box. The reaction position is in the sky-blue box at the end of (a) and (b) beam lines. After the reaction chamber, the anti-hydrogen beam line (c) as marked by green arrow is started from the switchyard to the anti-proton trap and currently the beam line is used to guide anti-proton beam to the anti-proton trap. The anti-proton trap (d) is in the end of the beam line which is a middle of the picture.

2.2 Anti-Proton Beam Line

As a unique anti-proton facility, Antiproton Decelerator (AD) at CERN provides 5.3 MeV anti-proton beam using the proton beam from PS. With commissioning of ELENA ring during last beam time which is an additional storage ring for further deceleration and cooling of the anti-proton beam from AD, it can provide

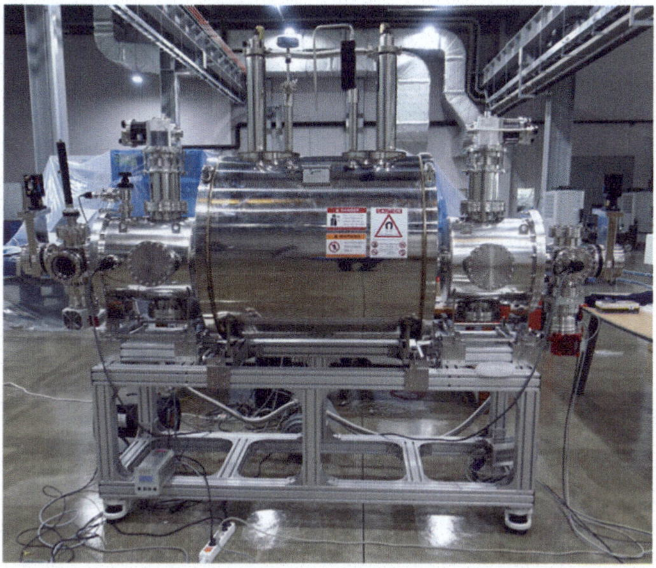

Fig. 3. The picture of the antiproton trap during basic test period in Korea

an anti-proton beam intensity about 5×10^6 with 100 keV kinetic energy to each experiment at the AD hall [14].

To use lower energy of anti-proton beam for the purpose of experiments, most experiments use degrading foil to reduce the anti-proton beam energy. Different with other experiments, the GBAR experiment has developed the anti-proton decelerator using electrostatic lens with fast switching. With the anti-proton decelerator, it's possible to achieve high efficiency of anti-proton deceleration and to have mono energetic anti-proton beam with possible tuning of 1 - 10 keV energy range [16]. The test of deceleration was succeeded and the commissioning of the anti-proton decelerator has been ongoing currently.

As shown in the schematic drawing in Fig. 1, there's a plan to install the anti-proton trap after the anti-proton decelerator finally. But to develop the anti-proton trap and to produce anti-hydrogen atom and anti-hydrogen ion at the same time, the anti-proton trap is installed at the end of the beam line and there are beam optic devices in the final position of the antiproton trap. The beam emittance is increased after the deceleration because only longitudinal momentum is decreased by the deceleration while the transverse momentum is not changed. The optic for proper guiding and focusing in the reaction target is chosen by the anti-proton beam simulation performed using SIMION program.

2.3 Anti-Proton Trap

The anti-proton trap is a newly added apparatus to achieve a better beam quality and stacking power for anti-proton beam. The picture of the anti-proton trap is in Fig. 3. The trap is a penning-malmberg trap with 7T superconducting magnet. The designed function of the trap is anti-proton beam trapping, cooling with radial compression, bunching with acceleration and stacking. Especially, the injection and extraction part of the beam was designed based on simulation study using WARP PIC simulation tool [15]. After basic operation check about temperature and vacuum condition in Korea, the device was moved to CERN and mounted at the end of the antihydrogen ion beam line to confirm the operation by anti-proton beam before the installation at the final position. For the anti-proton cooling, cooled electron by cyclotron radiation is used for sympathetic cooling for anti-protons. The electron loading and cooling have been tested and anti-proton beam test has been ongoing.

2.4 Collision Section

Double charge exchange process which is composed of the anti-hydrogen production between anti-proton and Ps and the anti-hydrogen ion production between anti-hydrogen and another Ps is needed to be happened sequentially. This is to avoid the loss of anti-hydrogen in flight because it's hard to focus and to control the neutral anti-hydrogen beam.

So, the same Ps cloud is used for two interactions and the production rate of the antihydrogen ion is related with square of the Ps density in the interaction point. To mix the anti-proton beam with bunch length 300ns(95%) and high density ortho-Positronium(o-Ps) ($10^{11-12} cm^{-3}$, τ = 142ns, K.E = 48 meV), the cavity shaped Ps target ($1 \times 1x20mm^3$ and $2 \times 2x20mm^3$) is developed. The Si_3N_4 film is used for the positron beam passage and the porous SiO_2 target(30% efficiency) is used for the o-Ps production [20]. The cavity has holes for passage of antiproton beam and other two sides are covered for o-Ps reflection. The generated o-Ps has almost 100% reflection efficiency in the target cavity wall and the the o-Ps cloud can be stayed in the cavity without loss except annihilation by lifetime and escaping to holes. Based on the cross-section calculation [21], a few anti-hydrogen ion is expected to be generated with and without Ps excitation.

Eventually, the antiproton beam needs to pass through the narrow passage. For this, the guiding by several optics (Einzel lenses, Steerer and Triplet quadrupole) has been studied by simulation and tested by real beam. After the interaction between anti-proton and Ps, all species in the anti-proton beam line, i.e. anti-proton, anti-hydrogen atom and anti-hydrogen ion beam stream together to downstream. The switchyard after the reaction chamber splits the beam electrostatically.

The basic method of the anti-hydrogen production and detection was demonstrated at 2018 [22,23] and the effort to produce anti-hydrogen atom and ion has been ongoing.

2.5 Anti-hydrogen Ion Trap and Free-Fall Detection

After guiding to the FFC with proper deceleration, the anti-proton ion is trapped at the RF linear trap by DC switching with high efficiency. After the catching, the anti-hydrogen ion is cooled down to 5∼100mK sympathetically by cooled Be^+ ion cloud [24]. To reduce the temperature less than the Doppler limit temperature, the anti-hydrogen ion is moved to the precision trap to form $Be^+/\bar{H}^+$ ion pair which is cooled by Raman side-band cooling [25]. By the method, the temperature of the anti-hydrogen ion is cooled down to ∼uK temperature range.

After the cooling, the anti-hydrogen ion losses one of the positron by the photo-detachment laser. The anti-hydrogen atom which loses the positron is kicked out from the trap by neutralization. The FFC is a chamber with expected size L = 10 cm and H = 30 cm which is enough to see the free-fall when the normal gravitational acceleration is assumed (the FFC has designed for better detection). After the anti-hydrogen's free-fall motion, the atom hits the FFC wall. Then the antihydrogen will be annihilated in the wall and the secondary particles mainly pions are generated.

The MicroMegas(MM) detector and TOF (time-of-Flight) scintillation detector as a secondary particle detector surround the FFC. The MM detector with 50cm×50cm size and three layers per each plane measures the track of charged secondary particles and can do a vertex reconstruction. The TOF detector [26] which covers 4 planes with 170 cm × 100 cm size per each side (each side composed of 10 scintillation bars($170 \times 10 \times 5\,\mathrm{cm}^3$) measures the arrival time of the secondary particles. The TOF detector can get a precise timing of the annihilation and also can veto the background cosmic-ray muon.

3 Current Status and Plan

The GBAR experiment is relatively new experiment in the anti-proton society and The first operation of electron linac at CERN was the end of 2017. Until now, the experiment has spent time for the machine development. Currently, the positron beam line and the anti-proton beam line without the anti-proton trap have been commissioned. With basic demonstration of anti-hydrogen production and detection technique, the GBAR experiment has tried to produce anti-hydrogen atom first and then anti-hydrogen ion.

With anti-hydrogen atom and ion production and the measurement of the cross-section, the GBAR experiment will aim to cool the anti-hydrogen ion and then do a free-fall experiment although there will be more studies to increase the beam intensity and anti-hydrogen ion amount.

Also, there's studies to reduce the possible uncertainty of the free-fall measurement and to achieve better result by adding some devices [27–29].

4 Conclusions

The GBAR experiment has been developed to measure the gravitational interaction of antihydrogen atom in the terrestrial gravitational field. The antiproton beam line and the positron beam line has been developed and it has been commissioned. With demonstration of the antihydrogen atom and ion production and detection technique at 2018, the GBAR experiment has tried to produce antihydrogen and antihydrogen ion. At the same time, there's efforts to develop the antiproton trap, design of the FFC and the free-fall detection system.

Acknowledgments. This work was supported by IBS-R016-Y1.

References

1. Nieto, M.M., Goldman, T.: Phys. Rep. **205**, 221–281 (1991)
2. Benoit-Levy, A., Chardin, G.: A&A 537, A78 (2012)
3. Villata, M.: Astrophys. Space Sci. **339**, 7 (2012)
4. Hajdukovic, D. S.: Int. J. Theor. Phys. **49**, 1023
5. Walz, J., Hansch, T.: Gen. Relativity Gravitation **36**, 561 (2004)
6. Borchert, M.J., et al.: Nature. **601**, 53–57 (2022)
7. Charlton, M., Eriksson, S., Shorem, G.M.: arxiv:2002.09348
8. Testera, G., et al.: Hyperfine Interact. **233**, 13 (2015)
9. Bertsche, W.A.: Phil. Trans. R. Soc. A **376**, 20170265 (2018)
10. Phillips, T.J.: EPJ web of conferences **181**, 01017 (2018)
11. Cassidy, D.B., Hogan, S.D.: Int. J. Mod. Phys. Conf. Ser. **30**, 1460259 (2014)
12. Chardin, G., et al.: GBAR, Proposal to Measure the Gravitational Behaviour of Antihydrogen at Rest, CERN-SPSC-P-342, 30/09/2011
13. Perez, P., et al.: Hyperfine Interact. **233**, 21–27 (2015)
14. Bartmann, W., et al.: Phil. Trans. R. Soc. A, **376**, 20170266
15. Yoo, K., et al.: JINST **17**, T10003 (2022)
16. Husson, A., et al.: Nucl. Instrum. Methods Phys. Res. A **1002**, 165245 (2021)
17. Blumer, P.: arxiv:2203.16174
18. Charlton, M., et al.: Nucl. Instrum. Methods Phys. Res. A, **985**, 164657 (2021)
19. Blumer, P., et al.: Nucl. Instrum. Methods Phys. Res. A, **1040**, 167263 (2022)
20. Crivelli, P., et al.: Phys. Rev. A **81**, 052703 (2010)
21. Comini, P., et al.: Hyperfine Interactions,**228**, 159–165 (2014)
22. Latacz, B.: arxiv:1905.06404
23. Charlton, M., et al.: CERN-SPSC-2019-005/SPSC-SR-244
24. Hilico, L., et al.: Int. J. Mod. Phys. Conf. Ser. **30**, 1460269 (2014)
25. Schnitzler, W., et al.: Phys. Rev. Lett. **102**, 070501 (2009)
26. Kim, B.H., et al.: Nucl. Instrum. Methods Phys. Res. A **973**, 164162 (2020)
27. Rousselle, O., Cladé, P., Guellati-Khélifa, S., Guérout, R., Reynaud, S.: Phys. Rev. A **105**, 022821 (2022)
28. Rousselle, O., Cladé, P., Guellati-Khélifa, S., Guérout, R., Reynaud, S.: New J. Phys. **24**, 033045 (2022)
29. Crépin, P.-P., Christen, C., Guérout, R., Nesvizhevsky, V.V., Voronin, A.Y., Reynaud, S.: Phys. Rev. A **99**, 042119 (2019)

Status of the SUB-Millicharge ExperimenT (SUBMET) at J-PARC

Hoyong Jeong(✉), Sungwoong Cho, Suyong Choi, Hyunki Moon, Eunil Won, and Jae Hyeok Yoo

Department of Physics, Korea University, Seoul 02841, Republic of Korea
hoyong5419@korea.ac.kr

Abstract. Electric charge quantization is a long-standing question in particle physics. While fractionally charged particles (millicharged particles hereafter) have typically been thought to preclude the possibility of Grand Unified Theories (GUTs), well-motivated dark-sector models have been proposed to predict the existence of millicharged particles while preserving the possibility for unification. Such models can contain a rich internal structure, providing candidate particles for dark matter. Several experiments have searched for millicharged particles (χs), but in the parameter space of the charge (Q) and mass (m_χ), the region of $m_\chi > 0.1$ GeV/c^2 and $Q < 10^{-3}e$ is largely unexplored. SUB-Millicharge ExperimenT (SUBMET) has been proposed to search for sub-millicharged particles using 30-GeV proton fixed-target collisions at J-PARC. The detector is composed of two layers of stacked scintillator bars and photo-multiplier tubes (PMTs) and is proposed to be installed 280 m distance from the target. The main background is expected to be a random coincidence between the two layers due to dark counts in PMTs, which can be reduced significantly using the timing of the proton beam. With $N_{\rm POT} = 5 \times 10^{21}$, the experiment provides sensitivity to χs with the charge down to $7 \times 10^{-5}e$ in $m_\chi < 0.2$ GeV/c^2 and $10^{-3}e$ in $m_\chi < 1.6$ GeV/c^2. This is the regime largely uncovered by the previous experiments. The status of this experiment is presented in this talk.

Keyword: millicharged particle, SUBMET, dark-sector model, J-PARC

1 Motivation

The quantization of electric charge is a long-unanswered question in particle physics. Whereas fractionally charged particles (millicharged particles hereafter) have typically been thought to preclude the possibility of Grand Unified Theories, well-motivated dark-sector models have been proposed to predict the existence of millicharged particles while preserving the possibility for unification. Such models may contain rich internal structures, which means they may be candidates for dark matter. Therefore, the existence of fractionally charged particles can be a stepping stone beyond the standard model (Fig 1).

H. J. Choi et al. (Eds.): APPC 2022, *Proceedings of the 15th Asia Pacific Physics Conference*, pp. 189–195, 2024.
https://doi.org/10.1007/978-981-96-0191-2_20

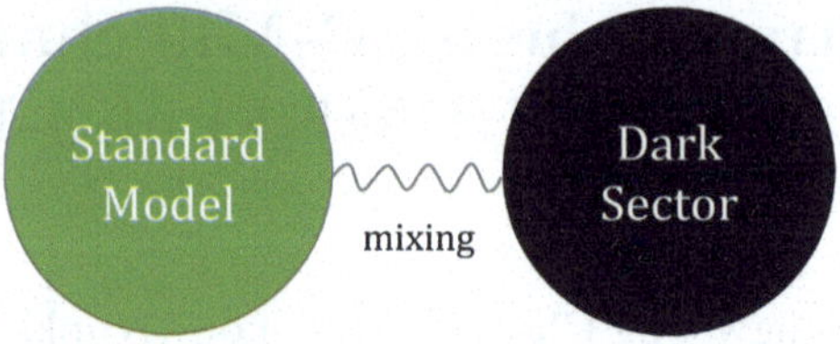

Fig. 1. Mixing between the standard model and the dark sector. See text for details.

One of the simplest extensions to the Standard Model is to introduce an extra $U(1)$ gauge in the dark sector. If there is a new $U(1)$ gauge boson (dark-photon A' hereafter), and if the dark-photon and the standard-model photon B are coupled [12][1], this assumption allows dark fermion (ψ) to interact with the B by a small electromagnetic interaction. The Lagrangian of this mechanism is written in

$$\begin{aligned}\mathcal{L}_{\text{dark sector}} = &-\frac{1}{4}A'_{\mu\nu}A'^{\mu\nu} \\ &+ i\bar{\psi}\left(\partial\!\!\!/ + ie'A\!\!\!/' + im_{\psi}\right)\psi \\ &- \frac{\kappa}{2}A'_{\mu\nu}B^{\mu\nu},\end{aligned} \tag{1}$$

where the last term shows that A' and B kinetically mix with the mixing parameter κ. After replacing $A\!\!\!/'$ with $(A\!\!\!/' + \kappa B_{\mu})$, the total Lagrangian would become

$$\begin{aligned}\mathcal{L} = \mathcal{L}_{\text{Standard Model}} &- \frac{1}{4}A'_{\mu\nu}A'^{\mu\nu} \\ + i\bar{\psi}\left(\partial\!\!\!/ + ie'A\!\!\!/' + i\kappa e'B\!\!\!/ + im_{\psi}\right)\psi.\end{aligned} \tag{2}$$

The charge of the dark fermion is proportional to κ. It makes dark fermion seen as fractional electric charge. In fact, our experiment is sensitive to any fractionally charged particles, not limited to this model.

2 Current Reach

There have been experiments searching for millicharged particles such as electron-fixed target experiments [1], proton-proton collider [2–4], proton-fixed target experiments [5] and neutrino experiments [6,7]. A comprehensive review is in Ref. [8]. Figure 2 shows the region in charge-mass parameter space excluded by the existing experiments. As shown in Fig. 2, the region of $m > 0.1$ GeV/c^2 and $Q < 10^{-3}e$ of charge (Q) and mass (m_χ), respectively, is still largely not probed at present. In Fig. 2, **SUB-Millicharge ExperimenT (SUBMET)** is

[1] Therefore, dark-photon in this model must be mass-less to mix with the standard-model photon kinetically.

the name of our new proposal. We aim to explore the unexplored region which is indicated by the red curves. In this proceeding, we cover the experimental design of the SUBMET detector and its status.

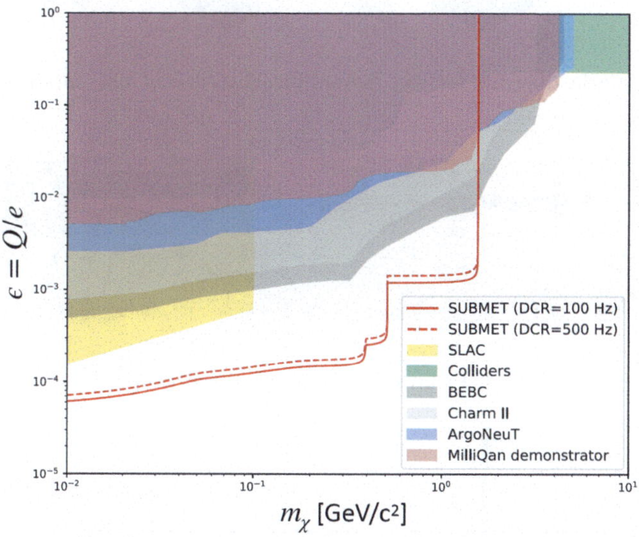

Fig. 2. The constraints from previous experiments are shown as shaded areas.

3 Design of a Single Module

Detection of a millichagred particle by the SUBMET is done by plastic a scintillator coupled with a photo-multiplier tube (PMT). A millichagred particle can generate scintillation inside the scintillator by electromagnetic interaction because the electric charge is nonzero. However, since the amount of charge is expected to be small, the number of photons generated by scintillation is also small. $\mathcal{O}(1)$ photons are generated for $Q < 10^{-3}e$. Therefore, it is necessary to verify whether even one photon can be detected by the PMT in high efficiency. The basic component of the detector design is a module, and this section describes what a module is, how it is designed, and how it is built.

The detailed design of the SUBMET detector is not completely finalized, but the design of a single module has been materialized to some extent. A module is defined as a combination of a PMT, a scintillator, and a support structure. Since the SUBMET detector has a structure in which horizontally placed modules are stacked up, there is a risk that the optical coupling between the PMT and the scintillator may be weakened by gravitation. To prevent this, a plastic support structure is added to the interface to maintain the optical coupling. A plastic support is attached to the opposite end of the scintillator for horizontal leveling.

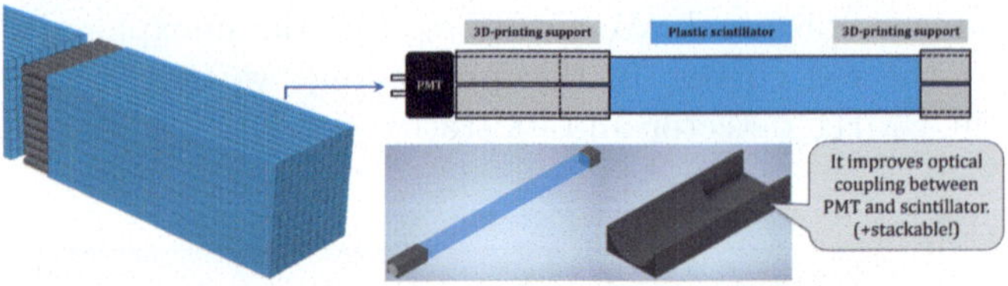

Fig. 3. A schematic diagram of the assembly structure of one module and a CAD drawing.

Fig. 4. Picture of the module prototype.

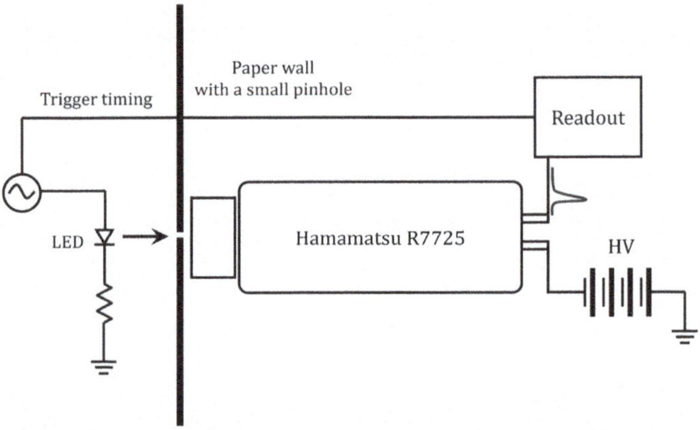

Fig. 5. A schematic diagram of the experimental setup for single-photoelectric signal detection test.

Figure 3 is a schematic diagram of the assembly structure of one module and a CAD drawing. Figure 4 shows the photo of a prototype made for testing the stability of optical coupling.

4 Study with Prototype

The main question about the detector is whether it can detect a single photon signal with high efficiency. The experimental setup to verify this is shown in Fig. 5. After putting the PMT in the dark box, a weak electric signal from a function generator is sent to the light-emitting diode (LED) to generate several

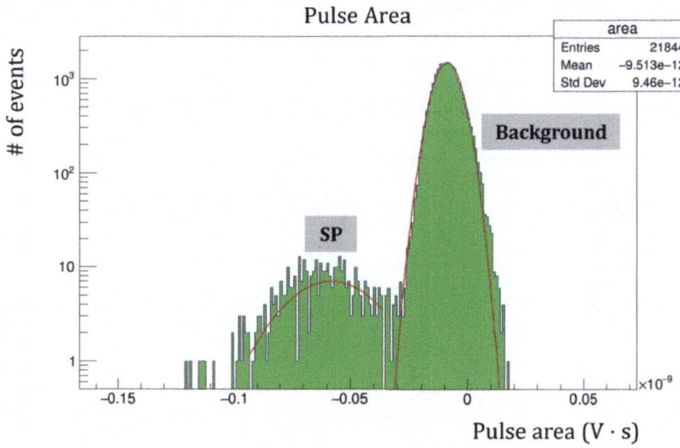

Fig. 6. SPE pulse area spectrum at HV = 1.2 kV. Red curves are the Gaussian fittings of each peak.

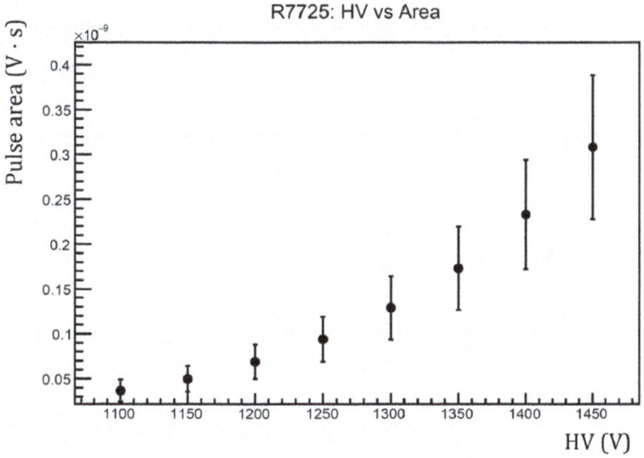

Fig. 7. SPE pulse area vs HV supplied to the PMT

photons. A needle hole is drilled in the hardboard paper and placed between the PMT and the LED to block most of the photons except a few of them. Then, the PMT signal is read and an analysis done to confirm that single photoelectric (SPE) signal can be detected. Figure 6 shows the peaks due to SPEs and the pedestal; the two distributions can be distinguished clearly. The size of the pulse as well as the level of separation depends on the high voltage (HV). This is shown in Fig. 7. The central value is the Gaussian mean and the error bars denote Gaussian sigma.

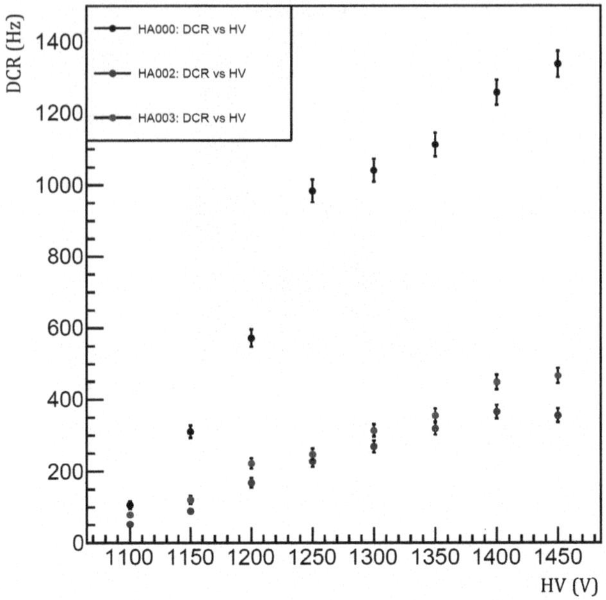

Fig. 8. DCR vs HV. Blue and purple dots represent typical aspect of DCR. Whereas black dots shows abnormal high DCR.

5 Dark Count Rate (DCR)

There are output pulses present even when no photon falls on the PMT. These are called dark current pulses. DCR is the count rate of the dark current pulses. The origin of the DCR is thermal electrons in the photocathode of the PMT. In order to generate high internal gain, typical photo-cathode of the PMT is made so that photoelectric effect happens easily. However, thermal electrons also can escape from the photo-cathode more as a side effect. Typical value of DCR for common PMT is a few hundred Hz.

The dark current pulses are expected to be the main background source for SUBMET because a random coincidence from such dark current pulses in different layers can be identified as a millicharge signal. Estimation of DCR and the coincidence rate becomes study of background source and signal sensitivity. For this, we measured the DCR of sample PMTs, *Hamamatsu Photonics* R7725 [9] and *ET Enterprises, Ltd.* 9814B [10]. Figure 8 is a plot of the DCR measurement result for several R7725 samples. Measured DCR values of R7725 and 9814B are not significantly different being in the range of typical value $\sim$ 500 Hz. Occasionally, significantly higher DCR was measured in some PMTs, irrespective of the model. (See "HA000" in Fig. 8.) DCR of all PMTs for the detector will be inspected. High-DCR PMTs won't be used for modules to prevent degradation of sensitivity to millichared particle detection.

6 Summary and Outlook

We proposed a new experiment for millicharged particle search [11]. This proposal was submitted to Japan Proton Accelerator Research Complex (J-PARC) and stage-I has been approved by J-PARC. We have already visited the J-PARC site in the spring in 2022 to understand the experimental environment. We aim to receive final approval in the upcoming winter. The prototype system is being tested and full system design is being finalized. In September 2022, module assembly started and is expected to be completed by the end of 2022. If the final approval is obtained from the J-PARC, the manufactured modules will be shipped and installed at the site.

Acknowledgments. This work was supported by the National Research Foundation of Korea (NRF) grant funded by the Korea government (MSIT) (No. 2021R1A4A1022977).

References

1. Prinz, A., et al.: Search for millicharged particles at SLAC. Phys. Rev. Lett. **81**, 1175 (1998)
2. Collaboration, C.M.S.: Search for fractionally charged particles in pp collisions at $\sqrt{s} = 7$ TeV. Phys. Rev. D **87**, 092008 (2013)
3. Collaboration, C.M.S.: Searches for long-lived charged particles in pp collisions at $\sqrt{s} = 7$ and 8 TeV. J. High Energy Phys. **07**, 122 (2013)
4. Ball, A., et al.: Search for millicharged particles in proton-proton collisions at $\sqrt{s} = 13$TeV. Phys. Rev. D **102**, 032002 (2020)
5. Marocco, G., Sarkara, S.: Blast from the past: constraints on the dark sector from the BEBC WA66 beam dump experiment. Sci. Post Phys. **10**, 043 (2021)
6. Davidson, S., Campbell, B., Bailey, D.C.: Limits on particles of small electric charge. Phys. Rev. D **43**, 2314 (1991)
7. Acciarri, R., et al.: Improved limits on millicharged particles using the argoneut experiment at fermilab. Phys. Rev. Lett. **124**, 131801 (2020)
8. Emken, T., Essig, R., Kouvaris, C., Sholapurkar, M.: Direct detection of strongly interacting sub-gev dark matter via electron recoils. J. Cosmol. Astropart. Phys. **09**, 070 (2019)
9. Photomultiplier tube - R7723, R7724, R7725
10. ET Enterprises, Ltd. 9814B series
11. Kim, J.H., Hwang, I.S., Yoo, J.H.: Search for sub-millicharged particles at J-PARC. J. High Energy Phys. **05**, 031 (2021)
12. Therefore, dark-photon in this model must be mass-less to mix with the standard-model photon kinetically

Physics Education

A Game of Bingo: A Secondary-School Inquiry Activity About Percolation

Yukio Terashima(✉)

Naruto University of Education,Naruto,Tokushima, Japan
yterashima@naruto-u.ac.jp

Abstract. Percolation theory refers to a geometrical and stochastic approach to describing whole macroscopic properties from local microscopic connections. Percolation is associated with many familiar phenomena; however, scholars have not sufficiently developed teaching materials and learning activities that allow students to experientially study its basic features. In connection with this, bingo is a familiar game that can be explained by percolation theory. Hence, this study conducted a simple inquiry activity for junior high school students to learn about percolation through bingo as an extracurricular task. The game's experimental scale and methods were designed in such a way that even younger students could easily perform the experiment and that the percolation threshold could be clearly identified from the results. Through the bingo game, students were able to understand a critical situation where many bingo patterns are suddenly completed. The teaching material and inquiry activity were found to be useful for teachers/learners to teach/learn the basic concept of percolation with low-cost preparations and easy operations.

Keywords: percolation · bingo game · inquiry activity

1 Introduction

Percolation theory is a geometrical and probabilistic concept that enables an understanding of the whole macroscopic properties of systems from microscopic connections among local elements. This theory was developed and widely introduced as a model to mathematically explain not only literal percolation phenomena such as liquid penetration through gaps in rocks but also natural phenomena such as the spread of forest fires, epidemics, the dilution of ferromagnetism through the mixing of nonmagnetic components, and metal–nonmetal transitions [1–3].

Scholars have widely applied some lattice percolation models to mathematically explain various natural phenomena. An element placed at a lattice point and its neighbors are considered to be connected. Element clusters are formed locally and grow larger as the number of elements increases. For infinite systems, when the ratio of elements (P) reaches a certain value, the clusters are connected to each other, and their size diverges to infinity. The fraction or probability of elements at this point is called the percolation threshold or the critical percolation probability (P_c). Typical lattice models include the site process, which focuses on whether an element occupies a lattice point,

H. J. Choi et al. (Eds.): APPC 2022, *Proceedings of the 15th Asia Pacific Physics Conference*, pp. 199–207, 2024.
https://doi.org/10.1007/978-981-96-0191-2_21

and the bond process, which focuses on the bonds connecting lattice points. The percolation behavior of elements depends on the system dimension and the lattice type, and the values of P_c have been reported as shown in Table 1 [3].

For example, familiar examples of a two-dimensional (2D) square lattice percolation include the Go game, which involves alternately placing black and white stones on a 19 × 19 square board, and bingo, which is played using 5 × 5 square cards. For general bingo games, many bingo patterns are suddenly completed when about 40 numbers are drawn from numbers 1–75 and when about 15 squares are marked on the card [3].

Although probabilistic and statistical concepts such as percolation models can explain many familiar phenomena, practitioners have yet to adequately develop teaching materials and learning activities to study the basic concepts of percolation, which has left few opportunities to learn about probabilistic phenomena and ideas in primary and secondary physics education in Japan. Therefore, this study planned and conducted a simple inquiry task to learn about the basic concept of percolation easily through a bingo game for lower secondary school students as an extracurricular activity. This paper reports the examined teaching materials and experimental methods as well as the results of the activity.

Table 1. Percolation thresholds (P_c) for site process lattice models

Dimension	Lattice model	P_c
2D	Square	0.5927454
2D	Honeycomb	0.6962
2D	Triangular	0.5
3D	Diamond	0.428
3D	Simple cubic	0.3117
3D	Body-centered cubic	0.2460
3D	Face-centered cubic	0.198

2 Preliminary Experiment

To determine the optimal number of bingo cards that would enable junior high school students to clearly and easily understand the percolation phenomenon, we conducted a preliminary bingo game experiment using different numbers of cards.

Bingo cards typically have 25 squares in total (5 × 5 = 25), with any five natural numbers from 1 to 15, 16 to 30, 46 to 60, and 61 to 75 randomly printed on the first, second, fourth, and fifth columns, respectively; the third column contains only four numbers from 31 to 45 because the center square is free. Bingo game toys are usually sold in toy stores and include bingo cards with different number sequences and lottery balls numbered 1–75. Recently, developers have created various bingo game apps that randomly draw natural numbers one by one. For the experiments in this study, commercial bingo cards and a bingo lottery were used.

We played three bingo games using sets of 30 cards with different number sequences (a, b, and c), the results of which are presented in Figures 1(a)-(c), respectively. Based on the percolation threshold for the site process of a 2D square lattice (P_c = 0.5927454), a steep increase in complete bingo cards was expected when approximately 15 (0.5927454 × 25) squares are marked on the card. However, for the first game (a), most bingo cards were completed with 11 marked squares. For the second game (b), most bingo cards were completed with 15 marked squares, which is close to the predicted value, while the third game (c) showed a large variation in the number of squares marked on complete bingo patterns. Since the experimental results were found to be different for each trial, an experiment with only 30 cards was deemed insufficient to obtain statistically reliable results with certainty.

The three experiments above resulted in a total of 90 cards, shown in Figure 1(d). Although most completed bingo patterns had 15 marked squares, the second peak at 11 marks could not be considered negligible. From these results, we found that a single percolation threshold might not be clearly observed in a smaller-scale experiment with approximately 100 or fewer cards in which each student in a single classroom of around 35 is assumed to have three or fewer cards.

Meanwhile, when students have a large number of cards in a bingo game, they may take a longer time than expected to check the numbers on their cards. If they miss the point at which their bingo cards are complete, accurate results cannot be recorded. The preliminary game conducted by the author with 30 bingo cards took about one hour or more to finish. Considering these issues, we planned a bingo game experiment where each student has four cards.

3 Inquiry Activity Procedure

In July 2017, we conducted an inquiry activity with the bingo game experiment involving 42 tenth graders at a public junior high school in Japan as an extracurricular class. This class was organized so that students could learn how to perform research activities efficiently by themselves during their summer vacation. In the first half of the class (50 minutes), the author, as a teacher, taught the students the basic process of scientific research including setting a research theme, conducting experiments, analyzing data, and summarizing a conclusion.

After the lecture, the teacher led the inquiry activity with a bingo game for 100 minutes guided by the following steps: (1) presenting the rules of the game, (2) setting the research question, (3) developing the hypothesis on the critical situation in bingo, (4) planning experiments to test the hypothesis, (5) conducting the bingo game experiments and recording data, (6) analyzing and interpreting results, and (7) concluding the activity.

First, the students and teacher established the rules of bingo. Second, the students shared their experiences of the game, such as having a few bingo cards at the beginning stage and then suddenly increasing the number of cards and accordingly the number of marked squares and connecting them vertically, horizontally, or diagonally. Based on these experiences, the students set two specific research questions: (a) How many numbers are drawn until the bingo cards suddenly increase? (b) How many marked

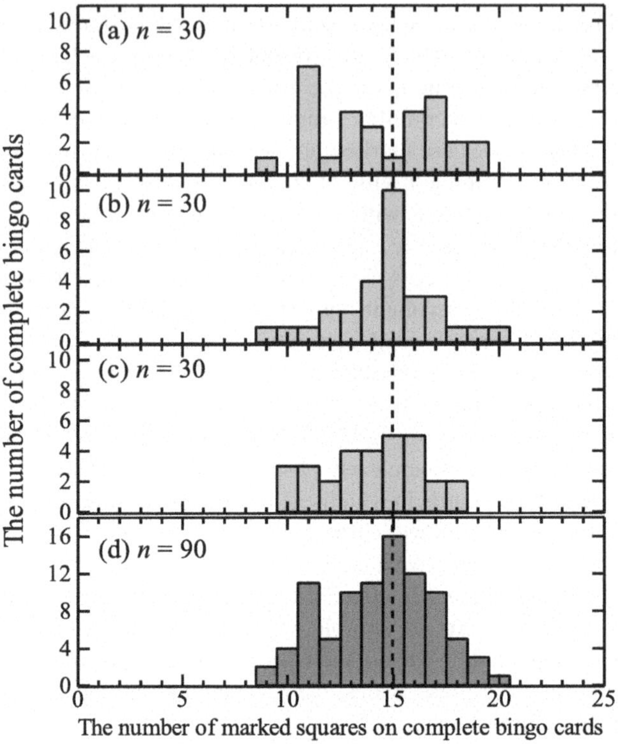

Fig. 1. Distribution of the number of marked squares on completed bingo cards in the preliminary experiments. The vertical dashed line corresponds to the percolation threshold for the bingo card. The results of each preliminary experiment using different sets of 30 cards are shown in (a), (b), and (c), and the combined results are shown in (d).

squares do the complete bingo cards have? Third, each student predicted answers to the two questions as their hypotheses. Fourth, the students organized a bingo game experiment to investigate the change in the total number of complete bingo cards with (a) increasing the number of drawn natural numbers and (b) increasing the number of marked squares in bingo cards with the teacher's guide. Fifth, all students participated in the bingo game, recording the data in a worksheet as will be discussed later. Sixth, after the game, the students analyzed and interpreted their results by plotting the data on a graph sheet, and they identified the critical phenomenon of percolation in the game. At the last step, the students arrived at a reasonable conclusion about the bingo game based on the experimental results.

Afterward, the author summarized the activity by briefly introducing to students the basic concepts of percolation and other related phenomena such as the spread of infectious diseases.

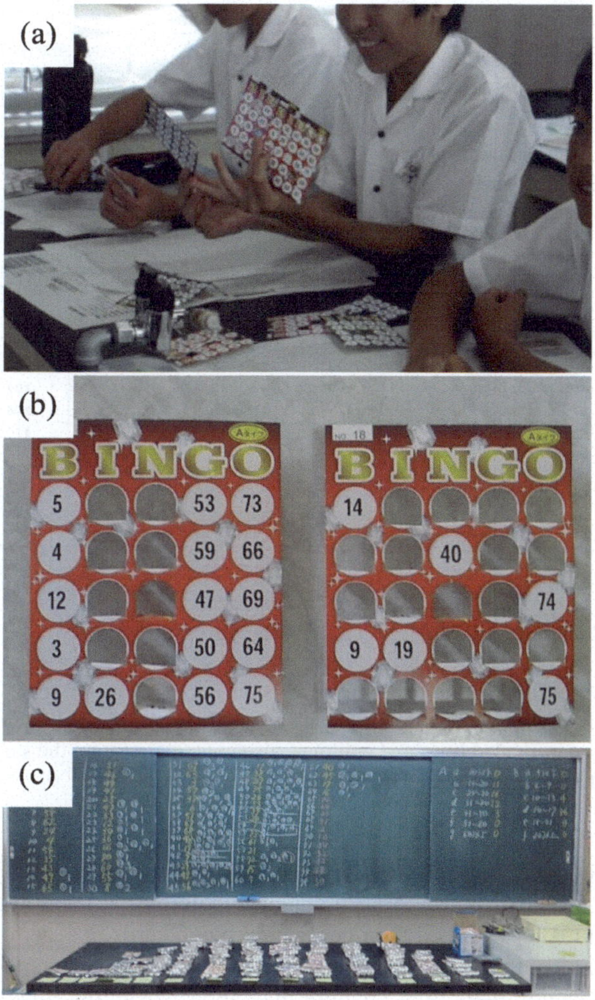

Fig. 2. (a) Students actively participating in the bingo experiment, (b) examples of completed bingo cards, (c) teacher's board notes for sharing game data and all bingo cards arranged on the teaching table for each number of marked squares.

4 Results of Student's Bingo Experiment

The bingo game experiment was conducted with 42 students with a total of 168 bingo cards (4 cards per student). Some scenes and results of this activity are illustrated in Fig. 2. The author acted as a facilitator, randomly selecting one natural number from 1 to 75 and writing each drawn number on the blackboard; meanwhile, the students checked each drawn number and the status of their own cards accordingly. When a bingo pattern was completed in a card, the students shared the results with all participants including the number of squares marked on the card. The teacher and each student then recorded

Drawing times	Drawn No.	# of marked squares on complete bingo cards 5	10	11	12	13	14	15	16	17	20	25	# of complete bingo cards	Total # of complete bingo cards
1	*23*												*0*	*0*
2	*28*												*0*	*0*
3	*59*												*0*	*0*
5	*5*	*1*											*1*	*1*
10	*4*												*0*	*1*
20	*25*												*0*	*4*
30	*8*			*2*									*2*	*25*
40	*48*		*1*	*1*	*1*	*3*	*3*	*3*					*12*	*72*
41	*19*						*2*	*2*	*1*				*5*	*77*
42	*9*					*2*		*2*					*4*	*81*
43	*29*					*1*	*1*	*1*	*2*				*6*	*87*
44	*42*							*1*	*1*				*2*	*89*
45	*56*				*1*	*3*	*1*		*2*				*7*	*96*
46	*11*					*1*	*1*	*1*	*3*				*6*	*102*
47	*38*								*1*	*2*			*4*	*106*
48	*27*					*1*	*1*		*1*				*4*	*110*
49	*75*							*2*	*2*	*2*			*6*	*116*
50	*26*					*1*		*1*					*5*	*121*
60	*7*												*2*	*161*
70	*10*												*0*	*168*
75	*50*												*0*	*168*
# of complete bingo cards		*1*	*3*	*11*	*16*	*20*	*20*	*22*	*22*	*19*	*5*	*0*		
Total # of complete bingo cards		*1*	*10*	*21*	*37*	*57*	*77*	*99*	*121*	*140*	*167*	*168*		

Fig. 3. The tally sheet used in the experiment for students. Because of space limitations, only some lines and columns are shown with students' example records in bold italics. In reality, the omitted parts, shown in shaded lines and columns, also contain data, and the sum of each number in this figure apparently does not match the total number of cards.

the results on the blackboard and a tally sheet. Figure 3 shows the tally sheet that the author prepared and distributed to each student, including an example of a student's recording. The students arranged their completed bingo cards in order of the number of marked squares in front of the teacher's table so that they could review the results later.

Theoretically, a complete bingo pattern may appear from the time the fourth natural number is drawn or later; five numbers are lined up either vertically, horizontally, or diagonally including the free square at the center. Figure 4 shows the number of complete bingo cards (a) and their cumulative count (b) obtained from the students' experiment as a function of the number of drawn natural numbers. The first bingo pattern actually appeared when the fifth natural number was drawn, the number of bingo patterns increased noticeably from around the 20th drawn number, and most of the bingo patterns emerged at around the 45th natural number drawn. The last card completed a bingo pattern when the 64th natural number was drawn. Figure 5 shows the number of complete bingo cards (a) and their cumulative count (b) as a function of the number of marked squares on the bingo cards. The minimum number of marks for a complete bingo pattern was five, the number of bingo patterns increased rapidly when the number of marks reached around 10, and most of the bingo patterns appeared when the number of marks was around 15. Theoretically, when five unmarked squares lined up on any diagonal lines in the bingo cards, a bingo pattern cannot be completed even

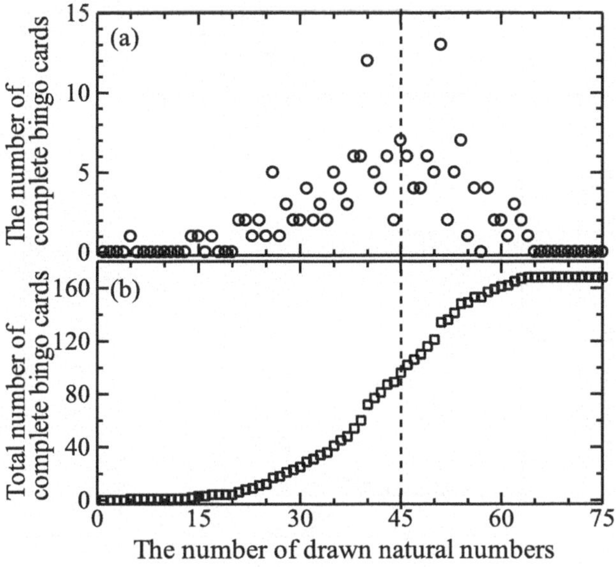

Fig. 4. (a) The number of bingo occurrences and (b) their cumulative number as a function of the number of drawn natural numbers. The dashed line corresponds to the percolation threshold (P_c) of the site process on a two-dimensional square lattice.

if the other 20 squares were marked. When the 21st square is marked on any diagonal spots, it is necessarily connected either vertically, horizontally or diagonally, resulting in a complete bingo pattern. In this game, such a card with the maximum 21 marked squares eventually emerged.

The percolation threshold, $P_c = 0.5927454$ [3], in the site process of the 2D square lattice model corresponds to approximately the 45th of the 75 natural numbers and approximately the 15th mark of the 25 squares, respectively. In the present experiment, the bingo patterns surged when the 40th to 50th natural numbers were drawn and when approximately 15 squares were marked on the cards, demonstrating a critical phenomenon of percolation on a 2D square lattice. The students were able to graphically examine the experimental results recorded as shown in Figures 3, 4, and 5, and conclude that bingo patterns appear rapidly when about 45 of 75 natural numbers were drawn and when 15 squares were marked on the card, which were equivalent to the percolation threshold. The students also found that the results of the bingo game were consistent with the percolation model's prediction.

Immediately after this activity, a brief interview was conducted to ask the participants about their opinions on the inquiry activity, and responses were obtained from 10 students. All 10 participants answered that they were very satisfied with this activity. The main reasons for their satisfaction were as follows: "I learned the importance of recording data in tables and using graphs." "It was very interesting and I discovered many new things on bingo." "The research contents were easy to understand and I learned a lot about percolation." "I could understand the tips for conducting research

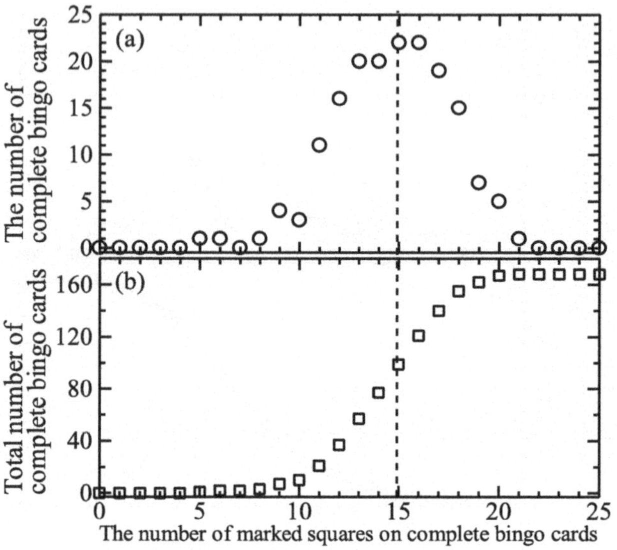

Fig. 5. (a) The number of complete bingo cards and (b) their cumulative number as a function of the number of marked squares. The dashed line corresponds to the percolation threshold (P_c) of the site process on a two-dimensional square lattice.

activities and how to summarize the results of experiments." From these students' responses, this activity with the bingo game was found to be easy and effective for junior high school students to learn about a simple example of percolation in addition to basic research skills.

5 Conclusions

This study conducted a simple inquiry activity on percolation using the familiar bingo game. The procedure and worksheet for this activity were designed in such a way that junior high school students could easily perform the experiment and clearly understand the critical situation with the percolation threshold in which many bingo patterns suddenly appeared. As a result, the participation of students in this activity was positive, and they were able to arrive at a reasonable conclusion; that is, the percolation model explains the critical phenomenon in the bingo game. This suggests that students may consider this activity as a useful hands-on experience when learning about the basic concept of percolation. In addition, the bingo game can be easily played by anyone, anytime, and anywhere with inexpensive, commercially available materials.

In the future, we plan to use this teaching material in a wide range of educational situations not only for junior high school students but also for high school and university students. We also aim to organize another activity involving the bingo game experiment focusing not only on the percolation threshold but also on the size of the clusters of connected squares and evaluate its educational effects more quantitatively.

We also intend to explore other simple and familiar games as teaching materials for physics experiments for different types of learners.

Acknowledgments. The authors would like to thank Ms. Nobuyo Mitsuhashi, science teacher at Tokushima Prefectural Jonouchi Junior High School, and her students for helping us conduct this research.

References

1. Stauffer, D., Aharony, A.: Basic Principles of Percolation. Talor and Francis, London (2001)
2. Odagaki, T.: The Science of Percolation. Shokabo, Tokyo (1993). in Japanese
3. Odagaki, T.: The Science of Connections: Percolation. Shokabo, Tokyo (2000). in Japanese

A Survey on Content Knowledge in Physics for Pre-service Teachers in Japan Using Science Problems of TIMSS 2015

Zehua Wang(✉), Yukio Terashima, and Keisuke Kita

Graduate School, Naruto University of Education, Naruto, Japan
wangzhehua2020@gmail.com

Abstract. Content knowledge is one of the important bases for teachers to acquire and improve their teaching skills. In Japan, although many pre-service teachers not majoring science are not confident to teach science, especially physics, because of a lack of content knowledge and experimental experience, there are few investigations to reveal their weak points specifically. In this study, we conducted a survey for undergraduates of a teacher training college in Japan to evaluate their content knowledge using science problems for 8th graders of TIMSS 2015. The results of the survey showed a positive correlation between the undergraduates and Japanese junior high school 2nd graders (8th graders) who participated in TIMSS 2015 in terms of the correct answer rate for each question. The undergraduates generally showed higher scores than the junior high school students, especially for some problems related to familiar phenomena in daily life, whereas some common weaknesses were found between them. Significant differences in the average percentage of correct answers were detected between students who studied "Advanced Physics" in high school, those who studied only "Basic Physics" and those who did not study physics at all. The experience of learning physics in high school may have influenced the pre-service teachers' understanding of physics contents. The analysis of undergraduates' wrong answers revealed a tendency of their mistakes. Even for questions about similar phenomena, the scores can depend on the question format. They tended to be good at problems given schematic diagrams with options, but poor at problems in which they are required to answer reasons with sentences. Further investigations on physics content knowledge for pre-service teachers can be useful not only for understanding their misconceptions more specifically but also for improving future physics education and teacher trainings more effectively based on their weaknesses.

Keywords: physics education · content knowledge · pre-service teachers · TIMSS2015

1 Introduction

The Organisation for Economic Co-operation and Development (OECD) compared the average science scores of Japanese secondary school students who completed compulsory education with those of other countries through the Programme for International

H. J. Choi et al. (Eds.): APPC 2022, *Proceedings of the 15th Asia Pacific Physics Conference*, pp. 208–215, 2024.
https://doi.org/10.1007/978-981-96-0191-2_22

Student Assessment in 2015 (PISA2015), and the ability of Japanese students was found to be relatively high among OECD countries [1]. In addition, The International Association for the Evaluation of Educational Achievement (IEA) surveyed the Trends in International Mathematics and Science Study in 2015 (TIMSS2015) for the 4 and 8th grades (Japanese primary school 4th grade and junior high school 2nd grade) students. According to the survey, Japanese students had achieved the best score in science in 1995 [2].

Shulman [3] pointed out that a teacher should have the following qualities: (1) content knowledge, (2) subject education knowledge, (3) curriculum knowledge, (4) pedagogical content knowledge (PCK), (5) knowledge related to learners' characteristics, (6) knowledge of educational context, and (7) knowledge of purpose, value, philosophy and social background. On this basis, he also pointed out that there are six stages in the process of classroom practice: 1) understanding, 2) adaptation, 3) guidance, 4) evaluation, 5) introspection, and 6) new understanding. Among them, the understanding of content knowledge is in the first stage as the basis of teaching. The quality of teaching activities can be controlled by the teachers' understanding of teaching contents. Therefore, teachers' content knowledge is very important for science teaching activities.

Liu and Sun [4] pointed out that teachers must be provided with supports to improve their professional abilities in order to improve the quality of education. Moreover, improving teachers' professional ability is also a necessary condition for education reform, so it is very necessary to investigate the professional ability of current and future primary and secondary school teachers. Related surveys have been carried out using questions from TIMSS2011 and the National Learning Ability and Learning Situation Survey in Japan so far, which had pointed out various characteristics concerning learning ability among pre-service teachers [5–7].

In this study, we conducted a survey on content knowledge in physics for first-year undergraduates in 2020 at a teacher training college in Japan who are the same age of the 8th grade students who participated in the TIMSS2015 survey by using some science questions including physics for the 8th graders from TIMSS2015. The purpose of this survey is to investigate the degree of understanding of content knowledge on the pre-service teachers after junior and senior high school physics course learning, and to find remaining learning difficulties among the undergraduates. By using the same science questions for the 8th graders from TIMSS 2015 to examine the content knowledge for the first-year undergraduates in 2020 who same age as the 8th grade participants in the TIMSS 2015 survey were, so that the performance between the two could be compared directly. This paper summarizes the results of this survey and analyzes the degree of content knowledge in physics and learning difficulties for the teacher training college students.

2 Methods

In October 2020, the survey on science content knowledge was conducted for 100 first-year undergraduates of Naruto University of Education, which is a teacher training college in Japan. Total 48 questions were selected for the survey from the eighth grade's science questions (Japanese translation version) of TIMSS2015 [2] including

14 physics, 14 chemistry, 11 biology and 9 earth science questions whose topics the Japanese eighth grade participants in TIMSS2015 had learned. The undergraduates answered the 48 science problems and a questionnaire on their leaning history of high school science for 45 min. In this paper, the results for the 14 physics questions are discussed.

From the perspective of cognitive fields, the physics questions were divided into the following three types: 1) "knowledge" in the field of cognitive science, information, concepts, tools, procedures and other related basic knowledge (3 questions), 2) "application" as knowledge and common sense applied directly to scenes, explanation of scientific concepts and principles of relevant information (6 questions) and 3) "reasoning" in order to make conclusions derived from scientific evidence (5 questions). From the point of view of the answer form, the questions can be divided into "choice" (7 questions) and "description" (7 questions).

Based on a publicly available scoring criterion [2], the students' answers to each question were examined to obtain the correct response rate for each question. The correct response rate of the participants was then compared with that of Japanese junior high school students of the eighth graders participated in TIMSS2015. The Mann-Whitney U test was used to check the significant difference in the correct response rate between the college students and junior high school students. The Kruskal-Wallis test was used to compare the correct response rate of the college students between the three academic skills and between the two answering styles classifications. Finally, wrong answers to some typical questions with lower correct answer rates for the college students were analyzed, and their specific misunderstandings in physics were detected.

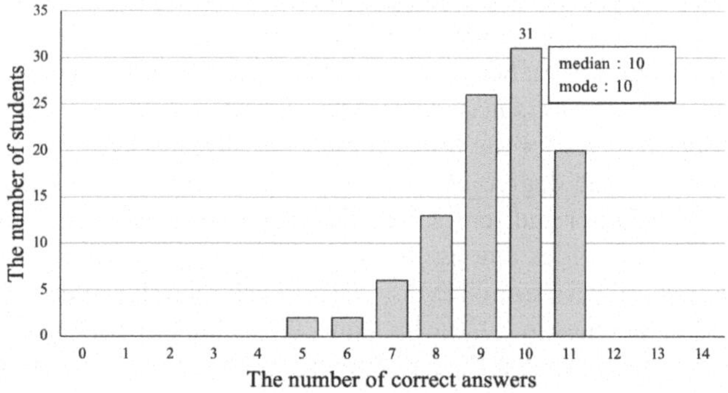

Fig. 1. Frequency distribution of the number of correct answers among the college students.

3 Results and Discussion

3.1 The Number of Questions Answered Correctly and Its Distribution

The distribution of the college students against the number of correct answers of 14 physics problems is shown in Fig. 1. The average number of correct answers (its correct answer rate) with the standard deviation was 9.3±1.3 (66.6±9.8%). The median was 10 questions (71.4%), and the mode was 10 questions (71.4%). The highest number of correct answers was 11 (78.6%) (20 respondents), while the lowest number of correct answers was 5 (35.7%) (2 respondents). No one answered less than 5 or more than 11 questions correctly. Even among the teacher training college students, no student was able to fully understand the entire contents of secondary school physics. The degree of understanding for the physics contents was limited at about 80%. The main reason for this might be related to the learning experience in high school as described later. Japanese high school students generally enter university after selectively studying two or three subjects from physics, chemistry, biology and earth science. Therefore, many undergraduates tend to forget knowledge in some areas that they did not study at high school. This may be one reason why the average correct answer rate did not reach nearly 100%.

The correct answer rate of college students and junior high school students in each category (%)

Category	College	Junior high school	Difference
knowledge	74.3	57.0	17.3
reasoning	78.6	56.4	22.2
application	88.2	63.3	24.9
choice	85.3	65.7	19.6
description	78.3	53.2	25.1
average	81.8	59.5	22.3

Mann-Whitney U: $p = 0.000$

Fig. 2. The correct answer rate of the college students and junior high school students in each classification.

3.2 The Correct Answer Rate of Each Classification

Figure 2 shows the average correct answer rates classified according to cognition, contents and form of question setting for the college students and Japanese junior high school students. In terms of overall average correct answer rate, there was a significant difference between the college students and junior high school students ($df = 1, p = 000 < 0.001$). The correct answer rate of the college students was significantly higher than that of the junior high school students. In the cognitive field, the correct answer rate of "application" was the highest (88.2%), followed by "reasoning" (78.6%) and "knowledge" (74.3%). There was no significant difference between the three terms ($df = 2$, $s = 3.52$, $p = 0.171 > 0.05$). There was no significant difference between "choice" and "description" ($df = 1$, $s = 15.5$, $p = 0.259 > 0.05$).

For the 14 physics questions used in this survey, the correct answer rate of junior high school students was taken as the horizontal axis, the correct answer rate of the

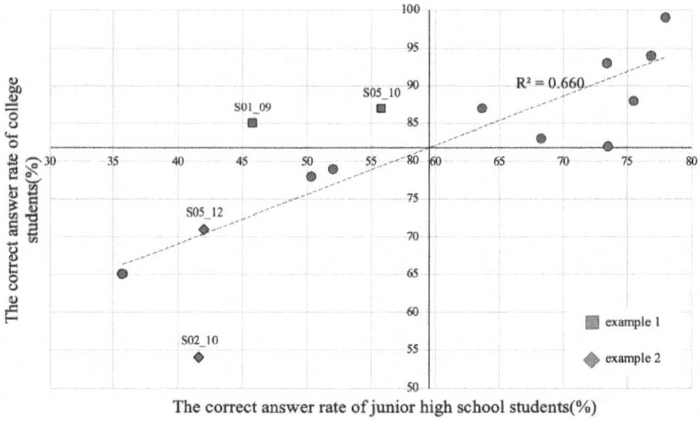

Fig. 3. Distribution of correct answer rate of each question.

	Physics (Total 14)
	Number
1st quadrant Both students are relatively good	S01_07 S03_11 S06_10 S06_12 S07_07 S07_09 S07_12 total (7)
2nd quadrant Junior high school students are relatively weak College students are relatively good	S01_09 S05_10 total (2)
3rd quadrant Both students are relatively weak	S02_08 S02_09 S02_10 S05_11 S05_12 total (5)
4th quadrant Junior high school students are relatively good College students are relatively weak	total (0)

2nd quadrant				Number of responses classified into each answer pattern code											
Number	contents	cognition	style	1	2	3	4	※10	※11	※12	20	70	71	79	99
S01_09	physics	application	description					85				1		13	1
S05_10	physics	knowledge	choice	0	6	※87	7								

3rd quadrant				Number of responses classified into each answer pattern code											
Number	contents	cognition	style	1	2	3	4	※10	※11	※12	20	70	71	79	99
S02_08	physics	reasoning	choice	1	2	32	※65								
S02_09	physics	reasoning	description					75						24	1
S02_10	physics	knowledge	description					56						40	6
S05_11	physics	reasoning	description					57	22			7		11	3
S05_12	physics	application	description					25	2	44			4	22	3

※ is the correct answers

Fig. 4. The distribution of the problem in each quadrant.

college students was taken as the vertical axis, and the intersection point of each average correct answer rate of all the questions was taken as the origin, as shown in Fig. 3. For the questions on the upper right (first quadrant) of the figure, the correct answer rates of both students are higher than their average correct answer rates, that is, both of them are good at these problems. Similarly, the upper left (second quadrant) of the figure is the problems that junior high school students are difficult to solve but the college students are good at; the lower left (third quadrant) of the figure is the problems that both students are difficult to solve; the lower right (fourth quadrant) is the problems that junior high school students are good at and the college students are not good at. As shown in Fig. 3 (the dashed line is the regression line), the correct answer rates of the college students and junior high school students were positively correlated ($R^2 = 0.660$). On the whole, the distribution of the questions that the college students and junior high school students were good/poor at was very clear. Figure 4 shows the problem numbers for the different areas classified in each quadrant. Examples 1 (S01_09, S05_10) and 2 (S05_12, S02_10), which are representative problems classified into the second and third quadrants, will be explained in detail later based on the analysis of wrong answers of the college students.

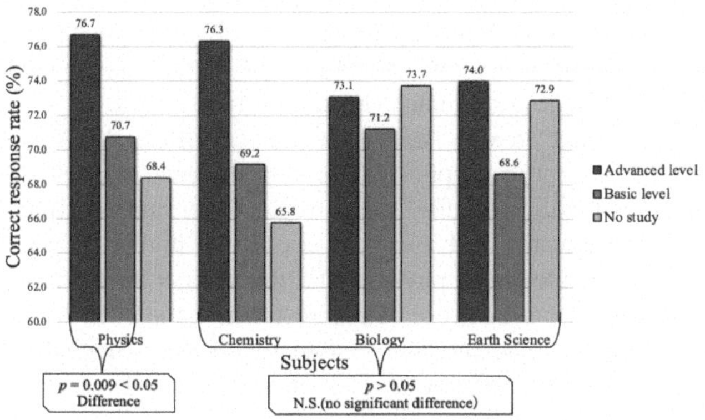

Fig. 5. Difference in the correct answer rate between the college students' science learning history in high school.

3.3 The Tendency of the College Students' Wrong Answers

For the example 1, as shown in Fig. 3, the correct answer rates of the college students were higher than those of junior high school students. The question of S01_09 was to write down the reason why a bottle filled with water was broken when it was frozen. The question of S05_10 was to choose how the volume and mass of water change after it freezes. According to the results of the chi-square test, students with correct/incorrect answers to S01_09 were found to tend to answer correctly/incorrectly to S05_10 ($p = 0.001 < 0.05$). Although matters generally become denser and thus smaller when changing from a liquid to a solid, water is an exception. The students' understanding on the thermal expansion of water may be better as their life experiences increase. The example 2 of S02_10 and S05_12 in Fig. 3 indicated typical difficult problems for both the college students and junior high school students. The question of S02_10 is about the path of light propagation and reflection in straight line. Students needed to answer the reason why an object hidden by the wall can be seen through a mirror that reflect light at an angle equal to the incidence angle. The 54% college students answered correctly, while the wrong answers included "Actually had such experience" and "The object has not been exposed to the angle of reflection/refraction of light." It can be seen that these college students mistook the meaning of reflection and refraction of light. The question of S05_12 is about the forces acting on a stationary person sitting on a wall. Although students of 71% answered correctly, we found incorrect answers such as "their feet hang under the wall due to gravity" and "they exist on Earth." Those students might not specifically understand about forces acting on objects and its balance because force is not visually seen in life, and it is hard for students to imagine.

3.4 The Relationship Between the Correct Answer Rate of the College Students and Their Learning Experience at High School

The comparison of the average correct answer rate of all questions between the college students' science electives at high school is shown in Fig. 5. The lower limit of the vertical axis of the graph is the average correct answer rate of the Japanese junior high school students participated in TIMSS2015 (about 60%), and the height of each graph indicates the improvement of the correct answer rate of the college students when the score of the junior high school students is taken as a benchmark.

In physics, the correct answer rates for students who took "Advanced Physics" "Basic Physics" and did not study physics (no study) were 76.7%, 70.7% and 68.4%, respectively. Compared with "no study" the correct answer rate for "Basic Physics" was higher by +2.3%, and that for "Advanced physics" was higher by +8.3%. Chemistry also showed the same tendency as physics, and the correct answer rate of each stage of study (the difference from the score of 65.8% for not studied) was 76.3% (+10.5%) for "Advanced Chemistry" and 69.2% (+3.4%) for "Basic Chemistry", respectively. For biology subjects, the correct answer rates (the difference from the score of 73.7% for not studied) were 73.1% (-0.6%) for "Advanced Biology" and 71.2% (-2.5%) for "Basic Biology" The students who did not study any biology subjects at high school were slightly better at physics than the other students who studied biology. For earth science subjects, the correct answer rate for not studied was 72.9%, which was slightly higher than that for "Basic Earth Science" (68.6%) and was approximately similar to that for "Advanced Earth Science" (74.0%). In the subjects of biology and earth science, the order of the correct answers rate was not consistent with the students' learning level at high school. The inconsistent reason might be the Japanese high school science curriculum in which most of students who did not study any biology or earth science subjects studied physics and chemistry at advanced level instead. As a result of multiple comparisons with the Kruskal-Wallis test, significant differences between "Advanced Physics" "Basic Physics" and "no study" were detected ($s = 9.336$, $p = 0.009 < 0.01$). Meanwhile, chemistry ($s = 2.365$, $p = 0.307 > 0.05$), biology ($s = 0.702$, $p = 0.704 > 0.05$), earth science ($s = 3.738$, $p = 0.154 > 0.05$) had no significant difference between the learning levels at high school. The results for chemistry, biology and earth science might be caused by the small sample size of this study. For physics, however, learning experience at high school can be a great effect on the college students' understanding of physics.

4 Conclusions

Using the scientific problems of TIMSS2015, the content knowledge in physics for the undergraduates at a teacher training college in Japan was evaluated, compared to the results for Japanese junior high school students who participated in TIMSS2015. The characteristics of the college students' understanding of physics were found and analyzed. Even for the college students, some poor understandings in physics similar to those for junior high school students and the effects of learning history at high school on their content knowledge in physics were detected.

Based on the results of this survey, the future research task is to develop more effective textbooks and learning methods for Japanese college students to complement physics content knowledge and scientific concepts. For example, it is necessary to investigate the physics textbooks used in junior and senior high schools in Japan and analyze their learning contents in detail with methods of structural evaluation analysis for textbooks, so as to specify the learning process of students.

Acknowledgments. The survey was conducted with the assistance of Professor Makoto Honda of Naruto University of Education and his students. In addition, the chair and participants of APPC15 made helpful suggestions for this study. We would like to thank all those who assisted our research.

References

1. National Institute for Education Policy Research, OECD Student Achievement Survey (PISA 2015) Key Points (2015). in Japanese
2. National Institute for Education Policy Research: International Comparison of Mathematics and Science Education TIMSS2015 International Mathematics and Physics Education Trends Survey 2015 Survey Report. Akashi Shoten, Tokyo (2017). in Japanese
3. Shulman, L.: Harvard. Educ. Rev. **57**, 1 (1987)
4. Liu, H., Sun, S.: J. Northeast Normal Univ. **5**, 113 (2003). in Chinese
5. Terashima, Y.: Res. Bull. Naruto Univ. Educ. **32**, 454 (2017). in Japanese
6. Terashima, Y.: J. Japan Assoc. Univ. Educ. **35**, 25 (2017). in Japanese
7. Terashima, Y.: Phys. Educ. Univ. **23**, 35 (2017). in Japanese

Plasma Physics

Power Balance in TOKASTAR-2 Small Tokamak for Feasibility Study on Light Source Application

A. Okamoto(✉), S. Kimata, K. Tsunoda, K. Kado, K. Yasuda, and T. Fujita

Graduate School of Engineering, Nagoya University, Nagoya 464-8603, Japan
a-okamoto@energy.nagoya-u.ac.jp

Abstract. An application of small tabletop tokamak as light source is considered. Temporal evolution of the electron temperature and density is measured using passive spectroscopy applicable for small low temperature tokamaks. Ion temperature measurement is also performed. Those are applied to the smallest tokamak, TOKASTAR-2, major radius of which is $R_0 \simeq 0.1\,\mathrm{m}$. The stored energy is determined from the temperature and density measured in the tokamak. Energy confinement time deduced from the temperature and density is comparable with a scaling law for larger tokamaks. Line radiation power is evaluated from direct measurement and a collisional radiative model calculation based on the measured temperature and density. These clarifies the power balance in TOKASTAR-2. Neutral depletion during tokamak operation is suggested. Resonant EUV line radiation is estimated from measurement data.

Keywords: tokamak · line radiation · collisional radiative model · neutral depletion

1 Introduction

While tokamaks lead the development of fusion energy, application of tokamak for light source will be interesting. It is expected to be high emissivity radiation source due to high density plasma confinement. A conventional tokamak was considered as a soft X-ray or extreme ultra violet (EUV) light source [1]. There are, however, few reports on light source application of small tokamaks.

Small (tabletop) tokamaks have been used for fundamental research on plasma confinement [2–8]. Recently, position controlling have been achieved using non-circular coils [9] or local helical coils [10]. Advantages of small tokamaks for light source application compared to conventional tokamaks are smaller occupation of the device, lower cost, suitable temperature range for visible or EUV radiation. Typical electron temperature of the small tokamaks is $T_\mathrm{e} \leq 100\,\mathrm{eV}$, which is suitable for those radiation application. On the other hand, the temperature is under the radiation barrier [11,12]. Then neutral particles will become a major radiator, density of which should be suitably controlled in terms of power balance and radiation efficiency.

H. J. Choi et al. (Eds.): APPC 2022, *Proceedings of the 15th Asia Pacific Physics Conference*, pp. 219–227, 2024.
https://doi.org/10.1007/978-981-96-0191-2_23

In order to clarify the power balance and confinement characteristics of small tokamak, the ion temperature measurement as well as electron temperature and density measurement is required. Then radiation power evaluation using suitable model is needed. However, measurement of the density and temperatures in small tokamak was difficult. Smaller dimensions and lower density restrict the applicable wavelength range of the interferometer than that for larger tokamaks. Low temperature results in difficulty in soft x-ray and diamagnetic measurements. Electrostatic probe is in a suitable range for temperature and density; however, the robustness of the probe for heat flux causes disturbance for core plasma. An available measurement method is passive spectroscopy in the visible wavelength range.

We have applied passive spectroscopy to the smallest tokamak. Then power balance has been clarified based on the density and temperatures measurement. EUV radiation power has been estimated using collisional radiative models. Neutral depletion during tokamak discharge has been observed.

2 Experimental Setup

Experiments were performed using the TOKASTAR-2 device in Nagoya University [13–15], in which typical major and minor radii were $R_0 = 0.12\,\mathrm{m}$ and $a = 0.05\,\mathrm{m}$. A toroidal magnetic field of $B \simeq 0.1\,\mathrm{T}$ was applied quasi-stationary during tokamak discharge using a slow time-constant capacitor bank. A microwave with a frequency of 2.45 GHz, which was utilized for pre-ionization, was launched with the toroidal field coil start-up and was terminated after the plasma current decreased to zero. Output of microwave power supply was 1.6 kW. X-mode injection was adopted. The microwave power absorption during tokamak formation was negligible due to higher electron density than the cut-off density. A tokamak with a plasma current up to $I_\mathrm{p} \simeq 2.5\,\mathrm{kA}$ was typically generated and sustained by induction of center solenoid coils. All the toroidal field (TF) coils, pulsed vertical field (PVF) coils, and center solenoid coils were installed inside a cylindrical vacuum vessel of 0.6 m in diameter and 0.6 m in length as shown in Fig. 1. Two pairs of one-turn voltage loop and poloidally aligned 16-pickup coils [16] were installed in the vacuum vessel for reconstruction of the cross section of the tokamak. Tokamak equilibrium was confirmed for typical discharges [17].

A tangential viewing chord was used for passive spectroscopy. The electron temperature and density were obtained from line intensities of atomic helium [18]. We measured temporal evolution of four He I emission lines (501.6, 667.8, 706.5, and 728.1 nm) using a low dispersion monochromator and a photomultiplier tube. A collisional radiative model for the helium atom [19] was applied. The electron temperature and density were deduced as the best fit value for reproducing line intensity ratios. Three ratios, I_{667}/I_{728}, I_{728}/I_{706}, and I_{501}/I_{706}, were used. On the other hand, absolute value of photon flux is utilized for neutral density estimation. For each line emission, five shots of tokamak discharges were accumulated. Verification of the temperature and density was performed by comparing poloidal beta β_p determined from equilibrium and that deduced from the temperature and density [17].

Ion temperature was obtained from Doppler broadening of He II line spectrum ($\lambda = 468.6\,\mathrm{nm}$), which was meareued at the same viewing chord with a high dispersion spectrometer [20]. The focal length of the collimating lens and the groove

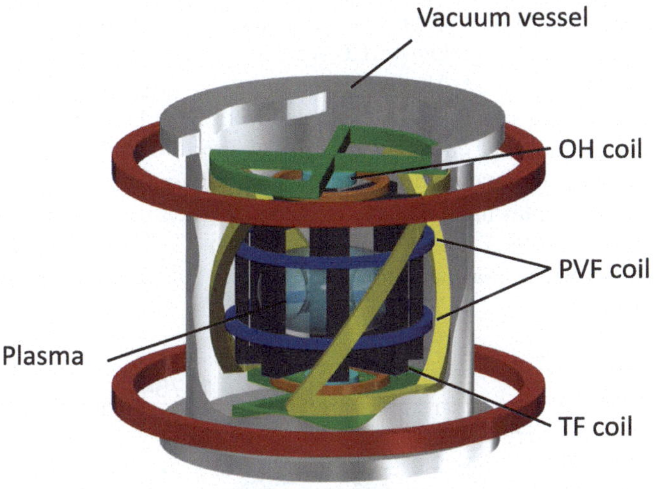

Fig. 1. (Color online) Schematic of TOKASTAR-2. Toroidal field (TF) coils, pulsed vertical field (PVF) coils, and center solenoid for Ohmic-heating (OH) coils are used.

density of the grating were 0.85 m and 2400 lines/mm. Reciprocal linear dispersion $\simeq 0.4\,\mathrm{nm/mm}$ was determined by two adjacent lines of He II ($\lambda = 468.6\,\mathrm{nm}$) and He I ($\lambda = 471.3\,\mathrm{nm}$) with a charge coupled device detector. A hollow cathode discharge tube [21] was used as a reference light source for instrumental width and dispersion calibration. In the tokamak experiment, a photomultiplier tube was attached to the spectrometer. Line emission spectrum was obtained with shot-by-shot wavelength scanning. For each wavelength, 5 – 20 shots of tokamak discharges were accumulated. The spectrum is fitted by superposed 13 Gaussian functions considering the fine structure of transition [22]. The total intensity and wavelength shift as well as the Gaussian width are determined by the fitting. The ion temperature is obtained by deconvolution from the Gaussian width and the instrumental width.

3 Results

Temporal evolution of He I emission lines are shown in Fig. 2(a) [23]. A tokamak sustained during $5.6\,\mathrm{ms} \leq t \leq 6.0\,\mathrm{ms}$ as plasma current I_p indicated in Fig. 2(b). The emission intensities monotonically decreasing during the tokamak discharge after rapid increasing at initial phase. Time when each intensity reaches maximum is $t \simeq 5.65 - 5.75\,\mathrm{ms}$, which is earlier than the peak time for plasma current, $t = 5.8\,\mathrm{ms}$.

Using three intensity ratios described in previous section, the electron temperature and density are determined as shown in Fig. 2(b). The electron density increases up to $n_\mathrm{e} = 3.2 \times 10^{18}\,\mathrm{m}^{-3}$ in the plasma current ramp-up phase, while the electron temperature decreases to $T_\mathrm{e} = 4.7\,\mathrm{eV}$. Due to increase of uncertainty in the emission intensities, larger uncertainties arise in the electron density and temperature in the plasma current ramp-down phase.

Line emission spectra of Helium ion were obtained for both the clockwise (CW) plasma current and counterclockwise (CCW) cases. Then, we have obtained the ion temperature $T_i = 0.77 \pm 0.24\,\mathrm{eV}$ (CW case) and $T_i = 0.63 \pm 0.24\,\mathrm{eV}$ (CCW case). Toroidal rotation velocity of a tokamak evaluated from Doppler shifts in both cases was negligiblly small. Because the emission intensity was too weak, the values shown above were obtained by time integration during tokamak discharge.

Reconstruction of plasma shape from magnetic measurement revealed that the major radius $R_0 = 0.105\,\mathrm{m}$ and the minor radius $a = 0.029\,\mathrm{m}$ at the time of the maximum plasma current. Plasma volume $V = 1.7 \times 10^{-3}\,\mathrm{m}^{-3}$ is also deduced from the reconstruction.

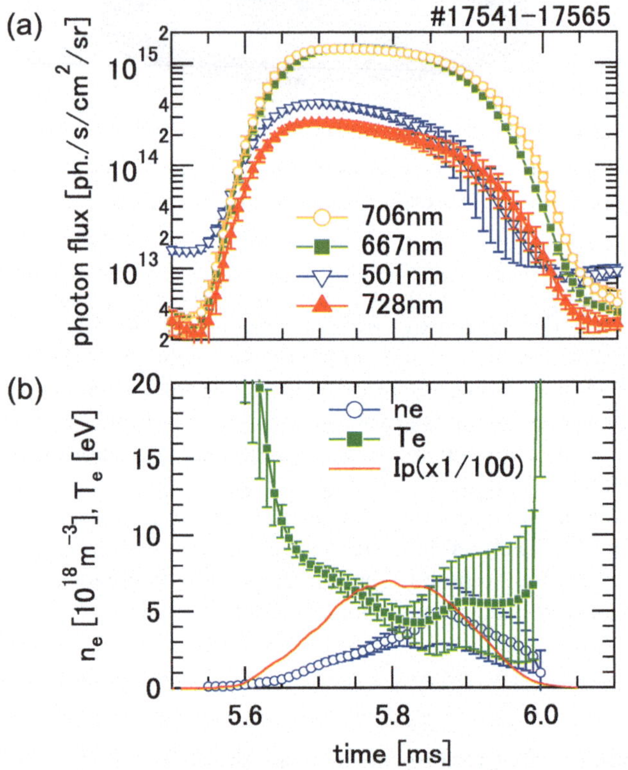

Fig. 2. (Color online) Time evolution of (a) He I line emission intensities and (b) electron density and temperature deduced from the line intensities. Plasma current averaged among the discharges is also shown in (b). S. Kimata, *et al.*, AIP Advances **12**, 045204 (2022); licensed under a Creative Commons Attribution (CC BY) license.

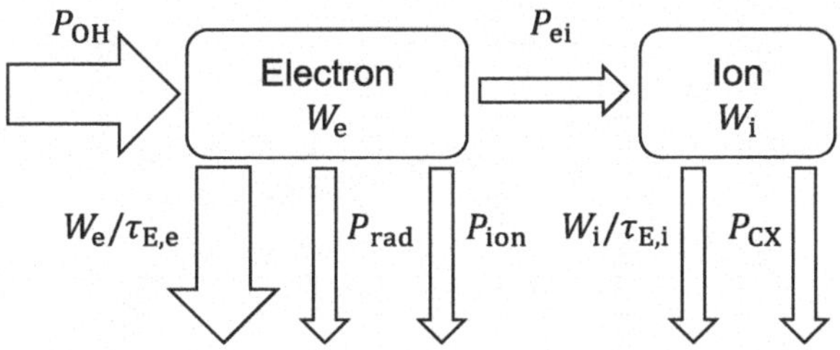

Fig. 3. Schematic of power balance. The description and value of each symbol are described in text. S. Kimata, *et al.*, AIP Advances **12**, 045204 (2022); licensed under a Creative Commons Attribution (CC BY) license.

4 Discussion

4.1 Power Balance

Power balance in the tokamak plasma schematically shown in Fig. 3 is discussed based on the experimental results described above. All power injected into the tokamak is via ohmic heating $P_{OH} = 3.2\,\text{kW}$. The ions receive power from the electrons via electron – ion collisional relaxation $P_{ei} = 15\,\text{W}$. Power loss channels for electrons are a conductive loss characterized by the energy confinement time $W_e/\tau_{E,e}$, a radiation loss $P_{rad} = 57\,\text{W}$, and an ionization loss $P_{ion} = 64\,\text{W}$. Power loss channels for ions are a conductive loss characterized by the energy confinement time $W_i/\tau_{E,i}$, and a charge-exchange loss $P_{CX} = 47\,\text{W}$. Also we can deduce electron and ion stored energies; $W_e = 6.1\,\text{mJ}$ and $W_i = 0.9\,\text{mJ}$. Then the electron energy confinement time $\tau_{E,e} = 2.0\,\mu\text{s}$ is obtained using the equation of power balance for electron: $P_{OH} = P_{rad} + P_{ion} + P_{ei} + W_e/\tau_{E,e}$. The main loss channel for electron stored energy is conduction, while the neutral density determined from gas pressure is one order higher than the electron density. Detail on neutral density is discussed in the following subsection.

On the power balance for ion, equation based on Fig. 3, $P_{ei} = P_{CX} + W_i/\tau_{E,i}$, is rather naive, because $P_{ei} < P_{CX}$ in straightforward calculation results in non-positive confinement time. A possible reason is variation of neutral density during the discharge, which is discussed in the following subsection.

When we consider the total stored energy, $\tau_E = (W_e + W_i)/(P_{OH} - P_{rad} - P_{ion} - P_{CX}) = 2.3\,\mu\text{s}$ is obtained. The confinement time τ_E is compared with a scaling law known as neo-Alcator scaling, which is determined from mid-sized ($R_0 \sim 1 - 2\,\text{m}$) ohmic heating tokamaks [24]. The energy confinement time ${\tau_E}^{SC}$ is a function of safety factor q, major radius R_0, minor radius a, and density $n_{e,20}$ in $10^{20}\,\text{m}^{-3}$ unit: ${\tau_E}^{SC} = 0.103\,q^{0.5}\,{R_0}^{2.04}\,a^{1.04}\,n_{e,20} = 1.5\,\mu\text{s}$. Then, the confinement time from the scaling law agrees with the confinement time directly determined from the experimental power balance. The result suggests that the scaling law supports design optimization of small tokamaks. However, the mass dependence and availability of the scaling law in the μs range should be studied with parameter scan experiments.

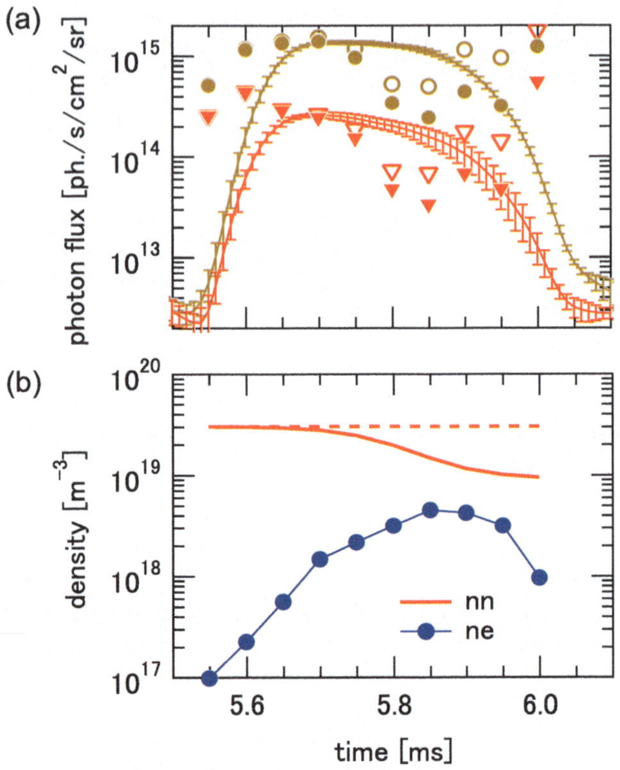

Fig. 4. (Color online) Time evolution of (a) He I line emission intensities and (b) electron and neutral densities. In panel (a), curves with error bar represent absolute photon fluxes directly determined from optical emission spectroscopy as the same as Fig. 2(a). Open symbols represent photon fluxes determined from electron density, temperature, and a constant neutral density shown in panel (b), while filled symbols that considered with neutral depletion. Circles and brown curve correspond to wavelength of 706 nm. Triangles and red curve, 728 nm. In panel (b), dashed line represents the constant neutral density; thick curve, neutral density with depletion; circle, electron density.

4.2 Neutral Density Depletion

Using the electron density and temperature, we can estimate absolute values of photon flux. In Fig. 4(a), comparison between directly measured and estimated photon fluxes is shown. In the estimation, we assume two temporal evolution model for neutral density. One is a constant density model; the neutral density unchanged during tokamak discharge is determined from initial neutral gas pressure. The other is a depletion model; the initial neutral density is the same as that of the constant density model, while it decreases as the plasma current increases and stays in lower density in current ramp-down phase. In Fig. 4(a), typical two line emissions are shown. In both emissions, estimated initial ($t = 5.65 - 5.7$ ms) photon fluxes well agree with those directly measured. Since it is in initial phase, neutral density difference between two models is negligible.

In both emissions, the same initial neutral density that is determined from experimental neutral gas pressure is applied. Therefore the agreement between the models estimation and direct measurement suggests consistency in gas pressure measurement, spectroscopic measurement, and calibration.

At the peak plasma current ($t \simeq 5.8\,\mathrm{ms}$), estimated photon flux underestimate the direct measurement. A possible reason is variation of radiation transport processes due to variation of neutral temperature. However, further investigation will be required.

After the peak plasma current similar electron temperature and density as the current ramp-up phase are obtained. The estimated photon flux with the constant neutral density model results in the similar photon flux as in the current ramp-up phase. On the other hand, directly measured photon fluxes decreases in the current ramp-down phase even with the similar electron temperature and density as the ramp-up phase. Therefore overestimation occurs in the constant neutral density model. Introducing temporal variation of neutral density in the estimation is essential to reproduce the direct measurement. In the neutral depletion model, following temporal variation of neutral density $n_\mathrm{n}(t)$ is assumed:

$$n_\mathrm{n}(t) = n_0\{1 - c_1 \tanh[(t - c_2)/c_3]\}, \tag{1}$$

where $n_0 = 3.1 \times 10^{19}\,\mathrm{m}^{-3}$ is the initial neutral density as the same as the constant model. Parameters $c_1 = 0.35$, $c_2 = 5.8\,\mathrm{ms}$, and $c_3 = 0.1\,\mathrm{ms}$ are used. Then the estimated photon fluxes in the neutral depletion model well agree with those directly measured. The result suggests that the neutral density decreases during the tokamak discharge down to about 1/3. While we have proposed a possible reason on inconsistency in ion power balance is decreasing neutral density [23], it is qualitatively verified. In terms of application on light source, lower degree of neutral depletion will be desired. However, the result also suggests that there is still 30% of initial neutral particle in the present experiment. In Fig. 4(b), electron density is also plotted. Neutral density is at least 3 times larger than the electron density. A picture of tokamak immersed in neutral gas is still valid.

4.3 EUV Radiation

The most valuable radiation is a resonant line radiation in terms of both photon energy and photon flux. For neutral helium atom, $1^1S - 2^1P$ transition (wavelength $\lambda = 58.4\,\mathrm{nm}$) is considered, while for helium ion, $1 - 2$ transition (wavelength $\lambda = 30.4\,\mathrm{nm}$) is considered. We can calculate the upper state of the transitions using CR models with experimentally obtained electron density, $n_\mathrm{e} = 3.2 \times 10^{18}\,\mathrm{m}^{-3}$, electron temperature, $T_\mathrm{e} = 4.7\,\mathrm{eV}$, and neutral density, $n_0 = 3.1 \times 10^{19}\,\mathrm{m}^{-3}$.

For neutral helium atom, CR model calculation results in population density of upper state $n(2^1P) = 1.2 \times 10^{14}\,\mathrm{m}^{-3}$ and corresponding radiation power density $P_\mathrm{rad}(2^1P) = 2.2 \times 10^4\,\mathrm{W/m^3}$. This is 2/3 of the total radiation power $P_\mathrm{rad} = 3.3 \times 10^4\,\mathrm{W/m^3}$. Assuming a part of torus (toroidal angle $\Delta\phi = 15$ deg.) as a volume source and a solid angle $\Omega/4\pi \simeq 1/2$ for which EUV photon available, 0.3 W or 1×10^{17} photons/s of photon flux is obtained. Considering tokamak duration time of $\tau_\mathrm{d} \simeq 0.1\,\mathrm{ms}$, available energy $\simeq 33\,\mu\mathrm{J}$ or available dose $\simeq 1 \times 10^{13}$ photons is obtained for single discharge shot. This is comparable to 1 s operation of a commercially available EUV photon source (40 nm, 5×10^{12} photons/s) based on higher harmonic generation of femto-second laser [25]. Increasing repetition rate and controlling neutral particle density will improve the photon flux. Small tokamak could be an alternative EUV light source after design optimization for these viewpoint based on the power balance and the energy confinement scaling.

5 Conclusions

Ion temperature as well as electron temperature and density is measured in the smallest tokamak, TOKASTAR-2, using passive spectroscopy. Power balance is clarified. Electron heat conduction dominates power loss. Neo Alcator scaling, describing electron energy confinement time for ohmic heating tokamak, is applied for the smallest tokamak. Neutral depletion during tokamak operation is suggested. Resonant EUV line radiation is estimated from measurement data. Small tokamak could be an alternative EUV light source after design optimization based on the power balance and the energy confinement scaling.

Acknowledgments. This work was supported by JSPS KAKENHI Grant Numbers JP19H01869 and JP22H01202 and NIFS collaboration research program, NIFS20KBAH031, NIFS20KLEP038, and NIFS22KIIP008.

References

1. Lee, P.: Tokamak as an X ray / XUV light source. In: Conference Proceedings of SPIE Technical Symposium on Optics and Optoelectronic Engineering (1986)
2. Robinson, D.: Nucl. Fusion **25**, 1101 (1985)
3. Vannucci, A., Nascimento, I., Caldas, I.: Plasma Phys. Control. Fusion **31**, 147 (1989)
4. Kojima, H., et al.: Electr. Eng. Jpn. **133**, 8 (2000)
5. Kojima, H., et al.: Czechoslovak J. Phys. **53**, 895 (2003)
6. Takamura, S., Kikuchi, Y., Uesugi, Y., Kobayashi, M.: Nucl. Fusion **43**, 393 (2003)
7. Kikuchi, Y., Uesugi, Y., Takamura, S., Elfimov, A.G.: Nucl. Fusion **44**, S28 (2004)
8. Majeski, R., et al.: Nucl. Fusion **45**, 519 (2005)
9. Naito, S., et al.: Nucl. Fusion **61**, 116035 (2021)
10. Yasuda, K., Fujita, T., Okamoto, A., et al.: Phys. Plasmas **28**, 082108 (2021)
11. Hawryluk, R.J., Schmidt, J.A.: Nucl. Fusion **16**, 775 (1976)
12. Yang, J.G., et al.: J. Korean Phys. Soc. **26**, 336 (1993)
13. Yamazaki, K., Taira, Y., Oishi, T., Arimoto, H., Shoji, T.: J. Plasma Fusion Res. SERIES **8**, 1044 (2009)
14. Oishi, T., Yamazaki, K., Okano, K., et al.: J. Plasma Fusion Res. SERIES **9**, 69 (2010)
15. Ueda, T., Arimoto, H., Fujita, T., et al.: Plasma Fusion Res. **10**, 3402065 (2015)
16. Yasuda, K., Arimoto, H., Okamoto, A., et al.: Plasma Fusion Res. **13**, 3402072 (2018)
17. Ikeda, R., Fujita, T., Arimoto, H., et al.: Plasma Fusion Res. **15**, 2402047 (2020)
18. Yokoyama, R., Okamoto, A., Fujita, T., et al.: Plasma Fusion Res. **13**, 3402047 (2018)
19. Goto, M.: J. Quant. Spectr. Rad. Transfer **76**, 331 (2003)
20. Okamoto, A., Yoshimura, S., Terasaka, K., Tanaka, M.: Plasma Fusion Res. **14**, 1201165 (2019)

21. Okamoto, A., Kobuchi, T., Kitajima, S., Sasao, M.: Plasma Fusion Res. **2**, 029 (2007)
22. Kado, S., Shikama, T.: J. Plasma Fusion Res. **79**, 841 (2003)
23. Kimata, S., et al.: AIP Adv. **12**, 045204 (2022)
24. Goldston, R.: Plasma Phys. Coltrol. Fusion **26**, 87 (1984)
25. KM Labs Inc., Pantheon 30nm EUV laser system

Parametric Instabilities of Arc-Polarized Alfvén Waves Revisited

Yasuhiro Nariyuki(✉)

Faculty of Education, Academic Assembly, University of Toyama, Toyama, Japan
nariyuki@edu.u-toyama.ac.jp

Abstract. We revisit the parametric decay instability of arc-polarized Alfvén waves. Three dimensional magnetohydrodynamic (MHD) simulations show that the growth of decay instabilities of "shear" component of arc-polarized waves is in good agreement with linear growth rates of circularly polarized waves with effective value (root-mean-squared) amplitude. Daughter waves of the decay instability including "compressible" components of arc-polarized waves, whose frequency is twice as high as the shear component frequency, is also observed.

Keywords: Alfvén wave · parametric instability · solar wind

1 Introduction

The finite amplitude Alfvén wave is an exact solution of the ideal MHD system [1,2] and unstable to parametric instabilities [3–5]. While monochromatic, circularly polarized waves are often assumed in theoretical studies [3–5], non-monochromatic, non-circularly polarized waves are frequently observed in the solar wind plasma [6,7]. Parametric instabilities of the arc-polarized Alfvén wave, which is one of exact Alfvén waves, have been addressed by using the ion hybrid simulation [8] and the MHD simulation [9,10]. Del Zanna [9] has shown that the decay instability of arc-polarized waves is qualitatively similar to that of circularly polarized waves. It has also been shown that growth rates measured in numerical simulations quantitatively disagrees with those predicted by the linear analysis for the circularly polarized waves [9,10]. In the present study, we revisit its quantitative disagreement between the decay instability of the arc-polarized wave and that of the circularly polarized wave by using the three dimensional MHD simulation. The influence of the way to give initial waves on the decay instability and the decay instability of the "compressible" component are also discussed.

2 Numerical Simulation and Result

Three dimensional MHD simulations are carried out to study the parametric instability of arc-polarized Alfvén waves. Initial Alfvén waves are given under the conditon [2,11–14]

$$|\mathbf{b}|^2 = B_0^2 + b_f^2 = const., \tag{1}$$

H. J. Choi et al. (Eds.): APPC 2022, *Proceedings of the 15th Asia Pacific Physics Conference*, pp. 228–233, 2024.
https://doi.org/10.1007/978-981-96-0191-2_24

where $\mathbf{b} = (b_x, b_y, b_z) = \mathbf{B}_0 + (\delta b_x, b_y, b_z)$ is the magnetic field and the uniform ambient magnetic field $\mathbf{B}_0 = (B_0(> 0), 0, 0)$, b_f^2 corresponds to the portion of the fluctuation in the magnetic energy. In this study, the wave number vector of initial parent Alfvén waves is set as $\mathbf{k}_0 = (k_{x0}(> 0), k_{y0}(> 0), 0)$. From the divergence free condition, magnetic fluctuation of initial parent Alfvén wave is given as $\mathbf{b}_0 = (\delta b_{x0}, b_{y0}, b_{z0}) = (-\frac{k_y}{k_x} b_{y0}, b_{y0}, b_{z0})$.

In this study, we give initial compressible components (δb_x, b_y) as the sinusoidal function [14]:

$$b_y = b_0 \cos(\omega_0 t - \mathbf{k}_0 \cdot \mathbf{x}), \tag{2}$$

where $b_0 > 0$. The shear component (b_{z0}) is given by using Eq. (1) as [14]

$$b_{z0} = \pm\sqrt{(B_0^2 + b_f^2) - (\delta b_{x0} + B_0)^2 - b_{y0}^2}, \tag{3}$$

Since $b_{z0}^2 \geq 0$, $(\delta b_{x0} + B_0)^2 + b_{y0}^2 \leq (b_0 + B_0)^2 + (-b_0)^2 \leq B_0^2 + b_f^2$. By using the case of the equality ($(b_0 + B_0)^2 + (-b_0)^2 = B_0^2 + b_f^2$), we find the quadratic equation of b_0 [14]. The wave form given by Eq. (3) with one sign in front of the square root has the asymmetric wave form, since δb_x and b_y are sinusoidal functions. In this study, in order to give the symmetric wave form, the sign in front of the square root in Eq. (3) is changed when $|b_{z0}|$ arrives at the minimum value, which theoretically equal to zero. The initial wave form is shown in Fig. 1. The numerical points before and after the change of the sign are linearly interpolated. Since the minimum value is usually close to zero, the wave form is smooth enough to give the stable propagation of Alfvén wave. While the shear component is given by sinusoidal function in the previous studies [2,8,9], these differences do not affect the decay instability of the shear component as shown below. The initial plasma bulk velocity is given to satisfy the condition of the Alfvén wave solution [1,2] in the ideal MHD system: $\mathbf{u}_0 = -\frac{C_A}{B_0}(\delta b_{x0}, b_{y0}, b_{z0})$. The number of grid points $(N_x, N_y, N_z) = (128, 128, 128)$ or $(256, 256, 256)$. In all the runs, the system size $(L_x, L_y, L_z) = (32, 32, 32)$. As shown below, it is confirmed that the difference of the number of grid points does not affect the growth of the daughter waves.

Table 1 shows $\tan\theta = \frac{k_y}{k_x}$ and initial amplitude of $\mathbf{b}$ in each run. In this study, the ideal MHD system with scalar pressure (p)

$$\frac{\partial \rho}{\partial t} = -\nabla \cdot (\rho \mathbf{u}), \tag{4}$$

$$\frac{\partial \rho \mathbf{u}}{\partial t} = -\nabla \cdot (\rho \mathbf{u}\mathbf{u}) - \nabla p + (\nabla \times \mathbf{b}) \times \mathbf{b}, \tag{5}$$

$$\frac{\partial \mathbf{b}}{\partial t} = \nabla \times (\mathbf{u} \times \mathbf{b}), \tag{6}$$

$$\nabla \cdot \mathbf{b} = 0, \tag{7}$$

is solved by using the rationalized Runge-Kutta method [15] for time integration and the pseudo-spectrum method for the spatial derivative with the periodic boundary. Density (ρ), velocity ($\mathbf{u}$) and magnetic field ($\mathbf{b}$) are normalized to the background uniform density (ρ_0), the background uniform magnetic field B_0, and the Alfvén speed C_A, respectively. The isothermal electron is assumed and $\beta_f = C_s^2/C_A^2 = 0.36$, where C_s is the ion sound speed.

Table 1. $\tan\theta = \frac{k_{y0}}{k_{x0}}$ and initial amplitude of **b** in each run. In Run A, $k_{x0} = \frac{2\pi\times 6}{32} = 1.18$ and $k_{x0} = \frac{2\pi\times 8}{32} = 1.57$ in Run B and C.

	Run A	Run B	Run C
$\tan\theta$	$\frac{1}{3}$	$\frac{1}{4}$	$\frac{1}{2}$
b_f^2	0.16	0.16	0.16
$\max[\delta b_{x0}]$	0.061	0.055	0.068
$\max[b_{y0}]$	0.18	0.22	0.14
$\max[b_{z0}]$	0.50	0.47	0.52

As shown in Fig. 1, since the initial amplitude of shear component b_z is larger than those of the other components, the decay instability of b_z is the dominant instability at the linear stage. Figure 2 shows the growth of the wave number mode of parallel propagating daughter density modes with maximum growth rates in (a)Run A, (b)Run B, and (c)Run C. As the decay instability of circularly polarized waves excites daughter waves [10, 16–19], that of the arc-polarized wave also excites the daughter waves with the different direction of the wave number vector from that of the parent wave [10].

It has been reported by the previous study [9] that growth rates of the decay instability of arc-polarized Alfvén waves are proportional to the parent wave frequency, while it does not agree with those of circularly polarized Alfvén waves in the limit of the small propagation angle. Two straight lines in each panel of Fig. 2 indicate the maximum growth predicted by the linear analysis of circularly polarized Alfvén wave [3,4] with $\max[b_{z0}]$ and $\max[b_{z0}]/\sqrt{2}$. It is clearly shown that the growth rates calculated with the effective value $\max[b_{z0}]/\sqrt{2}$ are in good agreement with numerical results. To the best of author's knowledge, this agreement between the model with effective values and the numerical results has not been reported yet.

The model with the effective value is also consistent with the result in Del Zanna [9]. In Run A and B of Del Zanna [9], the decay instability is observed with parameters $(k_{x0}, \max[b_{z0}], \beta_f, \gamma_{max}) = (4, 0.2, 0.1, 0.412)$ and $(4, 0.5, 0.5, 0.385)$, where γ_{max} is the maximum linear growth rate for the circularly polarized parent wave. When the effective value ($\max[b_{z0}]/\sqrt{2}$) is applied, $\gamma_{max} = 0.296$ (Run A of Del Zanna [9]) and 0.266 (Run B of Del Zanna [9]), respectively. Since $k_{x0} = 2$ for $tan\theta = \sqrt{3}$ in Del Zanna [9], the growth rates to be compared with the values measured in the numerical result are $\gamma_{max} = 0.148$ (Run A of Del Zanna [9]) and 0.133 (Run B of Del Zanna [9]), which are fairly close to the measure values [9] (0.135 and 0.113). On the other hand, the present model with the effective value does not agree with Run C of Del Zanna [9], in which the beat instability [5] is observed.

It is also reported in Del Zanna [9] that the decay instability of linearly polarized waves is similar to that of the arc-polarized wave. The present result quantitatively

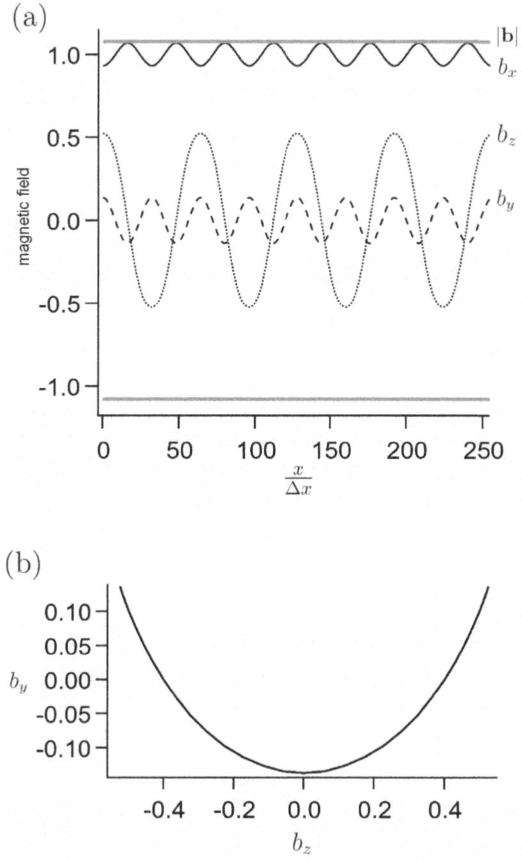

Fig. 1. (a) Initial wave forms of $\mathbf{b}(x, y)$ in Run C at $y = 0$. Solid, dashed, dotted black lines indicate b_x, b_y, and b_z, respectively. Gray lines show $|\mathbf{b}|$ and $-|\mathbf{b}|$. (b) The hodogram between b_y and b_z shown in (a).

supports the previous result. Namely, the good agreement with the linear growth rate with the effective value implies that the shear component of the arc-polarized wave can be regarded as the linear polarized wave and the influence of compressible components on the decay instability of the shear component is negligible.

On the other hand, compressible components are also unstable to the decay instability, although their growth is much weaker than that of the shear component because of their small amplitude. Figure 3 shows two-dimensional wave number spectra of ρ (left panel), b_y (middle panel), and b_z (right panel), which are averaged over z axis before carrying out the Fourier transform to x and y axis. Daughter waves excited through the decay instability of b_{z0} ($(k_{x0}/2, k_{y0}/2)$; (1) in Fig. 3) are observed in b_z ((4) in Fig. 3)) and ρ ((5) in Fig. 3)). As shown above, this decay instability is consistent with the one reported in the previous study [9], in which the shear component is given as the sinusoidal function. This is because the amplitude of nonmonochromatic component of b_{z0}

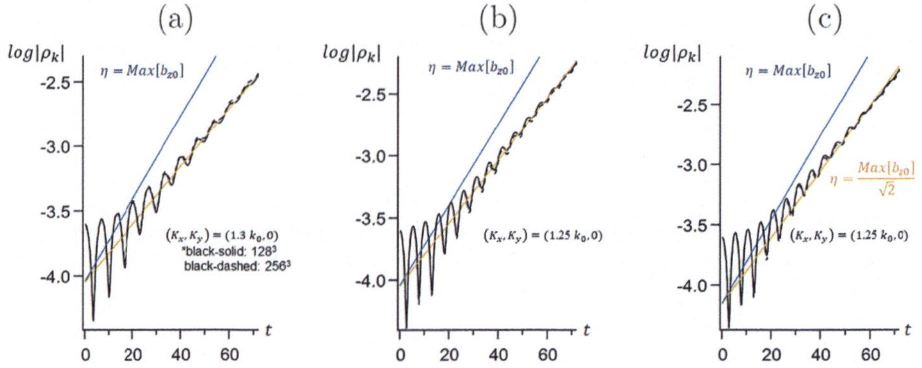

Fig. 2. (Color online) Growth of the wave number mode of parallel propagating daughter density modes with maximum growth rate in (a) Run A, (b) Run B, and (c) Run C. Black solid and dashed lines, which are almost overlapped, indicate runs with $(N_x, N_y, N_z) = (128, 128, 128)$ and $(256, 256, 256)$, respectively. Two straight lines in each panel indicate the maximum growth predicted by the linear analysis [3,4] with $\max[b_{z0}]$ (blue lines) and $\max[b_{z0}]/\sqrt{2}$ (orange lines), respectively. Note that the parent wave number of b_z is given as $k_{x0}/2$, since the wave length of b_{z0} is the double of b_{y0} (Fig. 1).

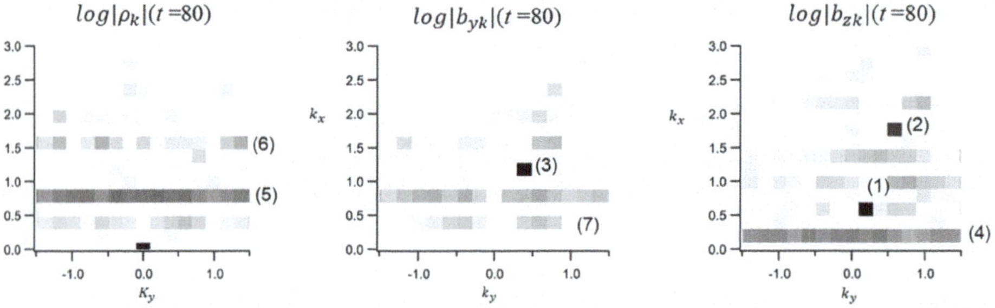

Fig. 3. Two dimensional wave number spectra of (left panel) density ρ, (middle panel) compressible component of magnetic field b_y, and (right panel) shear component of magnetic field b_z in Run A with $(N_x, N_y, N_z) = (128, 128, 128)$. (1)-(6) indicate (1)the dominant wave number mode of b_{z0}, (2)the nonmonochromatic component of b_{z0}, (3)the dominant wave number mode of b_{y0} (and b_{x0}, (4)dominant wave number modes of daughter waves of the decay instability of b_{z0}, (5)dominant wave number modes of daughter density waves of the decay instability of b_{z0}, (6)the second higher harmonic waves of (5), (7)daughter Alfvén waves of decay instability among (3) and (6).

((2) in Fig. 3) is much smaller than that of $(k_{x0}/2, k_{y0}/2)$. We can also find the daughter wave in b_y ((7) in Fig. 3) which is excited by the decay instability among the compressible component of the initial parent wave in b_y ((3) in Fig. 3)) and the daughter density mode ((6) in Fig. 3)), which corresponds to the second harmonic of the decay instability of b_{z0}.

3 Conclusions

In this study, the decay instability of the arc-polarized Alfvén wave is revisited. The growth of daughter waves excited by the instability of the shear component is similar to the one reported in the previous study, while the shear component of the initial arc-polarized waves in the present study is not sinusoidal. It is shown that the growth of the decay instability is consistent with the prediction of the linear analysis for circularly polarized wave, in which the effective value of the shear component is used as the parent wave amplitude. The decay instability of the compressible component of the initial parent wave is also observed, while it is much weaker than that of the shear component.

Acknowledgments. The computer simulation was partially performed on the KDK computer system at Research Institute for Sustainable Humanosphere, Kyoto University.

References

1. Alfvén, H.: Nature **150**, 405 (1942)
2. Barnes, A., Hollweg, J.V.: J. Geophys. Res. **79**, 2302 (1974)
3. Derby, N.F.: Astrophys. J. **224**, 1013 (1978)
4. Goldstein, M.L.: Astrophys. J. **219**(2), 700 (1978)
5. Jayanti, V., Hollweg, J.V.: J. Geophys. Res. **98**, 19049 (1993)
6. Bruno, R., Carbone, V.: Living Rev. Solar Phys. **10**, 2 (2013) [Online Article], http://www.livingreviews.org/lrsp-2013-2
7. Tsurutani, B.T., Lakhina, G.S., Sen, A., Hellinger, P., Glassmeier, K.H., Mannucci, A.J.: J. Geophys. Res. **123**, 2458–2492 (2018)
8. Vasquez, B.J., Hollweg, J.V.: J. Geophys. Res. **101**(A6), 13527–13540 (1996)
9. Del Zanna, L.: Geophys. Res. Lett. **28**(13), 2585–2588 (2001)
10. Del Zanna, L., Matteini, L., Landi, S., Verdini, A., Velli, M.: J. Plasma Phys. **81**, 325810102 (2015)
11. Roberts, D.A.: Phys. Rev. Lett. **109**, 231102 (2012)
12. Valentini, F., Malara, F., Sorriso-Valvo, L., Bruno, R., Primavera, L.: Astrophys. J. Lett. **881**, L5 (2019)
13. Primavera, L., Malara, F., Servidio, S., Nigro, G., Veltri, P.: Astrophys. J. **880**, 156 (2019)
14. Nariyuki, Y.: Rev. Mod. Plasma Phys. **6**(22) (2022)
15. Wambecq, A.: Computing **20**, 333 (1978)
16. Ghosh, S., Vinas, A.N., Goldstein, M.L.: J. Geophys. Res. **98**(A9), 15561 (1993)
17. Del Zanna, L., Velli, M., Londrillo, P., A&A.: **367**, 705 (2001)
18. Shoda, M., Yokoyama, T.: Astrophys. J. Lett. **859**, L17 (2018)
19. Nariyuki, Y.: J. Phys. Soc. Jpn. **89**, 045002 (2020)

Statistical Physics

A Spin Model for Global Flat-Foldability of Random Origami

Chihiro Nakajima(✉)

Tohoku Bunka Gakuen University, Sendai Miyagi 6-45-1, Japan
chihiro.nakajima@ait.tbgu.ac.jp

Abstract. We map the problem of determining flat-foldability of the origami diagram onto the ground-state search problem of spin glass model on random graphs. If the origami diagram is locally flat-foldable around each vertex, a pre-folded diagram, showing the planar-positional relationship of the facet, can be obtained. For remaining combinatorial problem on layer ordering of facets can be described as a spin model. A spin variable is assigned for the layer-ordering of each pair of facets which have an overlap in the pre-folded diagram. The interactions to prohibit the intrusion of each facet into the other component of the same origami diagram are introduced among two or four spins. The flat-foldability of the diagram is closely related to the (non-)existence of frustrated loops on the spin model with the interactions on the random (hyper)graph.

Keywords: Computational origami · Statistical mechanics · Flat-foldability · Random spin model

1 Introduction

Origami has its origin in traditional artwork of folding a piece of paper along straight crease lines into various shapes of animals, plants, or objects. By the extensive research in various fields, origami has found a number of interesting mathematical and engineering applications for today. In order to fold a sheet along straight crease lines into many kinds of geometric patterns, conformation of facets and creases should be incorporated into a sheet of paper in a well-planned manner. This sheet with planned facets and creases is called an origami diagram or a crease pattern. The design of the origami diagram has been one of research subjects in the field of computer seience.

The problem to determine whether a given origami diagram is actually foldable into a flat plane, is called the flat-foldability problem, which is one of the major topics in the computational origami [1]. On the computational complexity class, this problem is proven to be NP-hard [2] and it is known that to assign the global layer ordering of flaps, namely closed polygonal facets separated by creases, can be the central source of its hardness [2,4].

Analyses of foldability properties associated with spin models have been previously performed in somewhat limited ways, each with interesting results [3,4].

H. J. Choi et al. (Eds.): APPC 2022, *Proceedings of the 15th Asia Pacific Physics Conference*, pp. 237–249, 2024.
https://doi.org/10.1007/978-981-96-0191-2_25

Furthermore, there are also works dealing with the folding of the membrane, that is a physical reality of folding itself under the thermal fluctuation, with the treatment by spin models [5,6]. Previous statistical mechanics studies on origami have focused on the proximity of facets on origami diagrams, and have dealt with this by relating to the proximity of interactions on models. While, we consider the overlap for every pair of facets, including pairs for not close, or connected, on the origami diagram. We pay attention to whether two facets have the overlapped region while illustrating the positional relationship between two facets after the origami diagram is folded along the creases. With such manner we can obtain a complete correspondence with the model of random spin systems. In fact, it was previously pointed out that the global layer ordering of facets should in principle be translated into a statistical mechanics model [4]. However, by actually doing this, it is possible to highlight things that are not apparent from the proximity on the origami diagram. We can find the elements that give unfoldability in a more intuitive way on a hypergraph composed of interactions. In addition we can expect a reduction in the amount of computation in the flat-foldability problem by considering the structure of the hypergraph.

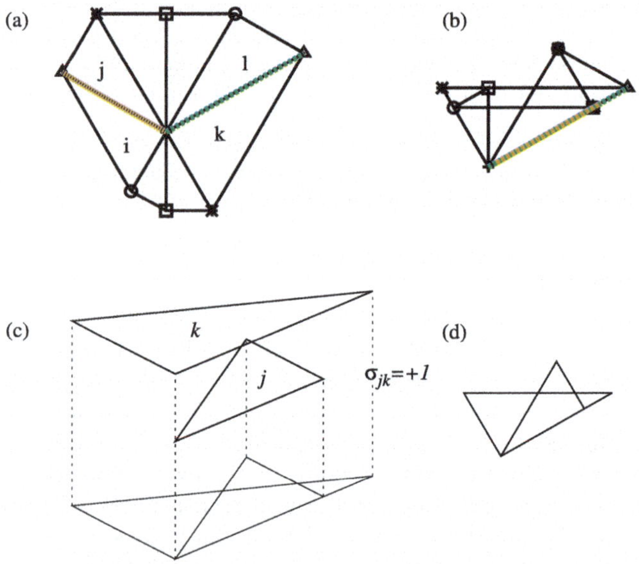

Fig. 1. (a)Example of origami diagram. Each edge in the figure represents a crease. In this figure there are no overlaps of facets. (b)Corresponding pre-folded diagram, which describes the overlaps of facets when the figure (a) is folded along the creases. Each mark corresponds to the vertex in Fig.(a) with the same mark. Unmarked cross-points in Fig.(b) are simply the intersection of two edges caused by the overlaps, not the vertex of a facet. (c)Schematic picture of introduction of the Ising variable to a local layer-ordering between the facets j and k. (d)The planar-positional relationship of the facets j and k corresponding to Fig.(c), also included in Fig.(b).

2 Models and Implementations

There are Kawasaki's theorem which gives necessary and sufficient conditions for the (local) flat-foldability of the origami diagrams which consist of one interior vertex and other surrounding exterior vertices [7]. According to the theorem, the number of facets clustered around the central vertex must be even and the alternating sum of the angles clustered around the vertex must be 0. For an origami diagram that satisfies the conditions from Kawasaki's theorem for every internal vertices, it is obtained a second diagram describing the planar-positional relationship of the facets after folding every creases. Here this second diagram is called pre-folded diagram. For example, Fig. 1(a) is an origami diagram which satisfies the condition with Kawasaki's theorem. The eight polygons (triangles) included in this figure are all facets, and the edges connecting the facets are creases. When the eight creases in Fig. 1(a) are folded, the shape becomes as Fig. 1(b). Folding causes the overlaps among facets. This Fig. 1(b) is the pre-folded diagram corresponding to the origmai diagram of 1(a). Hence, the (global) flat-foldability problem is as a problem of determining whether or not each facet can be stacked without breaking, that is, the intrusion of a facet of the origami paper into other component in the same paper.

Next, according to the pre-folded diagram, the layer-ordering of the facets should be determined. The local layer ordering, whether one facet of the pair is below- or above- side of the another, can be represented by the binary variable. Therefore the Ising spin variables $\sigma \in \{+1, -1\}$ are assigned for each pair of facets which have an overlap in the pre-folded diagram. For example, in the case of the figure in Fig. 1(c) $\sigma_{j,k}$ is defined to take its value $+1$ when the facet k is above the facet j in z-axis (the direction of height). Here, $\sigma_{j,k}$ is also defined to invert its value for the exchange of the order of indices, namely $\sigma_{j,k} = -\sigma_{k,j}$. From the set of $\{\sigma\}$ which describes the local layer orderings, the total layer ordering of facets is reconstructed. The variables $\sigma_{j,k}$ according to the above definition are convenient to give terms of the energy function described later, as Eqs. (7), (9), and (11), using spin variables from the geometry of facets via subscripts. However, from the point of view of statistical mechanics, it obscures the prospects for unambiguously defining the state space or the configuration space to use the variables with exchange of the order of subscripts or sign-reversal. To ensure better prospect, we introduce the variables $s_{j,k}$ and τ_{jk} as follows and use them to rewrite $\sigma_{j,k}$.

$$\tau_{jk} = \mathrm{sign}(k - j), \tag{1}$$

$$s_{j,k} = \sigma_{\min(j,k),\max(j,k)}. \tag{2}$$

With the variables $\sigma_{j,k}$, $s_{j,k}$ and τ_{jk}, the following relation holds,

$$\sigma_{j,k} = \tau_{jk} s_{j,k}. \tag{3}$$

The behavior of the variable for the exchange of the order of subscripts is, respectively,

$$\sigma_{j,k} = -\sigma_{k,j}, \tag{4}$$

$$\tau_{jk} = -\tau_{kj}, \tag{5}$$

$$s_{j,k} = s_{k,j}. \tag{6}$$

The variable $s_{j,k}$ also trivially takes two values +1 or −1. By considering the state space in terms of $s_{j,k}$, the state sum is taken on the state space of $\{s_{j,k}\}$, where $j < k$. Thus the bounds of the state space for the origami-diagram with N facets are limited to $2^{N(N-1)/2}$. After here, when the spin variable $s_{a,b}$ in which the former part of whose subscript is larger than the latter one appears, it shall be received according to Eq. (6). One might be concerned that different styles of facet labeling will have different signs for the above variables even for the same pre-folded diagram. In fact, however, we can label the facets in any order, for reasons based on gauge transformations described afterward.

By representing the problem with the combination of the local layer-ordering, the flat-foldability problem of the origami becomes the combinatorial optimization problem. Under the planar placement of each facet in the pre-folded diagram, depending on the choice of total layer-ordering, there are some cases in which an intrusion between facets may become unavoidable. To prohibit such ordering, some cost function terms are introduced. Constrains are introduced so that the energy takes a positive value for prohibited configurations. For the spin configurations that satisfy all constraints, they will be corresponding to the realizable layer-ordering.

As the first kind of constraints, we introduce a term that contains a product of two spin variables to prohibit an intru- sion of a facet into a crease. If one facet k has an overlap with other two facets i, j, connected to each other, in the pre-folded diagram, the former may be sandwiched between the latter so that it intrude into the junction of the two facets. Such an arrangement is not realizable. The case which the facet k is sandwiched between i and j is corresponding to combinations of spin that is $\{\sigma_{i,k}, \sigma_{k,j}\} = \{1, 1\}$ or $\{-1, -1\}$.

$$\begin{aligned} E^{(i)}_{ij,k} &= \frac{1}{2}\big(1 + \sigma_{i,k}\sigma_{k,j}\big) \\ &= \frac{1}{2}\big(1 - J_{(ik)(kj)} s_{i,k} s_{k,j}\big) \end{aligned} \tag{7}$$

$$J_{(ik)(kj)} = -\tau_{ik}\tau_{kj}. \tag{8}$$

The second kind of terms are required if the two creases are in geometrically coincidental position in the pre-folded diagram. Suppose that there are two creases that partially coincide with each other and pairs of facets, called i, j and k, l, each connected by the creases. The example is exhibited in Fig. 1(a) and (b). For such arrangements and connections of facets there are no intrusion even if both i and j are sandwiched between k and l, different from the cases considered in Eq. (7) and (8). However, if only one of i or j is sandwiched between k

and l, the creases will penetrate each other at the region where two connections coexist. In Fig. 2(a)-(c), the schematic picture of the local layer-ordering of the four facets i, j, k and l in the origami diagram and the pre-folded diagram of Fig. 1(a) and (b), an example of two pairs of facets connected by creases with partial coincidence in planar-positioning. The dotted lines of color orange and green in Fig. 2 are with respect corresponding to the edges connecting i and j, k and l in Fig. 1(b). The each pair of facets connected to the both side of each edge, i, j, k and l, is represented as solid line with the same color of the edge. Figure 2(a) and (b) represent the acceptable layer-orderings. Especially Fig. 2(b) is corresponding to the cases that both k and l are sandwiched between i and j. While, Fig. 2(c) is corresponding to the prohibited layer-ordering that only k is samdwitched between i and j. In these cases we need the product of four spin variables as the following form,

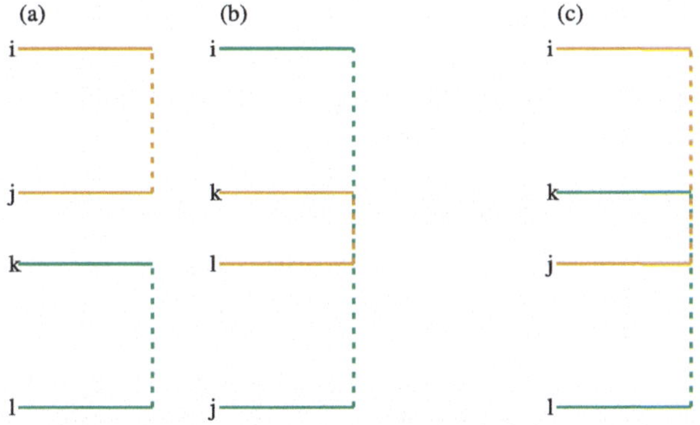

Fig. 2. (a),(b) Layer orderings of facets which are accepted under the coincident of two creases. (c) Layer ordering which is NOT accepted under the coincident of two creases. One connected pair of facets must penetrate the another connected part of the remanent two facets.

$$
\begin{aligned}
E_{ijkl}^{(q)} &= \frac{1}{2}\Big(1 - \sigma_{i,k}\sigma_{i,l}\sigma_{j,k}\sigma_{j,l}\Big) \\
&= \frac{1}{2}\Big(1 - K_{ijkl} s_{i,k} s_{i,l} s_{j,k} s_{j,l}\Big), \qquad (9) \\
K_{ijkl} &= \tau_{ik}\tau_{il}\tau_{jk}\tau_{jl}. \qquad (10)
\end{aligned}
$$

In addition to the above two kinds, the third kind of constraints is imposed to prohibit cyclic ordering of three facets. For three facets i, j, and k with $i < j < k$, cyclic ordering, for example the case represented with $\{\sigma_{i,j}, \sigma_{i,k}, \sigma_{i,k}\} = \{1, 1, -1\}$, can be considered in general. However, in the case these three facets have areas shared with each other in the pre-folded diagram, such ordering is unrealizable. To prohibit the cyclic ordering, the set of interaction terms of spins are introduced as follows,

$$
\begin{aligned}
E^{(c)}_{ijk} &= \frac{1}{4}\big(1 + \sigma_{i,j}\sigma_{j,k} + \sigma_{j,k}\sigma_{k,i} + \sigma_{k,i}\sigma_{i,j}\big) \\
&= \frac{1}{4}\big(1 - L_{(ij)(jk)} s_{i,j} s_{j,k} \\
&\qquad - L_{(jk)(ki)} s_{j,k} s_{k,i} - L_{(ki)(ij)} s_{k,i} s_{i,j}\big), \qquad (11) \\
L_{(ij)(jk)} &= -\tau_{ij}\tau_{jk} \qquad (12)
\end{aligned}
$$

Notice that there are such cases for 3 facets that no regions where all of them stack at the same time while any 2 pairs of them have each overlap. For these cases the set of terms like Eq. (11) is not to be introduced because the cyclic ordering is actually realizable without intrusion.

In the process of labeling indices into facets on an pre-folded diagram, any order can be employed. The sign of each interaction coefficient $J_{(ij)(jk)}$, K_{ijkl}, and $L_{(ij)(jk)}$ varies depending on the details of the index-labeling of facets. However, the changes of them due to the differences of the way of labeling can be rewritten back to the original Hamiltonians by the following transformations with gauge variables $u_{ij} \in \{1, -1\}$ such as

$$
\begin{aligned}
J_{(i'k')(k'j')} &\rightarrow \hat{J}_{('i'k)(k'j')} = u_{i'k'} u_{k'j'} J_{(i'k')(k'j')}, \qquad (13) \\
K_{i'j'k'l'} &\rightarrow \hat{K}'_{i'j'k'l'} = u_{ik} u_{i'l'} u_{j'k'} u_{j'l'} K_{i'j'k'l'}, \qquad (14) \\
L_{(i'k')(k'j')} &\rightarrow \hat{L}_{(i'k')(k'j')} = u_{i'k'} u_{k'j'} L_{(i'k')(k'j')}, \qquad (15)
\end{aligned}
$$

where the value of u_{ij} is given by the relation between the two different lookings $s_{i,j}$ and $s_{i',j'}$ which are caused by different ways of index-labeling of facets,

$$
u_{i'j'} s_{i',j'} = s_{i,j}. \qquad (16)
$$

Since the partition function is invariant under the gauge transformations, the same layer-ordering can be computed with any index-labeling of facets [8,9].

As an example of the above transformation, we see the origami-diagram in Fig. 3. In the origami-diagram in Fig. 3(a) the indices $i \in \{1, 2, 3, 4\}$ is allocated to each facet. Suppose that $i' = 4, 3, 1, 2$ is assigned to the each same facet (with $i = 1, 2, 3, 4$) by different way of indexing. A term which prohibits the intrusion of the facet $i = 4$ between the two facets $i = 2$ and 3, whose form is $(1 - J_{(24)(43)} s_{2,4} s_{3,4})/2$ with $J_{(24)(43)} = 1$, is included in the terms in the energy function. Within the way of indexing with $i' = 4, 3, 1, 2$, the same prohibition term is described as the form $(1 - J_{(32)(21)} s_{2,3} s_{1,2})/2$ with $J_{(32)(21)} = -1$. Note that in these two descriptions $s_{2,4} = 1$ within $i = 1, 2, 3, 4$ and $s_{2,3} = -1$ within $i' = 4, 3, 1, 2$ represent the same local layer-ordering in the sense of geometry. Thus, the signs of the interaction coefficients differ each other depending on the way of indexing, even though the same in the meaning of the geometric content is prohibited. However, when we apply the above transformation with the gauge variable u_{ij} as

$$u_{14} = -1, \tag{17}$$
$$u_{34} = -1, \tag{18}$$
$$u_{24} = -1, \tag{19}$$
$$u_{13} = -1, \tag{20}$$
$$u_{23} = -1, \tag{21}$$
$$u_{12} = 1, \tag{22}$$

the interaction coefficients are transformed into,

$$\hat{J}_{(32)(21)} = (+1)(-1)J_{(32)(21)} = -J_{(32)(21)}, \tag{23}$$

and the term $(1 - \hat{J}_{(32)(21)}s_{2,3}s_{1,2})/2$ within $i' = 4, 3, 1, 2$ becomes the same as $(1 - J_{(24)(43)}s_{2,4}s_{3,4})/2$ within $i = 1, 2, 3, 4$. Looking at the same thing as this rewrite for the whole intrusion-prohibiting terms, at first the terms with $i = 1, 2, 3, 4$ is described as,

$$\sum E^{(i)}_{(ij,k)} = \frac{1}{2}\Big(6 - s_{1,3}s_{1,2} - s_{1,4}s_{1,2} - s_{1,3}s_{2,3} + s_{3,4}s_{2,3} - s_{1,4}s_{2,4} - s_{3,4}s_{2,4}\Big) \tag{24}$$

Similarly, the same contents within $i' = 4, 3, 1, 2$ is described as,

$$\sum E^{(i)}_{(i'j',k')} = \frac{1}{2}\Big(6 - s_{1,4}s_{3,4} - s_{2,4}s_{3,4} - s_{1,4}s_{1,3} - s_{1,2}s_{1,3} - s_{2,4}s_{2,3} + s_{1,2}s_{2,3}\Big), \tag{25}$$

where the order in which the terms are arranged is the order of the same geometric content, namely the prohibition for same subset of facets, as above. By the transformation with the gauge variables (17)-(22)the form of the prohibition term (25) is written back to the same as Eq. (24), as exhibited that

$$\sum \hat{E}^{(i)}_{(i'j',k')} = \frac{1}{2}\Big(6 - s_{1,4}s_{3,4} - s_{2,4}s_{3,4} - s_{1,4}s_{1,3} + s_{1,2}s_{1,3} - s_{2,4}s_{2,3} - s_{1,2}s_{2,3}\Big). \tag{26}$$

3 Details of Energy Function for Some Illustrative Examples

In the example of Fig. 3, each facet located outside, 1, 3, 4, cannot be sandwiched between every creases which consist of other two facets. Therefore in the energy function for this pre-folded diagram 6 terms of first kind are included. In addition there are 4 terms of third kind. Totally, the form of the energy function for this diagram is as follows,

$$H(\{s\}) = \sum E^{(i)}_{(ij,k)} + \sum E^{(c)}_{(ijk)} \tag{27}$$

$$\begin{aligned}\sum E^{(i)}_{(ij,k)} = \frac{1}{2}\Big(&6 - s_{1,3}s_{1,2} - s_{1,4}s_{1,2} - s_{1,3}s_{2,3} \\ &+s_{3,4}s_{2,3} - s_{1,4}s_{2,4} - s_{3,4}s_{2,4}\Big)\end{aligned} \tag{28}$$

$$\begin{aligned}\sum E^{(c)}_{(ijk)} = \frac{1}{4}\Big(&1 + s_{1,2}s_{2,3} - s_{2,3}s_{1,3} - s_{1,3}s_{1,2} \\ &+1 + s_{1,2}s_{2,4} - s_{2,4}s_{1,4} - s_{1,4}s_{1,2} \\ &+1 + s_{1,3}s_{3,4} - s_{1,4}s_{3,4} - s_{1,3}s_{1,4} \\ &+1 + s_{2,3}s_{3,4} - s_{3,4}s_{2,4} - s_{2,4}s_{2,3}\Big),\end{aligned} \tag{29}$$

where the interaction coefficients are expressed numerically. The terms in Eq. (28) come from the prohibition for intrusive layer-orderings. These form a cycle.

Similarly, the terms in Eq. (29) are given from the prohibition of cyclic orderings.

In this energy function, the chain of 2-spin interactions in the first 2 rows, given from $\{J_{(ij)(jk)}\}$, forms a closed cycle in which the product of $J_{(ij)(jk)}$s involved is negative, that is,

$$\begin{aligned}&\prod_{(ik)(kj)}^{\text{cycle}} J_{(ij)(kl)} \\ &\quad = J_{(31)(12)}J_{(41)(12)}J_{(13)(32)}J_{(43)(32)}J_{(14)(42)}J_{(34)(42)} \\ &\quad = -1.\end{aligned} \tag{30}$$

In the cycle the 6 spin variables are involved and then there are 64 possible binary combinations. When the sign of the product of $J_{(ij)(jk)}$s involved is negative, none of the combinations can give the value $\sum E^{(i)} = 0$ of the local energy corresponding to this cycle.

When 4-body interactions are involved, frustration is defined not for a single cycle, but for a composite structure of multiple cycles combined through 4-body interactions. In the example of the origami and pre-folded diagram in Fig. 4, the two creases each connecting facets 1, 6 and 7, 8 in (a) are coincident in (b). Hence, the Hamiltonian corresponding to this pre-folded diagram contains four-body interactions involving facets 1, 6, 7, and 8, whose interaction coefficient is represented as K_{1678}. In the Hamiltonian, totally there are 42 terms with the form same as Eq. (7), 84 terms with the form same as Eq. (11), and a term with the form same as Eq. (9). Then, 11 terms in $\sum E^{(i)} + \sum E^{(q)}$ form a composite of 2 cycles in which the sign of the product of interaction coefficients involved is negative as follows,

$$\begin{aligned}\prod_{(ij)(jk)}^{\text{cycle}} J_{(ij)(jk)} \prod_{(mnop)}^{\text{cycle}} K_{mnop} \\ &= J_{(1,7)(7,2)} J_{(7,2)(2,1)} J_{(1,2)(2,6)} J_{(2,6)(6,1)} J_{(1,6)(6,7)} \\ &\quad \times J_{(1,8)(8,7)} J_{(7,8)(8,9)} J_{(8,9)(9,10)} J_{(9,10)(10,7)} \\ &\quad \times J_{(7,10)(10,1)} J_{(1,10)(10,6)} J_{(10,6)(6,8)} \\ &\quad \times K_{1678} \\ &= -1. \end{aligned} \tag{31}$$

This diagram is also unfoldable.

In a model in which all interaction coefficients are given only by binary variables $+1$ or -1, it generally holds that there is no spin configurations that satisfy all the 2- (or 4-) spin interactions in the cases which the subgraph forms a cycle (or a composite of them) and the product of the signs of the interaction coefficients $J_{(ij)(jk)}$ (or K_{ijkl}) involved is negative. Therefore, if the model graph contains frustration, for example as illustrated here, we can conclude that the configurations with $\sum E^{(i)}_{ij,k} + \sum E^{(q)}_{ijkl} = 0$ does not exist. Same as the partition function, the product of interaction coefficients $J_{(ij,k)}$ or K_{ijkl} contained in the closed cycle or its complex, a quantity called frustration, is also invariant to the gauge transformation shown in Eq. (16). If we rewrite the expressions of Eq. (30) or Eq. (31) with τ_{jk}s according to Eq. (8) and (10), we see that each τ_{jk} corresponding to the same pair of the facet-indices always appears an even number of times. Therefore changes in the values of $J_{(ij,k)}$s and K_{ijkl}s accompanying the changes in the labeling order of facets are canceled in the cycle. Hence, the detection of the presence of frustration can be used to determine unfoldability.

This section focuses on frustration caused by the intrusion prohibition. In the case that the frustration caused by them is included in the Hamiltonian, no ground states with $H = 0$ exist. Hence, the searching for such frustration is

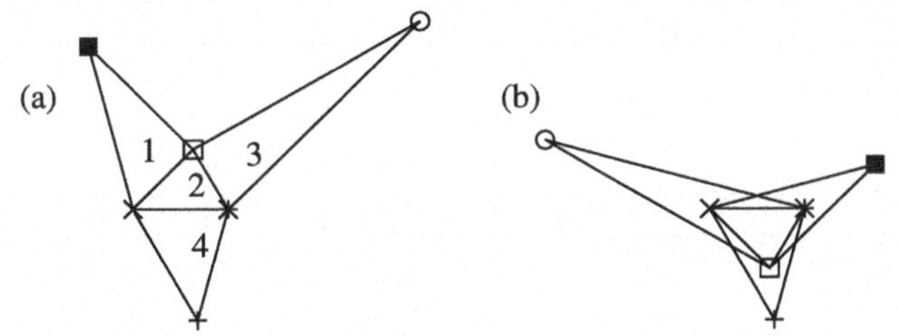

Fig. 3. (a)An example of the origami diagram which is flat-unfoldable. The numbers in the figure represent the indices of the facets. The pairs of numbers appearing in the subscripts of Eqs. (28) and (29) and (30) in the text are given corresponding to them. (b)Pre-folded diagram corresponding to the Fig.(a). Each mark corresponds to the vertex in Fig.(a) with the same mark. Unmarked cross-points in Fig.(b) are simply the intersection of two edges caused by the overlaps, not the vertex of a facet.

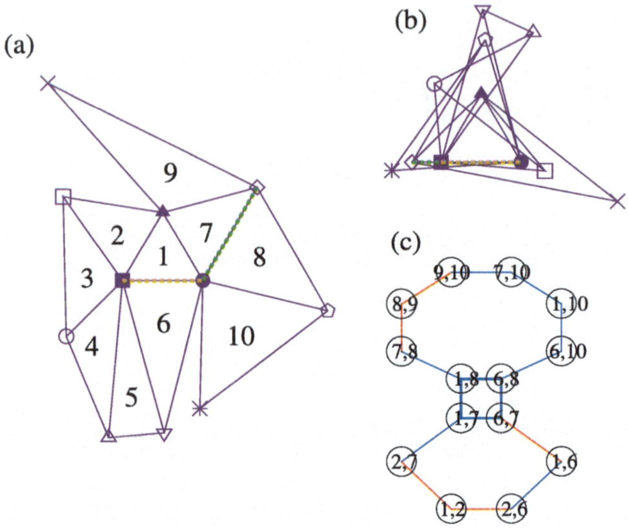

Fig. 4. (a)Another example of the origami diagram which is flat-unfoldable. (b)Prefolded diagram corresponding to the Fig.(a). Each mark corresponds to the vertex in Fig.(a) with the same mark. Unmarked cross-points in Fig.(b) are simply the intersection of two edges caused by the overlaps, not the vertex of a facet. (c)Subgraph with $J_{(ij,k)}$ and K_{ijkl}, which is frustrated. The numbers i, j inside the circles in the figure represent the subscripts of two facets whose layer-orderings are described by the spin variables $s_{i,j}$. For example, numbers 9, 10 in upper-left circle represents $s_{9,10}$. The solid blue line represents the ferromagnetic interaction, and the dashed red line represents the antiferromagnetic interaction.

certainly a part of discriminating the unfoldability. The cyclic prohibition terms do not prohibit any total layer ordering that does not include cyclic ordering. In this sense, it can be said that there is no non-trivial prohibition caused by cyclic prohibition terms. While, it is possible situation that only the layer orderings which contain cyclic order are allowed, namely be the ground states, in the Hamiltonians without frustration. Such situations are, in present, in need of further consideration.

4 Further Consideration on the Hypergraph Obtained from Spin Model

Hypergraphs formed by interactions obtained by rewriting to the spin model often contain branches extending from the core. The spins included in these branches can take values depending on those on the root side adjacent, as long as the branches have open extremities. Since the branches does not contribute to the frustration, we are allowed to consider the hypergraph with their branches removed, once after given according to the procedure in the previous section, for the purpose of detection of the frustration.

In addition, performing this removal in combination with the connected component analysis leads further reduction of computational effort. This is because there are cases that some arbitrary of folding way appear as a connected component which is independent, at least with respect to $J_{(ij,k)}$ and K_{ijkl}, from the other ones in the hypergraph. Both removal of the branches and connected component analysis can be done in polynomial time for the system size, *i.e.* the number of spins.

The problem of detecting frustration in partial hypergraphs still seems to face a combinatorial optimization problem. So it is true that the approaches presented in the present and the previous section may not provide significant speedups to the problem, such as reducing the exponential dependence to the polynomial dependence on system size. However, reducing the effective system size to the number of spins which is involved in the core of the largest connected component also multiplicatively reduces the amount of computation. Therefore we can realistically expect to reduce the amount of computation by rewriting the problem of detecting frustration in the spin model.

By using methods such as the Monte Carlo method and a thermodynamic integration or a reweighted histogram method, it is possible to detect the unfoldability and evaluate the total number of layering orders at the same time through the evaluation the number of states. This method allows the analysis of the energy landscape based on the spin model for each single instance of origami diagram. Although the Monte Carlo methods do not necessarily mean speeding up the algorithm, they are undoubtedly important as a realistic and highly versatile numerical method.

5 Conclusions

In summary, in this proceeding we show that by translating the origami diagram into a spin model it is possible to describe the unfoldability criterion in a natural way, as frustration, and also expectable to reduce the computational effort with focusing on the cycles and connected components of the hypergraph obtained from the model.

Formulating the problem by spin model is the first step to apply various statistical mechanics methods for random systems, including the replica method, to the flat-foldability determination problem. One of the advantages of the statistical-mechanical treatment is extracting information about the average-case properties and its parameter dependence through the analysis with replica method. It is expected that the whole picture of the flat-foldability problem of origami will be further clarified, for example, the existence or the location of the phase transition which divides the parameter-region into one whose most instances are flat-foldable and one whose most instances are unfoldable, by examining the average properties for sets of many instances. It is expected that the whole picture of the flat-foldability problem of origami will be further clarified by examining the average-case properties for sets of many instances.

Among them, the average-case computational complexity is an important topic, which is expected to be obtained through the analysis of replica symmetry breaking [11–13]. In the sense of the computational complexity class, which is given based on the worst-case complexity, the problem dealt with in this research has already proven with the polynomial reduction to be NP-hard, as mentioned in the introduction. However, the analysis of the average-case computational complexity has not progressed relatively to that of the worst-case ones.

However, in order to apply the replica method it is necessary to give some probabilistic model for the distribution or the probability of the realization of each instance, that is, each origami diagram in this problem. Probabilistic generation of locally flat-foldable origami diagrams is a non-trivial problem, and this problem must first be overcome in order to carry out the replica calculation. Alternatively, if we can build the model that is a randomized version of the origami diagram which gives an upper bound on the free energy, or mathematically find a polynomial reduction from the random spin model to the origami diagram, they may provide the gateway which leads to the replica analysis.

The spin-model formulation explained in this proceeding has also some direct extendability to other computational problems related to origami. For example, there is a map folding problem; the folding of the region tessellated into uniformly rectangular facets [14]. Variants of this problem range from those known to be solved with polynomial-time-algorithms to those known to be in the class NP-complete. In the map folding problem, including the variants, the manner of folding, mountain or valley, is assigned for each crease. This problem can be translated with the procedure explained in this proceeding into the Ising-spin model consisting of 4-body interactions and a random field for each spin. For some of the problems in the range of the extension, including the above example, it is expected to be relatively easy to obtain the average-case properties.

References

1. Hull, T.C.: Origametry: Mathematical Methods in Paper Folding (Cambridge University Press, Cambridge, 2020), T. C. Hull, *The combinatorics of flat folds: a survey*, in *Origami3* (Natick, MA: A. K. Peters, 2001), pp. 29–38
2. Bern, M., Hayes, B.: In: Proceedings of the Seventh Annual ACM-SIAM Symposium on Discrete Algorithms, (Society for Industrial and Applied Mathematics, Philadelphia, 1996), pp. 175–183
3. Ginepro, J., Hull, T.C.: J. Int. Seq. **17**, 14108 (2014)
4. Assis, M.: Phys. Rev. E **98**, 032112 (2018)
5. Di Francesco, P., Guitter, E., Mori, S.: Phys. Rev. E **55**, 237 (1997)
6. Di Francesco, P., Guitter, E.: Phys. Rep. **415**, 1–88 (2005)
7. Kawasaki, T.: In: Proceedings of the First International Meeting of Origami Science and Technology, (Universita di Padova, Padova, p. 229–237 (1989), Justin, J.: In: Proceedings of the First International Meeting of Origami Science and Technology, (Universita di Padova, Padova, p. 263–277 (1989). Kawasaki, T.: British Origami, p. 28–30 (1986)
8. Nishimori, H.: Prog. Theor. Phys. **66**, 1169 (1981)
9. Nishimori, H.: Statistical physics of spin glasses and information processing: an introduction. Oxford University Press, Oxford (2001)

10. Justin, J.: In: Proceedings of the Second International Meeting of Origami Science and Scientific Origami, Seian University of Art and Design, Otsu, p. 15–29 (1997)
11. Monasson, R., Zecchina, R., Kirkpatrick, S., Selman, B., Troyansky, L.: Nature **400**, 133 (1999)
12. Weigt, M., Hartmann, A.K.: Phys. Rev. Lett. **86**, 1658 (2001)
13. Krzakala, F., Montanari, A., Ricci-Tersenghi, F., Semerjian, G., Zdeborova, L.: Proc. Natl. Acad. Sci. U.S.A. **104**, 10318 (2007)
14. Arkin, E.M., et. al.: Comput. Geom. Theory Appl. **29**, p.23-46 (2004). Morgan, T.D.: Map Folding, PhD Thesis, (MIT 2012)

Dynamical Properties of Directed Transport in a Lattice System at the Nanoscale

Takaaki Kawaguchi(✉)

Department of Physics, Toho University, Ota-ku 143-8540, Japan
tkawaguchi@med.toho-u.ac.jp

Abstract. The dynamical properties of the directed transport in nanoscale systems have been studied using computer simulations based on a lattice model. Directed transport of atoms or molecules are an important phenomenon in nanotechnology because it has been considered as an effective method for driving nanomachines. In this study, using a spatiotemporal modulation method, the directed transport of atomic lattices are induced under certain conditions. We investigate the dynamical properties of the directed transport using numerical simulations based on a lattice model where impurity-type disorder of a substrate is taken into account. We clarify the effect of interfacial disorder on the directed transport on the basis of the concept of impurity pinning.

Keywords: directed transport · pinning · nanomachine

1 Introduction

Nanoscale mechanical systems such as nanomachines [1,2] are often needed to move or slide as a whole in a certain direction. Such a dynamical phenomenon is often called directed transport. The directed transport is a key feature for several nanomechanical systems because some types of mechanical motion of nanomachines might be realized by the directed transport of machine components. Basically, in such systems, atomic lattices or molecules may be considered as machine components. Hence, the dynamics and its control of the directed transport of atomic lattices or molecules will play important roles in nanomechanics and related fields.

In this study we investigate the physical properties of directed transport of the atomic lattice on a disordered substrate. using a simple model of atomic lattices as a component of nanomachines. To realize the directed motion of the lattice, we employ a dynamical method which is based on the spatiotemporal modulation of atomic lattices [3–5]. We assume that the atomic lattices interact with solid substrate lattices which possess impurity-type disorder because in realistic nanoscale systems there might exist a sort of disorder at the interface of systems. In the following, using computer simulations based on a lattice model, we investigate the effects of the disorder on the directed transport of the atomic lattice subject to the spatiotemporal modulation.

H. J. Choi et al. (Eds.): APPC 2022, *Proceedings of the 15th Asia Pacific Physics Conference*, pp. 250–259, 2024.
https://doi.org/10.1007/978-981-96-0191-2_26

2 Model and Numerical Method

We consider a simple lattice system, which consists of an atomic lattice on a solid substrate. The atomic lattice corresponds to an essential component of nanomachines on the substrate. The schematic diagram of the model is shown in Fig. 1. This is an extended version of the Frenkel-Kontorova (FK) model [6,7] in one-dimension. The original FK model consists of an atomic lattice interacting with each other via a harmonic force on a periodic substrate. The atomic lattice is described by a discrete elastic lattice. The static and dynamical behaviors of the FK and its related models have been studied in the context of friction and lubrication at the nanoscale [7–16]. The FK model is useful to study elementary processes of interfacial phenomena at nanoscale.

For the model used in this study two lattices are assumed, i.e., the atomic lattice and the substrate lattice. The substrate lattice is taken into consideration as the substrate potential, which is chosen here to be $U_{\mathrm{s}}(x) = -(U_{\mathrm{p}}/2)\exp\left[-4\left(x/c_s\right)^2\right]$, where U_{p} is the strength of the substrate potential and c_s the atomic spacing of the substrate lattice. In the model without impurities all the atoms of the substrate are fixed periodically at the regular sites. The total interlattice force $F_{\mathrm{s}}(x)$ which acts on the i-th atom of the atomic lattice is obtained by summing up the contribution from all the atoms of the substrate.

We introduce here impurity-type disorder in the substrate [13]. As shown in Fig. 1, the potential acting on the atomic lattice consists of two parts, i.e., the periodic substrate potential and the impurity potential which is distributed randomly on sites of the substrate. We assume that the impurity potential has the same form as the periodic potential except for the potential strength. The impurity potential U_{i} is written as $U_{\mathrm{i}}(x) = -(U_{\mathrm{imp}}/2)\exp\left[-4\left(x/c_s\right)^2\right]$, where U_{imp} is the strength of the impurity potential. The total interfacial potential is given by

$$U(x) = \sum_{i}^{\text{regular sites}} U_{\mathrm{s}}(x - x_i^{\mathrm{s}}) + \sum_{j}^{N_{\mathrm{i}}} U_{\mathrm{i}}(x - x_j^{\mathrm{d}}), \tag{1}$$

where U_{s} represents the original inter-atomic potential, U_{i} the impurity potential, and N_{i} is the number of impurities. The coordinates x_i^{s} and x_j^{d} are that of the $i-$th regular

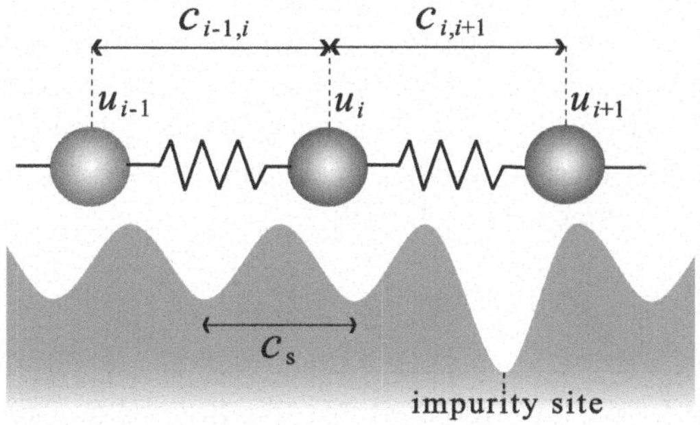

Fig. 1. Schematic diagram of the nanomachine which consists of an atomic lattice subject to the spatiotemporal modulation.

site of the substrate and that of the $j-$th impurity site, respectively. The interfacial force which works on the lattice is given by $F_{\rm I}(x) = -dU(x)/dx$. For convenience of later discussion we denote the ratio $U_{\rm imp}/U_{\rm p}$ as W.

To induce the directed motion of the atomic lattice we adopt a dynamical method, by which no external driving force is needed for the occurrence of the directed motion. We introduce spatiotemporal modulation of the natural length between adjacent atoms [3–5, 17–19]. The modulation has a sinusoidal wave form so that the natural length between i-th and $(i+1)$-th atoms $c_{i,i+1}$ is given by

$$c_{i,i+1} = c_0 + \alpha \sin(kic_0 - \omega t), \qquad (2)$$

where c_0 is the natural length between adjacent atoms in the absence of the modulation, and k is the wave-number. The modulation is controlled by changing the values of the amplitude α, the time period $T = 2\pi/\omega$ and the wave length $\lambda = 2\pi/k$. The effectiveness of this method on the occurrence of the directed motion has been reported in several studies [17–19] . The method have relationship with the driving mechanism of molecular machines which utilize photochromic processes [20].

The equation of motion of the i-th atom in the lattice is given by

$$m\ddot{u}_i + \gamma\dot{u}_i = K\{(u_{i+1} - u_i - c_{i,i+1}) - (u_i - u_{i-1} - c_{i-1,i})\} + F_{\rm I}(u_i) + F, \quad (3)$$

where u_i is the one-dimensional coordinate of the $i-$th atom, m, γ and K are the mass of an atom, dissipation coefficient, and the elastic constant for the inter-atomic force, respectively.

We consider a small system in which $N_a = 30$ and $N_s = 50$. Under the periodic boundary condition the ratio c_0/c_s is equal to N_s/N_a, where N_a and N_s are the numbers of atoms in the atomic lattice and substrate lattice, respectively. In the following, some parameters are fixed as $c_0 = 1.66\cdots$, $c_s = 1$, and $\gamma = m = K = 1$. The strength of the impurity potential, U_i, is changed in a certain range. We set the parameters used for the substrate potential and spatiotemporal modulation as follows: $U_p = 0.5$, $\alpha/c_0 = 0.60$ and $\lambda/c_s = 2.5$. The angular frequency ω is set as $2\pi/40$ or 0. The number of impurity sites is set as $N_{\rm i} = 5$.

Using a computer simulation method, the equations of motion are solved numerically under the periodic boundary condition by assuming overdamped motion of atoms. In the following, the velocity v is defined as the temporally averaged velocity of the center of mass of the atomic lattice. There is a unit velocity, which is characteristic for the system and is given by $v_0 = c_s/T$. For the directed motion of the system under a certain condition, it has been reported that v is locked to a discrete value which is given by $v = \pm nv_0$ where n is integer [3–5, 17–19].

3 Results

Using the present model we have reported that the directed transport of the lattice can occur under a certain condition on system parameters. In this study the system parameters are chosen so that the directed transport occurs in the absence of impurities when the spatiotemporal modulation is applied to the lattice. In the following we investigate the effect of disorder due to impurities on the dynamics of the system and the occurrence of the directed transport.

First we investigate how the directed motion is affected by the external force in order to reveals the effect of interfacial disorder due to impurities. The effect of the spatiotemporal modulation on the static friction force is also clarified. To clarify the effect of temporal modulation we consider two cases: $\omega = 2\pi/40$ and $\omega = 0$, where there remains a spatial static modulation of the lattice, which is defined by the wave number k in eq. (2). We calculate spatially and temporally averaged velocity v for two different strengths of the impurity potential. In the case of a weak impurity potential $W = 1.4$, Fig. 2(a) and 2(b) show the $v - F$ characteristics for $\omega = 0$ and $\omega = 2\pi/40$,

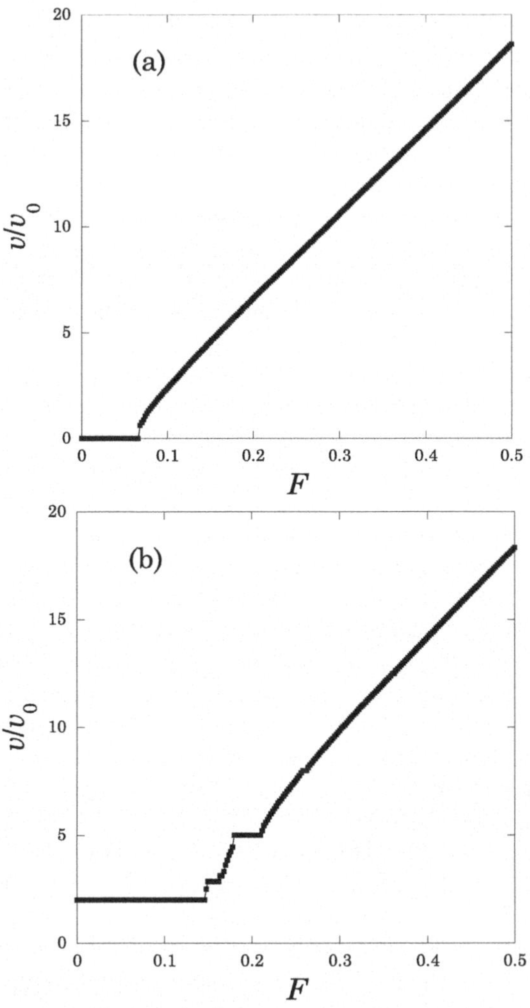

Fig. 2. Velocity-external force ($v - F$) characteristics for $W = 1.4$, where (a) $\omega = 0$ and (b) $\omega = 2\pi/40$.

respectively. where the velocity is normalized with the unit velocity v_0. If the directed transport occurs, the velocity at $F = 0$ is finite. That is the lattice shows directional motion even in the absence of the external force F. As shown in Fig. 2(a), $v = 0$ at $F = 0$ for $\omega = 0$. This means that the directed transport of the lattice can not occur in the absence of the temporal modulation. Note here that there exists a finite threshold force F_c between $v = 0$ and $v \neq 0$ regimes. The threshold behavior is relevant to pinning effects of the lattice, which is caused by the two kinds of underlying potentials. The lattice can be pinned by both the impurity and the periodic substrate potentials. The depinning of the lattice occurs at $F = F_c$, above which the lattice driven by F moves at a finite velocity.

For $\omega = 2\pi/40$, as shown in Fig. 2(b), v is finite at $F = 0$ and also in a small F regime, where the normalized velocity is locked at $v/v_0 = 2$. Many velocity locking regimes are observed complicatedly in the curve. The velocity locking is due to the spatiotemporal modulation of the lattice, and is characteristics of the directed transport induced by the modulation. It should be noted that the periodic potential works as a pinning potential for no temporal modulation case $\omega = 0$. For $\omega \neq 0$ case, however, the periodic potential assists the lattice subject to the spatiotemporal modulation to move as the directed transport. This reflects the barrier-assisted nature of the directed transport [17–19]. Thus, in the presence of the spatiotemporal modulation, the lattice on a weakly disordered substrate shows the directed transport even for $F = 0$. In other words, the static friction force (or pinning force) of the lattice subject to the spatiotemporal modulation is zero. It shows the robustness of the directed transport against the weak disorder in the substrate. In the large F regime ($F > 0.3$) both the curves in Fig, 2(a) and 2(b) become almost the same. This means that the velocity locking effect of the spatiotemporal modulation is smeared out by the strong driving due to large external force.

Next we investigate the driven dynamics for a strong impurity potential. For $W = 3$, the $v - F$ characteristics for $\omega = 0$ and $\omega = 2\pi/40$ are shown in Figs. 3(a) and 3(b), respectively. For both values of ω, we find that $v = 0$ at $F = 0$, i.e., no directed transport occurs, and there appear threshold behaviors. The lattice can not move in a certain direction when F is less than the threshold force. By comparing with the results in Fig. 2(a), it is found that the threshold force for $\omega = 0$ in Fig, 3(a) is greater than that in Fig, 2(a). This is due to the strengthening of impurity pinning. Furthermore, note that, in Figs. 3(a) and 3(b), the threshold force for $\omega = 2\pi/40$ is smaller than that for $\omega = 0$. Significant reduction of the threshold force is caused by the temporal modulation. When F is greater than the threshold force, both curves in Figs. 3(a) and 3(b) show similar behaviors, as observed also in Figs. 2(a) and 2(b). The velocity locking behavior is commonly observed in finite ω cases.

To clarify the reduction effect due to finite ω on the threshold force, we investigate the impurity strength W dependence of the threshold forces F_c. Figure 4 shows the $W - F_c$ characteristics for $\omega = 0$ and $\omega = 2\pi/40$. In all W regimes, the values of F_c for $\omega = 2\pi/40$ is significantly smaller than those for $\omega = 0$. This means that the reduction of F_c due to the modulation is effective regardless of the strength of the impurity potential. In a weak impurity potential regime around $1 < W < 2.2$, F_c is zero. This regime corresponds to the robust regime of the directed transport against the

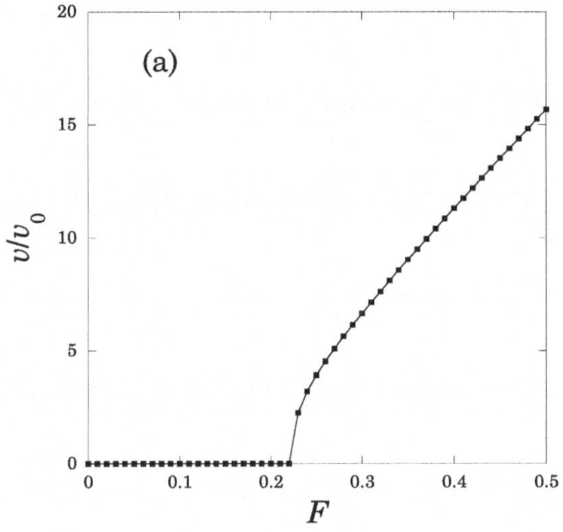

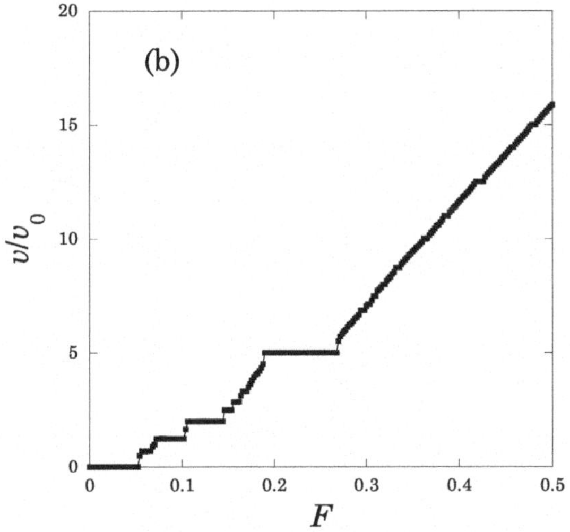

Fig. 3. Velocity-external force ($v - F$) characteristics for $W = 3$, where (a) $\omega = 0$ and (b) $\omega = 2\pi/40$.

interfacial disorder. Since the threshold force F_c is vanishing there, the lattice shows the directed transport in the absence of the external driving force F.

In the $W - F_c$ curve for $\omega = 0$ two sorts of scaling behaviors are observed. The scaling relations employed here are derived from the impurity pinning theory [21,22], which are effectively utilized in several fields concerning density waves, vortex lattices and interfacial friction [13,15]. Figure 5 shows again the $W - F_c$ curve for $\omega = 0$. The data shown in Fig. 5 are the same as those for $\omega = 0$ in Fig. 4. The dotted and the solid lines indicate scaling relations $F_c \propto W$ and $F_c \propto W^{\frac{4}{3}}$, respectively. According to the collective impurity pinning theory in one-dimension, we get these scaling relations of the threshold force F_c. On the basis of the collective impurity pinning theory, the

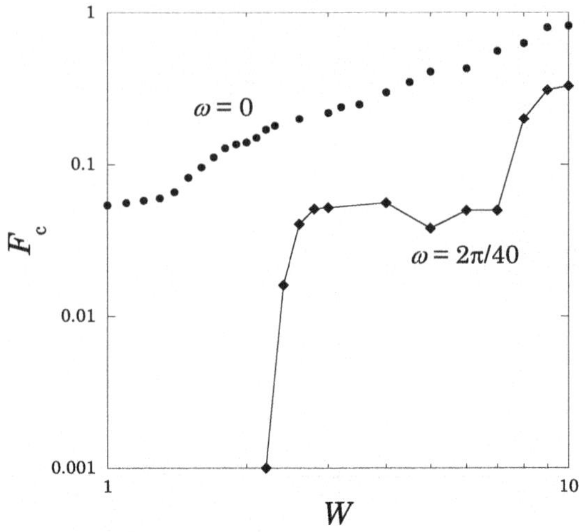

Fig. 4. Impurity strength- threshold force ($W - F_c$) characteristics for $\omega = 0$ and $\omega = 2\pi/40$.

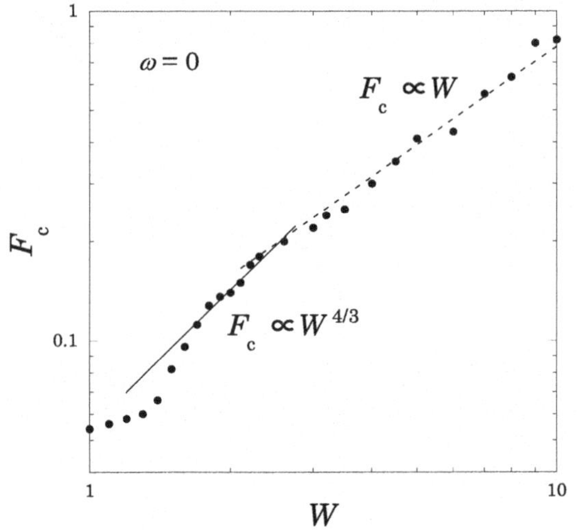

Fig. 5. Scaling behaviors in the $W - F_c$ curve for $\omega = 0$.

following two characteristic pinning mechanisms are understood. When the impurity pinning is weak, the lattice is collectively pinned by many impurity potentials, and then the scaling relation $F_c \propto W^{\frac{4}{3}}$ is derived from the pinning theory. The pinning described by this relation is called weak pinning. On the other hand, when the impurity pinning is strong, the lattice is pinned at each impurity potential site. In such a pinned state, collective behavior is absent and a simple scaling relation $F_c \propto W$ is obtained. Such pinning is called strong pinning.

As shown in Fig. 5, in the large W regime ($W > 3$), the scaling relation for the strong pinning, $F_c \propto W$, holds well. In a narrow regime around $W = 2$, there appears a scaling region for the weak pinning described as the relation, $F_c \propto W^{\frac{4}{3}}$. The fitting of the scaling relation is somewhat unclear, but the deviation from the strong pinning behavior is confirmed obviously. There appears crossover between the two characteristic impurity pinning regimes around $W \sim 3$. The scaling analyses, however, does not take into account the effect of the underlying periodic substrate potential. In the small W regime ($W < 2$) the scaling relation is broken obviously. This is because the effect of the periodic substrate potential is dominant there and the weak pinning state is destroyed by the pinning due to the substrate potential. It is a sort of commensurate pinning because the interface between the lattice and the substrate has some commensurability. A pinning effect due to the commensurability exists, which causes finite threshold force F_c for no temporal modulation.

By comparing between the results in Fig. 4 and Fig. 5, it is found that the threshold force F_c for $\omega = 2\pi/40$ in Fig. 4 is finite in the two scaling regimes of weak and strong pinning shown in Fig. 5. For $1 \leq W < 2$, which is outside the two scaling regimes of impurity pinning, F_c for $\omega = 2\pi/40$ in Fig. 4 is zero. Thus, the directed transport under zero external force ($F = 0$) can occur only outside the scaling regimes, where the effect of the periodic substrate potential is dominant for the interfacial interaction and the commensurate pinning takes place for $\omega = 0$. These results reflects that the periodic potential due to the substrate is essentially important for the occurrence of the directed transport. On the other hand, in the impurity pinning regime ($W > 2$) where the scaling behaviors of impurity pinning are observed, the directed transport under $F = 0$ is not possible any more because the impurity pinning effect is dominant for the interfacial interaction. In such pinning states, a finite external force is needed to induce the directed motion of the lattice even for finite ω.

The results obtained above are summarized in a table shown in Fig. 6. Several characteristic regimes concerning pinning and directed transport are indicated here. The overall picture of the phenomena can be understood from the table.

We make some remarks on the modulation method and the properties of directed transport discussed above. Different types of spatiotemporal modulation for which the directed transport occurs are possible as long as a certain dynamical condition is satisfied. In any case, modulation with respect to time and space is necessary for

no modulation ($\omega = 0$)	commensurate pinning	weak pinning	strong pinning
	periodic potential is dominant	impurity potential is dominant	
	finite F_c		
spatiotemporal modulation (finite ω)	directed transport ($F_c = 0$)	no directed transport ($F_c > 0$)	
	vanishing or significant reduction of F_c		

W →

Fig. 6. Characteristic behaviors versus W. Dotted lines indicate crossover between characteristic regimes. The horizontal arrow indicates an increase in W.

the directed transport. In particular, a condition for quasi-adiabatic cases is given by Popov [5], which is related to the commensurability between the atomic lattice and the substrate lattice. For finite ω, dynamical effects such as the relaxation of the atomic lattice becomes important to the occurrence of the directed transport [17–19].

In general, for interfacial systems, periodicity and disorder exist in their substrates. The results obtained in the present study show essentially universal behaviors and properties of the directed transport in many interfacial systems with a competing effect between periodicity and disorder. In such systems, it is expected that the type of disorder is irrelevant to qualitative results.

A candidate real-world system for the model used in the present study is considered to be a photochromic molecule, which undergoes a reversible structural change by external electromagnetic waves. The material design based on such a molecular system is a topic for future research.

When considering a novel application of the directed transport to nanotechnology, traction capability would be useful. The lattice in directed motion is expected to have a traction force of finite magnitude. Using the traction force the lattice may carry other nanoscale objects. The specific method and properties of the traction will be investigated in future studies.

4 Conclusions

We investigated the directed transport of the atomic lattice subject to spatiotemporal modulation on disordered substrates. The effect of the substrate potential and interfacial disorder on the directed transport was clarified on the basis of the concept of impurity pinning. There exists a robust regime against weak interfacial disorder in the substrate. In such a regime, the commensurate pinning effect of the periodic substrate potential is dominant and the atomic lattice can occur the directed motion even for zero external force conditions. Significant reduction of the threshold force for depinning of the atomic lattice was observed in the presence of the spatiotemporal modulation. Dynamical effects caused by the modulation play important roles in the threshold behavior.

References

1. Bhushan, B.: Nanotribology and Nanomechanics, p. 311. Springer, Berlin (2005)
2. Browne, W.R., Feringa, B.L.: Nature nanotechnol. **1**, 25 (2006)
3. Porto, M., Urbakh, M., Klafter, J.: Phys. Rev. Lett. **84**, 6058 (2000)
4. Zheng, Z., Cross, M.C., Hu, G.: Phys. Rev. Lett. **89**, 154102 (2002)
5. Popov, V.L.: Phys. Rev. E **68**, 026608 (2003)
6. Frenkel, Y., Kontorova, T.: Zh. Eksp. Teor. Fiz., **89**, 1340 (1938). **89**, 1349 (1938)
7. Braun, O.M., Kivshar, Y.S.: The Frenkel-Kontorova Model: Concepts, Methods, and Applications.(Springer 2004)
8. Persson, B.N.J.: Sliding Friction, 2nd edn. Springer, Berlin (2000)
9. Fundamentals of Friction and Wear on the Nanoscale. Gnecco, E., Meyer, E., (eds.), (Springer, Berlin 2007)
10. Physics of sliding friction. Persson, B.N.J., Tosatti, E., (eds.), (Kluwer Academic Publishers, Dordrecht 1996)

11. Hirano, M., Shinjo, K.: Phys. Rev. B **41**, 11837 (1990)
12. Shinjo, K., Hirano, M.: Surf. Sci. **283**, 473 (1993)
13. Kawaguchi, T., Matsukawa, H.: Phys. Rev. B **56**, 13932 (1997)
14. Kawaguchi, T., Matsukawa, H.: Phys. Rev. B **58**, 15866 (1998)
15. Kawaguchi, T., Matsukawa, H.: Phys. Rev. B **61**, R16366 (2000)
16. Hirano, M.: Surf. Sci. Rep. **60**, 159 (2006)
17. Kawaguchi, T., Matsukawa, H.: AIP Conf. Proc. **1518**, 737 (2013)
18. Kawaguchi, T., Matsukawa, H., Conf, J.P.S.: Proc. **1**, 012059 (2014)
19. Kawaguchi, T., Conf, J.P.S.: Proc. **5**, 011025 (2015)
20. Kay, E.R., Leigh, D.A., Zerbetto, F.: Angew. Chem. Int. Ed. **46**, 72 (2007)
21. Fukuyama, H.: J. Phys. Soc. Jpn. **41**, 513 (1976)
22. Fukuyama, H., Lee, P.A.: Phys. Rev. B **17**, 535 (1978)

Author Index

H. J. Choi et al. (Eds.): APPC 2022, *Proceedings of the 15th Asia Pacific Physics Conference*, pp. 261–262, 2024.
https://doi.org/10.1007/978-981-96-0191-2